Big Data Analytics for Smart and Connected Cities

Nilanjan Dey
Techno India College of Technology, India

Sharvari Tamane
Jawaharlal Nehru Engineering College, India

A volume in the Advances in Civil and Industrial Engineering (ACIE) Book Series

IGI Global
DISSEMINATOR OF KNOWLEDGE

Published in the United States of America by
IGI Global
Engineering Science Reference (an imprint of IGI Global)
701 E. Chocolate Avenue
Hershey PA, USA 17033
Tel: 717-533-8845
Fax: 717-533-8661
E-mail: cust@igi-global.com
Web site: http://www.igi-global.com

Library of Congress Cataloging-in-Publication Data

Names: Dey, Nilanjan, 1984- editor. | Tamane, Sharvari, 1973- editor.
Title: Big data analytics for smart and connected cities / Nilanjan Dey and Sharvari Tamane, editors.
Description: Hershey, PA : Engineering Science Reference, 2018. | Includes bibliographical references.
Identifiers: LCCN 2018003720| ISBN 9781522562078 (h/c) | ISBN 9781522562085 (ebk)
Subjects: LCSH: Municipal engineering--Data processing. | Municipal services--Data processing. | Smart cities. | Internet of things. | Big data. | Intelligent control systems--Handbooks, manuals, etc. | Community development, Urban--Handbooks, manuals, etc.
Classification: LCC TD159.4 .H36 2018 | DDC 628.0285/57--dc23 LC record available at https://lccn.loc.gov/2018003720

This book is published in the IGI Global book series Advances in Civil and Industrial Engineering (ACIE) (ISSN: 2326-6139; eISSN: 2326-6155)

British Cataloguing in Publication Data
A Cataloguing in Publication record for this book is available from the British Library.

For electronic access to this publication, please contact: eresources@igi-global.com.

Advances in Civil and Industrial Engineering (ACIE) Book Series

ISSN:2326-6139
EISSN:2326-6155

Editor-in-Chief: Ioan Constantin Dima, University Valahia of Târgovişte, Romania

MISSION

Private and public sector infrastructures begin to age, or require change in the face of developing technologies, the fields of civil and industrial engineering have become increasingly important as a method to mitigate and manage these changes. As governments and the public at large begin to grapple with climate change and growing populations, civil engineering has become more interdisciplinary and the need for publications that discuss the rapid changes and advancements in the field have become more in-demand. Additionally, private corporations and companies are facing similar changes and challenges, with the pressure for new and innovative methods being placed on those involved in industrial engineering.

The **Advances in Civil and Industrial Engineering (ACIE) Book Series** aims to present research and methodology that will provide solutions and discussions to meet such needs. The latest methodologies, applications, tools, and analysis will be published through the books included in **ACIE** in order to keep the available research in civil and industrial engineering as current and timely as possible.

COVERAGE

- Transportation Engineering
- Earthquake engineering
- Materials Management
- Coastal Engineering
- Operations research
- Hydraulic Engineering
- Optimization Techniques
- Structural Engineering
- Quality Engineering
- Production Planning and Control

IGI Global is currently accepting manuscripts for publication within this series. To submit a proposal for a volume in this series, please contact our Acquisition Editors at Acquisitions@igi-global.com or visit: http://www.igi-global.com/publish/.

The Advances in Civil and Industrial Engineering (ACIE) Book Series (ISSN 2326-6139) is published by IGI Global, 701 E. Chocolate Avenue, Hershey, PA 17033-1240, USA, www.igi-global.com. This series is composed of titles available for purchase individually; each title is edited to be contextually exclusive from any other title within the series. For pricing and ordering information please visit http://www.igi-global.com/book-series/advances-civil-industrial-engineering/73673. Postmaster: Send all address changes to above address.

Titles in this Series

For a list of additional titles in this series, please visit:
https://www.igi-global.com/book-series/advances-civil-industrial-engineering/73673

Contemporary Strategies and Approaches in 3-D Information Modeling
Bimal Kumar (Glasgow Caledonian University, UK)
Engineering Science Reference • ©2018 • 313pp • H/C (ISBN: 9781522556251) • US $205.00

New Approaches, Methods, and Tools in Urban E-Planning
Carlos Nunes Silva (University of Lisbon, Portugal)
Engineering Science Reference • ©2018 • 407pp • H/C (ISBN: 9781522559993) • US $205.00

Designing Grid Cities for Optimized Urban Development and Planning
Guiseppe Carlone (Independent Researcher, Italy) Nicola Martinelli (Politecnico di Bari, Italy) and Francesco Rotondo (Politecnico di Bari, Italy)
Engineering Science Reference • ©2018 • 305pp • H/C (ISBN: 9781522536130) • US $210.00

Design Solutions for nZEB Retrofit Buildings
Elzbieta Rynska (Warsaw University of Technology, Poland) Urszula Kozminska (Warsaw University of Technology, Poland) Kinga Zinowiec-Cieplik (Warsaw University of Technology, Poland) Joanna Rucinska (Warsaw University of Technology, Poland) and Barbara Szybinska-Matusiak (Norwegian University of Science and Technology, Norway)
Engineering Science Reference • ©2018 • 362pp • H/C (ISBN: 9781522541059) • US $225.00

Handbook of Research on Perception-Driven Approaches to Urban Assessment and Design
Francesco Aletta (University of Sheffield, UK) and Jieling Xiao (Birmingham City University, UK)
Engineering Science Reference • ©2018 • 641pp • H/C (ISBN: 9781522536376) • US $295.00

Convective Heat and Mass Transfer in the Free Atmosphere Emerging Research...
Kaliyeva Kulyash (Lorraine University, France)
Engineering Science Reference • ©2018 • 127pp • H/C (ISBN: 9781522530503) • US $135.00

For an entire list of titles in this series, please visit:
https://www.igi-global.com/book-series/advances-civil-industrial-engineering/73673

IGI Global
DISSEMINATOR OF KNOWLEDGE

701 East Chocolate Avenue, Hershey, PA 17033, USA
Tel: 717-533-8845 x100 • Fax: 717-533-8661
E-Mail: cust@igi-global.com • www.igi-global.com

Table of Contents

Section 5
The Case Studies: Urban Areas in the Conext of Smart City Infrastructures

Detailed Table of Contents

The world is growing and energy conservation is a very important challenge for the engineering domain. The emergence of smart cities is one possible solution for the same, as it claims that energy and resources are saved in the smart city infrastructure. This chapter is divided into five sections. Section 1 gives the past, present, and future of the living style. It gives the representation from rural, urban, to smart city. Section 2 gives the explanations of four pillars of big data, and through grid, a big data analysis is presented in the chapter. Section 3 started with the case study on smart grid. It comprises traffic congestion and their prospective solution through big data analytics. Section 4 starts from the mobile crowd sensing. It discusses a good elaboration on crowd sensing whereas Section 5 discusses the smart city approach. Important issues like lighting, parking, and traffic were taken into consideration.

Preface

This book explores the concepts of big data, IoT and smart city along with recent research developments. It also includes various real time/offline applications and case studies in the field of engineering, computer science, IoT and smart cities with modern tools and technologies.

As a population grows, the resources become scarcer, therefore, the efficient usage of these limited goods becomes more important. Smart cities are a key factor in the consumption of materials and resources. Built on and integrating with big data, the cities of the future are becoming a realization today.

The integration of big data and interconnected technology along with the increasing population will lead to the necessary creation of smart cities. To continue providing people with safe, comfortable, and affordable places to live, cities must incorporate techniques and technologies to bring them into the future. We are looking forward to seeing the advances that will come to our cities in the near future.

OBJECTIVE OF THE BOOK

The main objective of this book is to cover development in urban areas about Internet of Things (IoT), Information and Communication Technology (ICT) and the technologies used to manage city's assets in a secure manner. IoT has developed into our daily lives from few days back. It represents a general concept for the ability of network devices to sense and collect data from the world around us, and then share that data across the internet where it can be processed and utilized for various interesting purposes. IoT can do following things for us:

- Build a system specific application using microcontroller,
- Receive alert messages on our smart electronic devices (mobile phones, smart watches etc.) in case of any danger detected at nearby places,
- Self-parking automobiles,

- Automatic ordering of groceries and other home supplies,
- Automatic tracking of exercise habits and other day-to-day personal activity including goal tracking and regular progress reports.

ICT is used to improve the quality, performance and interactivity in the services provided by cities. It also utilizes resources while using the services and improves interaction between a common man and government. City's assets may include various departments' information systems, physical infrastructure, institutes (schools/ colleges, etc.), infirmaries, smart meters, smart water, power plants, public libraries, community services, etc.

Other objectives may include various real time/offline applications and case studies in the field of engineering, computer science, information security, interdisciplinary tools and cases of IoT, big data and smart city with modern tools and technologies used.

Smart cities are using information technologies to build the strong physical infrastructure, to open/process new innovative activities and to enable citizens to face new challenges to solve problems in an intelligent way. The reason for this growth is increase in communication through electronic devices, sensors, web logs, global positioning system (GPS) data, mobile data, IoT etc. The most challenging part in this is to discover new things in a city through technology. It requires to take new initiatives and government support. And this is possible through new technologies which involves cloud-based services, IoT, real-world user interfaces, use of smart phones, smart meters, networks of sensors and RFIDs, and more accurate communication based on the semantic web. We also need newer information management tools and technologies to deal with real time large volume and large variety of data.

This book is mainly focusing on IoT and technologies, big data and future horizon, smart city and future luxuries, the interdisciplinary tools, cases of IoT, big data and smart city.

ORGANIZATION OF THE BOOK

The book consists of 12 chapters that are organized in five sections as shown below. The first section consists of Chapters 1 to 5 which relates to information about next generation, a collaborative approach, a secure, distributed and reliable cloud-based reference architecture towards IoT, big data and smart city. It has also provided role of cloud, IoT and it's impacts in computing intelligence.

The second section elaborated on IoT and industry expectations. Chapter 6 from this section explained Industry 4.0: The New Industrial Revolution in detail. The third section of the book encloses two chapters that introduced the safety and security

Table 1. Organization of the book

<table>
<tr><td colspan="6">Section 1: Next Generation: IoT, Big Data, and Smart City</td></tr>
<tr><td colspan="3">Chapter 1
Smart and Connected Cities</td><td colspan="3">Chapter 2
A Collaborative Approach: IoT, Big Data, and Smart City</td></tr>
<tr><td colspan="2">Chapter 3
Towards a Secure, Distributed, and Reliable Cloud-Based Reference Architecture for Big Data in Smart Cities</td><td colspan="2">Chapter 4
Cloud- and IoT-Powered Smart and Connected Cities</td><td colspan="2">Chapter 5
Internet of Things and Its Impacts in Computing Intelligence: A Comprehensive Review</td></tr>
<tr><td colspan="6">Section 2: The IoT and Industry Expectations</td></tr>
<tr><td colspan="6">Chapter 6
Industry 4.0: The New Industrial Revolution</td></tr>
<tr><td colspan="6">Section 3: Safety and Security Systems</td></tr>
<tr><td colspan="6">Chapter 7
An Integrated Digital Authentication Mechanism for Intrusion Detection System</td></tr>
<tr><td colspan="6">Section 4: Applications in Electronic and Digital Technologies</td></tr>
<tr><td colspan="2">Chapter 8
Smart Traffic System Operations</td><td colspan="2">Chapter 9
Towards a Better and Smarter Healthcare and IoT</td><td colspan="2">Chapter 10
Smart Waste Management System</td></tr>
<tr><td colspan="6">Section 5: The Case Studies: Urban Area in Context of Smart City Infrastructure</td></tr>
<tr><td colspan="3">Chapter 11
Big Data Analytics and Internet of Things for Urban Transportation: A Case of Pune City, Maharashtra, India</td><td colspan="3">Chapter 12
Songdo Smart City: An Aerotropolis and a Ubiquitous City</td></tr>
</table>

system. Chapter 7 from this section has provided a sample real time application process of IDS which involves an access control for specified area forming a vital link in a security. It has also specified the digital authentication based secure system which can be adopted at home/organisations or any specified areas by the authorities.

The fourth section focuses on applications in electronics and digital technologies like smart traffic system operations, smarter healthcare and IoT and smart waste management system. The fifth section included case studies in urban areas in context of smart city infrastructure. Chapter 11 and 12 from this section provided case studies for big data analytics and IoT for urban transportation: a case of Pune city, Maharashtra, and Songdo smart city-urban analytics resp.

Section1: Next Generation – IoT, Big Data, and Smart City

The rapid growth in the urban population increases proper utilization of resources, services and infrastructure. These needs can be fulfilled by connecting and communicating various IoT devices with each other over the internet to establish smart systems.

This section elaborates the fundamental concepts of the smart and connected cities. The intention of Chapter 1 is to provide the information about: the needs of remembering the past (preservation and revitalization), the needs of living in the present (livability), the needs of planning for the future (attainability), the vision of smart and connected cities to improve livability, preservation, revitalization, and attainability of a community, how IoT is used to provide a ubiquitous network of connected devices and smart sensors for Smart and connected cities?, how IoT, machine to machine, big data and smart cities linkages can help in doing predictive analytics?, real potential of using big data analytics in smart cities, smart grid and traffic congestion management where big data analytics can be useful for decision makers and city planner.

The second chapter has provided a collaborative approach towards: IoT, big data and smart city. It has also focused on how big data helps decision-makers to plan for any expansion in either smart city services, resources, or areas along with the challenges involved in using big data and IoT in smart city. Chapter 3 outlined the development towards a secure, distributed, and reliable cloud-based reference architecture for the management of big data in a smart city. The focus of this chapter is on the conceptualization and implementation of an infrastructure to prove its feasibility.

In Chapter 4 smart communities through big data analytics and cloud-IoT are highlighted. This chapter covers the requirement of cloud computing, its elements, design issues and architecture model. In addition to this various design issues, overview of smart city communities through big data analytics are also described. Chapter 5 promotes the main concept of data growth for next generation IoT devices along with the discussion regarding how to handle the big data. The chapter has been concluded by providing various big data analytical tools that can be used effectively for faster processing and extraction of different significant information.

Section 2: The IoT and Industry Expectations

Use of IoT in industry with minimum investment is getting good results. This attracted various organizations to make use of IoT and gain insights of data to make important decisions. These decisions are benefited to improve productivity, work effectiveness, marketing capability etc. for growing their businesses.

Chapter 6 from this section provided the concept of industry 4.0 which can be considered as the 21st Century Industrial Revolution and will soon be the new form of manufacturing delight from the previous three industrial revolutions characterized by machines, electrical power, and information technology. The definitive customer would experience manufacturing requests determined by artificial intelligence, machine learning, and automated technologies linked with data science support for gauging customer necessities. This new industrial revolution would possess systems with transformative technologies for managing interconnected systems between its physical assets and computational capabilities.

Section 3: Safety and Security Systems

As population grows, use of smart equipment's grows, which increases generation of data rapidly. In next 2-3 years it may increase at least 20 times as it is available right now. IoT is also contributing some percentage in expanding this data. Organizations are spending a massive amount of money on security; therefore, security is playing a most important role in big data these days.

Chapter 7 has demonstrated an application on human intrusion detection in secure space using wireless sensor networks. It is able to identify intrusion, send alerts and then track the object of interest. The system comprises of Infra-Red (IR) transceiver for continuous surveillance of the specified location. The main scope of this chapter is that the information technology world, organizations move their business to outsource through cloud. This can be used at specified locations of the defense, army and societal need locations such as hospitals, Houses, organizations etc. The military and environmental surveillance based on vibration sensing using an array of micro machined accelerometers or microphones has attracted the attention of several groups.

Section 4: Applications in Electronic and Digital Technologies

This section discusses various applications of IoT, big data and smart cities. Chapter 8 focuses on use of technology in Traffic management. It has discussed about the use of data coming from different sources in real-time and processing of this data to take immediate decisions. These decisions are the keys to control traffic management successfully in our cities. It is the need of the hour to leverage the enormous amount of data around us and create a more meaningful and smooth living for us.

With billions of smart devices already in use, the IoT has seen a rapid acceleration. IoT is also proving to play a pivotal role in health monitoring due to its low-cost infrastructure, internet connectivity, and by allowing the ease of access to health data from anywhere. Chapter 9 has sought to first highlight the complexity and challenges in healthcare systems. The chapter then proposed cloud-based solutions to build an enterprise data platform that integrates with IoT devices. Finally, big data solutions and smart applications are presented for both healthcare systems and patients. The chapter concludes by discussing great potentials of IoT. It also illustrates new research challenges faced by rapid growth of multitude of smart devices and the potentials for big data analytics with potential integration with healthcare data.

Chapter 10 presented the design and development of Smart Waste Management System that is one of the primary components of a smart city. The Web Portal application that applies the MQTT and CoAP protocol for communicating data from the bins is presented. The usage of IoT Cloud, Data analytics and Visualization makes the decision making effective. This system avoids human intervention, saves human time and effort and provides a smart technology for waste management that leads to a healthy and pollution free environment. This system can be implemented in a smart city, where the public can be benefitted.

Section 5: The Case Studies – Urban Area in Context of Smart City Infrastructure

Chapter 11, primarily focused upon transportation sector in Pune and issues surrounding therein, wraps up with the finding that the city has a long way to go with respect to perking up mobility. Transportation is one of the few areas where the city lags behind compared to its urban counterparts in the country.

Chapter 12 discusses and explores South Korea's Songdo smart city. This chapter gives a 360-degree view of Songdo smart city project, specifically with regards to planning, design, deployment of technologies, and especially, urban analytics. The chapter's major areas of focus are: looking into Songdo from the perspectives of its being an Aerotropolis and a ubiquitous city. Songdo international business district is a new ubiquitous city built from scratch on 600 hectares of reclaimed land along Incheon's waterfront. With all the modern amenities and LEED-certified buildings, Songdo will also survive in the future.

Sharvari Chandrashekhar Tamane
Jawaharlal Nehru Engineering College, India

Nilanjan Dey
Techno India College of Technology, India

Acknowledgment

We can't always control what happens in our lives - things will go well, things will go poorly - but what we can control is our response to those events. —— Kenneth H. Blanchard

We would like to say thank you to the Almighty and our parents for the endless support, guidance and love through all our life stages. We are thankful to our beloved family members for standing beside us throughout our career, to move our career forward through editing this book. Our great thanks to our students, who have put in their time and effort to support and contribute in some manner. We dedicate this book to all of them.

We would like to express our gratitude towards all who supported, shared, talked things over, read, wrote, offered comments, allowed us to quote their remarks and assisted in editing, proofreading and design; through this book journey.

We believe that the team of authors provides the perfect blend of knowledge and skills that went into authoring this book. We thank each of the authors for devoting their time, patience, perseverance and effort towards this book; we think that it will be a great asset to the all researchers in this field!

We are grateful to the IGI-publisher team, who showed us the ropes to creating this book. Without that knowledge we would not have ventured into starting this book, which ultimately led to this. Their trusting in us, their guidance and providing the necessary time and resources, gave us the freedom to manage this book.

Last, but definitely not least, we'd like to thank our readers, who gave us their trust and we hope our work inspires and guides them.

Take a minute: look at your goals, look at your performance, see if your behavior matches your goals. —— Kenneth H. Blanchard

The Editors

Sharvari Chandrashekhar Tamane
Jawaharlal Nehru Engineering College, India

Nilanjan Dey
Techno India College of Technology, India

Section 1

Next Generation:
Smart City, IoT, and Big Data

Chapter 1
Smart and Connected Cities Through Technologies

Vrushali Gajanan Kadam
Jawaharlal Nehru Engineering College, India

Sharvari Chandrashekhar Tamane
Jawaharlal Nehru Engineering College, India

Vijender Kumar Solanki
CMR Institute of Technology Hyderabad, India

ABSTRACT

The world is growing and energy conservation is a very important challenge for the engineering domain. The emergence of smart cities is one possible solution for the same, as it claims that energy and resources are saved in the smart city infrastructure. This chapter is divided into five sections. Section 1 gives the past, present, and future of the living style. It gives the representation from rural, urban, to smart city. Section 2 gives the explanations of four pillars of big data, and through grid, a big data analysis is presented in the chapter. Section 3 started with the case study on smart grid. It comprises traffic congestion and their prospective solution through big data analytics. Section 4 starts from the mobile crowd sensing. It discusses a good elaboration on crowd sensing whereas Section 5 discusses the smart city approach. Important issues like lighting, parking, and traffic were taken into consideration.

DOI: 10.4018/978-1-5225-6207-8.ch001

INTRODUCTION: PAST, PRESENT AND FUTURE LIVING STYLE

The Needs of Remembering the Past (Preservation and Revitalization)

The history of human civilization can be traced back to the Indus valley and Mesopotamia settlements. With the civilization the concept of towns, cities and villages evolved. Whether in prehistoric or historic times, human settlements were always found near the sources of water. Cities are developed on the river banks around 3000 BC, when some of the first settlements happened in Mesopotamia, the Indus valley civilization, the Egypt's Nile river civilization and along rivers in china. As the settlements developed growing food and storing it for all seasons became a practice. This gave rise to an organized agriculture system using labor. Labor divisions led to the rise of an upper class and also the development of cities, which provided the foundation for civilization (Sun et al, 2016).

A major technological and cultural transition to technical or machine age began approximately in the 1500 century in Western Europe, and from this beginning new approaches to science and technology spread rapidly around the world, which also included earlier cultures into the industrial and technological age in to the present times.

In the later times due to industrialization a lot of rural population shifted to the urban areas or cities. Considering the current situation 50% or more population is dwelling in the city area. With reference to the various calculations by the United Nations, by 2030 the probability is that 70 percent of the population might live in the cities which expectedly is going to bring a lot of challenges regarding the energy usage and sustainability which would also disturb the social and economic balance in the communities. Hence appropriate use of the technological resources like integrated networking systems, advanced sensing & communication devices, data sources are being put in to use for proper traffic management, delivery systems, crime tracking, reducing pollution and help local authorities to govern the cities in an efficient manner. This concept constitutes the Smart and Connected Cities (SCC) which should also implement the actual motives of the wordplay Smart and Connected "Communities".

SCC believes in the concept of our need of recalling the past(preservation and revitalization),the importance and inevitability of living in the present-day(livability), and the requirements of forecasting for our future needs (sustainability).The goals of SCC are to live in present, plan for the future and remember the past.

NON-COMPUTING TO COMPUTING LIFE STYLE

Preservation and Revitalization

Considering India where our ancient civilization like the Harappa which were the well-planned towns to our cities now with excessive population and improper facilities we have been degrading our heritage. Our ancient civilization and heritages are being destroyed naturally by the decay and chemical mechanisms as well as by the changing social and economic conditions. In addition to cultural significant landscapes which is a source of identity and bonding in the communities, natural heritage the biodiversity is also being ruined. We need to preserve the traditions that come with the heritage. In modern times with the number of advanced technologies we can always revitalize the ways our cities operate. Demolishing the entire structure and building a new one would be a very expensive project and say would not be a feasible idea but working on the cities in already existing conditions to make them better with advanced facilities would implement the basic idea of revitalization. Making the cities smart and connected in terms of management of the Buildings, travel activities, Electricity, water supply which before was not monitored in terms of supply, consumption of resources, availability of resources is the aim.

Living in the Present (Livability)

We are living in an urbanizing world. Cities are accumulating importance as main business centers and hubs for socio-economic investments. For cities with such purpose, livability rankings and awards can give global recognition and can be a proper advertisement for attracting business investments.

As per certain definitions Livability is the sum of the factors that add up to a community's quality of life—including the built and natural environments, economic prosperity, social stability and equity, educational opportunity, as well as cultural, entertainment and recreation possibilities (Sun et al, 2016; "Public Policy Institute", n.d.).

A city with all the facilities at the click of the button and technologies best implemented is a great place to live. It is very progressives as well, with competitive social, economic and environmental advantages. By means of our definition, a livable city is also a healthy city, promoting liveliness, comfort, wellbeing and progress.

Livability index can be better if following facilities or developments are made:

- Restoration and Reuse of historic buildings
- Provision of more environment friendly transportation choices
- More schools and colleges in remote areas with digital education

- All of the streets, public places, junctions covered through surveillance systems
- Affordable housing facilities
- Management of policies and power investments.
- Support and value existing communities.

For example let us consider the cities like New York. With being densely populated to being the hub of investment and recreational activities it has a lot of facilities which cities like Mumbai don't have. Still their livability index is low compared to Melbourne in Australia. The reason being the New York City has lot of congestion, public transport problems, and high levels of crimes.

Hence for a city to be high with livability index it needs to organized and technologically developed in every aspect. Technologies like the mobile crowd sensing can be used to manage the traffic and get real time data. They have also implemented Adaptive Control Decision Support System (ACDSS). The system augments traffic signal patterns in reaction to varying traffic volumes. This improves traffic flow. Hence solutions to the problems like this in a way which uses the advanced technologies is the central idea of smart and connected cities which thus improve the quality of life and the livability index

The Needs of Planning for the Future (Attainability)

Our planning for future should be in a way that the existing needs of our society are fulfilled without compromising the ability of the future generations to meet their needs.

The factor of attainability rests on making and maintaining the community's economic and environmental health and promoting social justice. In order to have healthy communities we need clean air, natural resources and a non-toxic environment. The day by day increasing effects of global warming on climate and its future adverse effects have given an insight in to conditions that might arrive if we do not develop our environment and use the resources sustainably (Sharma et al, 2017).

Ignoring the issues of sustainable development has many possible penalties, such as rising sea levels, extreme droughts, erosion and loss of forests, increases in slum populations, species extinctions and loss of flora and fauna which damages the ecosystems.

And hence even though we aim at development using advanced machineries and technologies for the smart cities our approach should be how we use the resources for this sustainably without exhausting or causing any harm to the already existing resources.

How Do We Define Smart City?

Smart city can mean different things to different people. The concept of smart city can be different for different nation, cities depending on the development path the country follows or the resources available with them, the needs of their citizens and their will to have advanced smart technologies bringing a change in the old ways. A smart city is an altered concept for our country than to our neighboring countries like China. When we talk about the smart city in India we cannot give one particular definition. It depends on the cities and the variety of landscapes, culture and population they have. To consider what a smart city can mean to an urban dweller it will be scenario with huge infrastructures, advanced transport systems, education systems, and water management systems and so on. The city planners aim at creating an entire urban eco system which include economic, social, infrastructure, education and physical developments ("Smart City Mission", n.d.)

Smart city in its implementation would be using technologies for say online management of certain systems like the water usage meters or smart parking systems. Hence it would be using a lot of digital information to correspond with the services for better quality and performance.

Smart City concept can effectively implement cost reduction and help monitor the usage of resources sustainably. The applications for smart city are established with the aim of better management of the cities and for immediate solving of the real time problem encounters in the city. Various applications of smart city are shown in Figure 1.

Figure 1. Applications of Smart City

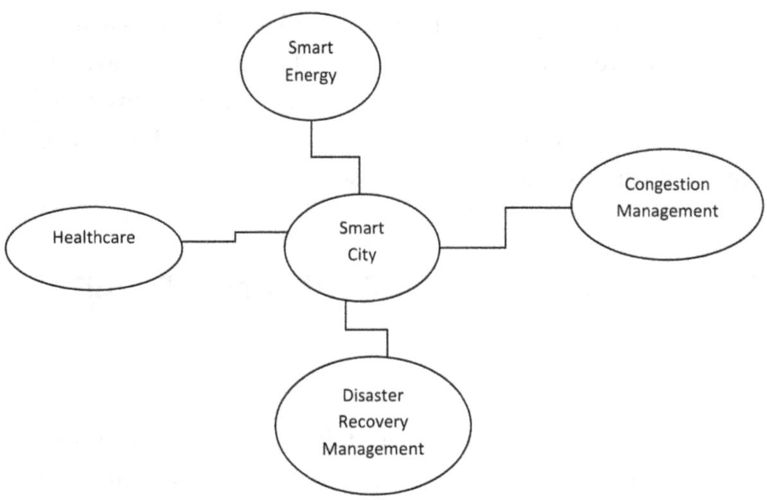

As per the on-date records in India 31% of the population resides in the city. These cities constitute63% of the nation's economic activity. These statistics are taking a huge surge with the estimations that by 2030 almost 70% of population will live in the city. Smart Cities focus on to improve quality of life for the city residents as of now and also the future. The Smart Cities Mission is a new initiative by the Government of India and it cites "To drive economic growth and make a considerable improvement in the livability index of people by enabling development locally and connecting technology as a way to create smart outcomes for citizens" ("Smart City Mission", n.d.; "Urbanisation in India", n.d.).

Smart city will always be based on the advanced software and the technologies to improve the quality of life and for proper management, considering this enormous quantity of data will be produced via different sources like sensors, social networking, video and audio, log files, and transactional data and so on. In our times majority of the people use the smart phones. Usage of social networking applications is at the highest. In 2014, according to Mary Meeker's annual *Internet Trends report,* people uploaded an average of 1.8 billion digital images every single day. That's 657 billion photos per year. Another way to think about it: Every two minutes, humans take more photos than ever existed in total 150 years ago. This is just the photos we are talking about. There are other forms of data existing like audios, videos, chats, word and PowerPoint files. The statistics of the data generated are in Exabyte (1000^6 bytes). Hence the older techniques of data storage and operations are not useful here. Hence the use of BIG data. The data if handled properly can be very useful. Not all data can be useful we need to extract the one we need with the variety of tools which help in sorting the useful data.

The whole new concept of big data analysis deals with the sorting of data and later use it for useful research. According to a source on the internet Big data analytics is the process of examining large and varied data sets i.e., big data -- to uncover hidden patterns, unknown correlations, market trends, customer preferences and other useful information that can help organizations make more-informed business decisions.

Hence, here we need to study the technical concepts like big data and IoT here to understand how the data can be put to best use.

BIG DATA, INTERNET OF THINGS AND SMART GRID

Big Data

As per the definitions Big Data, it is a "huge volume of both structured and unstructured data which is so large and complex that it is difficult to manage and process using traditional database and software tools." As per Gartner's definition, "Big Data is high-volume, -velocity and -variety information assets that demand cost-effective, innovative forms of information processing for enhanced insight and decision making".

The Four V of Big Data

The first V is *Volume*: Volume of the data or the quantity of data.(it can be anything from terabytes ($1000^{4)}$ to Yottabytes($1000^{8)}$).

The second V is *Variety*: Variety of data mean the data coming from varied sources which can be structured as well as unstructured. Due to the number of applications today data sources can be heterogeneous. Gartner used to refer to it as the dark data. As the dark matter in physics, dark data cannot be seen openly or in a straight away manner, and still is a bulk in the organizational universe.(this refers to the variety of files like audio, video, images, power point files, chats, IoT devices data).

The third V is *Velocity*: Velocity is the rate at which data flows in coming from the heterogeneous sources. The Big Data current is huge as well endless. As discussed earlier due to variety of social media applications and the increasing number of smart phones and IoT devices are bringing huge quantities of data in exa bytes per minute and hence the velocity of data is high.

The fourth V is *Veracity*: Veracity discusses the uncertainty of the data. It relates to the quality of data as well as its credibility and accuracy. It refers to the noise and abnormality in the data ("Big Data", n.d.).

Internet of Things

As per the definitions on the web, The Internet of Things (IoT) is the network of physical objects or "things" embedded with electronics, software, sensors, and network connectivity, which enables these objects to collect and exchange data. The Internet of things concept enables the physical objects in our vicinity like the temperature, pollution level to be detected and communicate with the existing network and collaborate the physical objects with the computer network system which results in getting real time information of the physical world as well as the accurate information for a better development of the technologies. The IoT concept interprets that the physical objects will be linked to the web and will be able to bond to devices which are computerized. A company named Cisco has predicted that the IoT will by 2020 have 50 billion of devices connected to the internet ("What is IoT?", n.d.).

Smart Grid

A grid can be synonymously defined as a matrix or a network. A grid can be of computers networks or roads and so on. Here we are going to study about the grid referring to the frame of electricity that holds electricity from the plant and then is distributed for consumption. A grid has components like switches, poles, substation, transformers and wires. Smart Grid however is more than just a network. Like a smart phone is more than just a calling device as it shows a calendar, calculator and has a camera and can be used for storage of documents. In short it acts like a mini computer. Similarly, a smart grid would mean computerizing the electricity network ("What is The Smart Grid?", n, d.). Figure 2 shows various aspects of a Smart Grid.

The smart grid was developed with the aim of overcoming the cons of the existing electricity supply system by using smart meters. It would help in monitoring various aspects like the voltage, faults in the network, power consumption etc. By means of the sensors on the network these data would be collected and later used for analysis. For example, the power consumption of the city changes according to the seasons, as well as when it is excessively consumed and if or not the consumption is as anticipated or is there any foul play going on can all be found out in smart grid. In short it will be intelligent and digitized system combining with the power technologies and with information systems. It is a transparent system which would cut down the losses and increase the reliability and the efficiency of the network. Maximum percentage of the power generated is provided to the users and thus it would also be an optimal use of the resources.

Advanced metering infrastructure (AMI) is a concept implemented in the smart grid. This irrespective of letting us know the consumers usage pattern or their personal information benefits the customers by giving them the power to control and monitor their consumption ("What is The Smart Grid?", n.d.; "The Smart Grid", n.d.)

Figure 2. Aspects of a smart grid

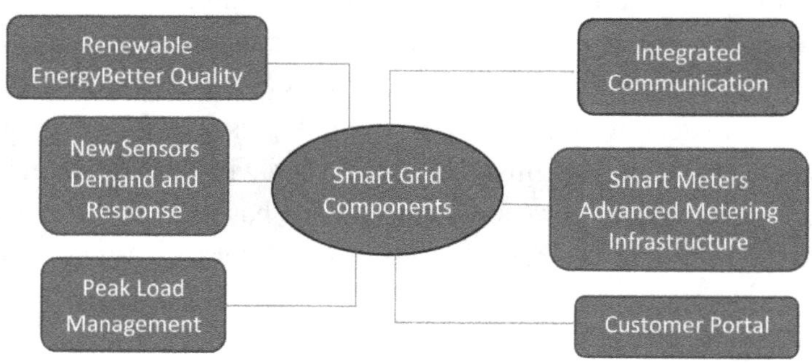

Variable pricing is another concept which also motivates the customers to save the electricity which in turn profits us by the sustainable use of the resources and avoid wastage of resources.

Due to a system which continuously 0monitors the usage transmission losses are prevented and reliable, robust grid is created.

This method is also said to reduce the carbon footprint. Hence indirectly this leads to decrease in pollution.

Big Data Analytics in Smart Grid

As discussed earlier, smart grid helps in the real time monitoring of the energy consumption and implements a powerful, intelligent electricity supply system. It will help in reduction of transmission losses, response management and in delivering quality services. It will be a key factor in enabling advanced technologies like the electric vehicles, with this we should also acknowledge the fact that smart grid will generate huge amounts of data since it is connected to a lot of devices and it gathers a lot of personal information as well as the energy production as well as usage data. The devices can be smart meters, sensing devices, measuring devices etc. ("Big Data Analytics", 2017).

This data which is collected can be put the optimum use. Other than that the phenomenal thing here to discuss why we use the big data concept is because the data collected is in huge quantities. It can be used for analysis which can give us the predictions about the future usage and requirements.

Roughly a few of the pilot projects have been implemented. With the Big Data Analytics which has huge amount of data coming via devices can be put to following use.

- **Power Outage:** Due to certain problems in the grid when the power is lost instead of the customers informing the shutdown, due to grid management the suppliers would know this beforehand. The Smart Grid in real time get the report of the outage and can document it as well as work for the immediate solutions ("Big Data Analytics", 2017; Daki, 2017).
- **Demand and Supply:** The data generated form the grid would help in analyzing the demand of the resources by the consumer and the capacity of the supplier. It would help to know the demand in peak hours and also manage the supply to be efficient in the peak hours without any load.
- **Transmission Losses:** Data collected from the sensor in the grid will help to know where the loss is taking place in the supplying process. The supply companies can then identify these losses and work to reduce the losses.

- **Forecasting:** The consumption data will be different in the different times of the years for e.g. in winter devices like AC and fan are less used and hence power consumption which will be the opposite in summer season. Similarly, it can be different for different regions. This can give a forecast on the demand of the power supply. Companies can thus work in advance and avoid any disruptions.
- **Resources Management:** With Big Data Analytics the companies can give an insight in to how the costly resources involved in the grid may fail and what at the worst the damage may cost economically. And therefore, it can design the best strategy for the maintenance. It can also help in designing strategies for managing the natural calamities and make the grid system more robust.
- **Disaster Management:** The visual analytics systems with the help of big data can be placed in the grid which can be of decision makers help to observe and examine the situations in the crisis.
- **Mobile Workforce Management:** Decision makers can better mobilize their workforce by doing real time monitoring of the systems in place on the Smart Grid. Managers can get accurate information on Fault location, real time load. People can be mobilized accordingly to fix the issue.
- **Power Theft:** The real-time metering data helps to discover the more than usual or required consumption and where energy is being stolen from.
- **Behavior Analysis of Customers:** The usage of electricity varies from season to season, day to night and different during various festivals and different across various regions. The huge amount of data available on the usage helps to know the customer behavior and prepare accordingly
- **Birds Eye View:** Due to predictive analysis complete idea of the city places and their respective features or problems are revealed. This can help for better planning of the connections which will result in the end user satisfaction.
- **Green Energy:** Since producers already have an idea of the consumption they can start pilot projects for alternative and ecofriendly energy resources which can later be expanded relating to the usage ("Big Data Analytics", 2017; Daki, 2017).

Case Study on Smart Grid Using Big Data Analytics

POWERGRID has implemented an initiative to develop Smart Grid Pilot Project at Puducherry with the Electricity Department of Govt. of Puducherry.Net metering, electric vehicle with charging through renewables, monthly tariff design and efficient use of technology are the few of many things implemented (Jha et al, 2014).

Under this project various Smart Grid presently, more than 1600 smart meters at customers place with Data Concentrator Units (DCU) & Meter Data Management System (MDMS) are implemented at one center i.e. Smart Grid Control Centre in Puducherry.

Real time monitoring of electricity usage and various alert devices with it etc. have been made possible with AMI system installed. Meters with various communication technologies have been installed. Narrow band, GPRS, broad band PLC, RF-2.4GHz, RF-865 MHz are some of the technologies.

Real time Distribution Transformer (DT) wise energy audit is conducted to minimize the blackout duration and assure efficient supply of energy to the consumers. Outage Management System (OMS) having Distribution Transformer Monitoring Unit (DTMU) and Fault Passage Indicators (FPI) have been installed and are associated with Smart Grid Control Center. The monitoring unit keeps an eye on various calculations of distribution transformers (DT) like oil level, oil temperature, load current, voltage, harmonics, palm temperature etc. Daily average Loading and temperature profile are monitored through a DTMU.A model of demand response has also been implemented It allows the user to receive the information on its consumption and it also provides facility to the customer to respond for demand management.

With this 126 street lights have been automated at Puducherry which has reduced the energy consumption of the streetlights by 57%.Renewable energy generation is the important aspect of this project. Rooftop solar generation has enabled "every consumer to become a "Prosumer" (energy producer and consumer)".Because of the rooftop solar energy customers are saving an appreciating amount on the electricity bill.

Electric vehicle which charge with Solar Photo Voltaic system which is implemented at Puducherry. Smart security, Micro grid controllers are also installed. Puducherry Electricity Department in the project area has been gaining great benefits from the successful implementation of Smart Grid Pilot. Some benefits reported by PED in the initial phase of implementation are as follows:

1. The Billing efficiency increased in response to the proper metering facility. In percentage it can be given as 14%.
2. Remote meter reading through AMI can cover huge customer base. Therefore, efficiently the billing cycles are modified for proper collection and flow of cash.
3. AMI has helped in detection of abnormal customer behavior in real time:
    a. In one case usage was observed very high by the customer. On investigating it was found that there was additional usage for lights and other sound systems for a family function.

b. Bypassing the meter by a customer immediately came to light.
c. The activity of moving the location of meter for construction was found. The customers were advised to let the center know before any such step.
d. Damaged wire sat user's place were also detected. The user was asked to change the wires.
e. Meters showing more readings than the actual consumption were detected. Such faulty meters were replaced (Jha et al, 2014).

Traffic Congestion Management Using Big Data Analytics

For Smart Cities where there are Sensors on signals, radio announcements on traffic updates, GPS trackers, mobile phones and social media posts huge amount of data is generated which is the main elements of traffic management. Predictive analysis from the data can be used by the cities to figure out how the traffic should be managed and how the congestion can be avoided. Drivers can get the updates from social media and can avoid traffic. City planners can analyze data and figure out where the transportation facilities are most needed. For now, it can be considered that the BDA is in its initial state for its complete implementation in the traffic management.

By getting the data in real time and monitoring the city can help in solving several problems like where to allow parking of vehicles, which roads are wide so that the heavy vehicles and buses can pass through? Is there any way to allow a separate path for bicycles?

Through the Big Data analytics the traffic management team can have in advance have the information of the locations and they can pass on this information to drivers and they can change their way which can also avoid accidents due to traffic congestion.

It can also save the fuel costs as well as save time for people.

Due to analytics the pattern of traffic during the day and night as well as during certain festivals can be configured

Data gathered from various sources can help measure the speed of traffic as well.

Patterns of traffic incidents can be configured through the historical as well as current data.

Sensor from the roads, the video camera can show which roads need repairing.

This all enables immediate detection of problems as well as pave a way for the effective solutions. The word "Connected" cities implies a proper co-ordination amongst the various authorities like traffic departments and the people travelling on roads which is possible through this system and covers the principle behind smart and connected cities.

Case Study on Traffic Congestion Management Using BDA

Zhejiang, China

This city in china has over 1000 installed cameras at various checkpoints for the real time visuals of the traffic. All the data time, place, vehicle information, pictures, videos are in storage at a centralized data center.

The traffic supervision team uses this data to for the real time analysis of the traffic congestions, incidents related to the traffic where the speed limits are being exceeded.

The system has following benefits:

- Due to the cameras the traffic violations information is stored in the system. The traffic department can easily know the violators vehicle number, his license information, the color model of the car also the history of the driver and his driving routes and his identity.
- Traffic law enforcement agency can without any difficulty get car plate numbers and the route of the vehicle from 2.4 billion records in the system.
- In 10 seconds the information of any vehicle can be gathered which require various investigation like from which checkpoint the vehicle passes and so on.

IOT TECHNOLOGIES

Mobile Crowd Sensing

What is crowd-sensing?

Crowd-sensing, also defined as mobile crowd-sensing is a method where a group of individuals who own mobile devices capable of sensing and computing, share data and that data is used to extract information to analyse and predict, measure or map any process which would benefit everyone around. It can also be called as crowdsourcing of sensor data from mobile devices (Ganti et al, 2011; "What is Mobile Crowdsourcing", n.d.)

These days most smartphones can sense light, noise, movement and many such things. These sensors collect huge quantities of data that are very useful in many different ways. For e.g. microphone data can be used to measure noise pollution and GPS and accelerometer can be used to locate the potholes on the roads.

The definition of mobile crowd-sensing was devised by Raghu Ganti, Fan Ye, and Hui Lei in 2011. Mobile crowd-sensing has 3 main categories: environmental (checking contamination), infrastructure (finding buildings or places feared to be collapsed or problems with roads like potholes), and social (tracking the data of how many people exercise).

On the type of involvement by the users crowd-sensing is divided in 2 types:

1. **Participatory Crowd-Sensing:** The users knowingly and at their own free will participate in contributing information.
2. **Opportunistic or Community Crowd-Sensing:** Here the data is sensed, collected and shared automatically without the knowledge of the user and without any of the user's contribution.

Process of Crowd Sensing

There are 3 main steps to Mobile crowd sensing: Data collection, Data storage, Data upload.

Data collection uses sensors available over the IoT kit. There are 3 key methods for collecting of the data.

- Data is collected manually by the user of the device. For e.g. by clicking the pictures or by using certain applications.
- Some data is collected automatically like when the user uses the mobile application.
- Data sensing which is caused by certain situation which has already been defined like when a user is in a certain place at a certain time.

MCS Applications

MCS applications are classified into3 categories. These are divided on the basis of phenomena which are being charted. This includes: Environment, Infrastructure and Social.

In environmental MCS applications, the singularities are those of natural atmosphere like quantifying pollution ranks of the cities and the water level. These applications sallow the mapping of numerous ecological occurrences by the participation of the public. An instance for the application is Common Sense uses specialized handheld air quality sensing devices that communicate with mobile phones (using Bluetooth) to measure various air pollutants (e.g. CO2, NOx). These devices when implemented across a large population, measures the air quality of a community or covers a larger habitat. Similarly, one can use microphones on mobile phones to monitor noise levels in environments.

The other model is Creek Watch. This is developed by IBM Almaden Research Center. It checks the water levels and quality in creeks by collecting reports from people through the images taken by them when they visit certain locations along the creek, or text messages or articles about the amount of garbage. This information can be used by the water control boards for keeping a check on the pollution levels in water resources.

Infrastructure applications include calculation of occurrences related to public infrastructure of which instances are assessing traffic jamming, road settings, parking availability, outages of public and real-time transit tracking.

The MCS application of this includes MIT's Cartel and Microsoft Research's Nericell. Cartel uses specialized devices installed in cars to measure the location and speed of cars. It then transmits the measured values using public Wi-Fi hotspots to a central server. This information stored at the central server is to be inquired to provide information such as the roads with minimum traffic.

Nericell employs cell phones to conclude average speed or the delays occurring due to the traffic, and also the honking levels which can be a measure of pollution or the potholes on the roads.

Other illustration is Park Net. It is an application that identifies the available parking places in cities. This is done by the ultrasonic sensing devices which are mounted on cars having smart phones.

The next track is that of social applications, here people share sensed information in their groups. For example, an individual can share his or her exercise data and compare their exercise levels with the rest of the people in the group. This can help the to be competent and improve their exercise structures. Examples of such applications are Bike Net and Diet Sense.

In Bike Net, individuals measure location and bike route quality and combine the data to obtain best route for biking. Diet Sense individual stake pictures of what they eat and share it within their group to compare their eating habits. This can be extremely useful for the people detected with diabetics.

Case Study

Mobile crowd-sensing is a very important concept in smart city. All the data gathered can be used for a better city planning. Crowd sensed information show urban areas where e a lot of people spend their free time and also the types of physical activities they do as well as the physical environment which can be useful to the urban planners to establish certain business and recreational centers according to the user's mobility in environment.

Figure 3.

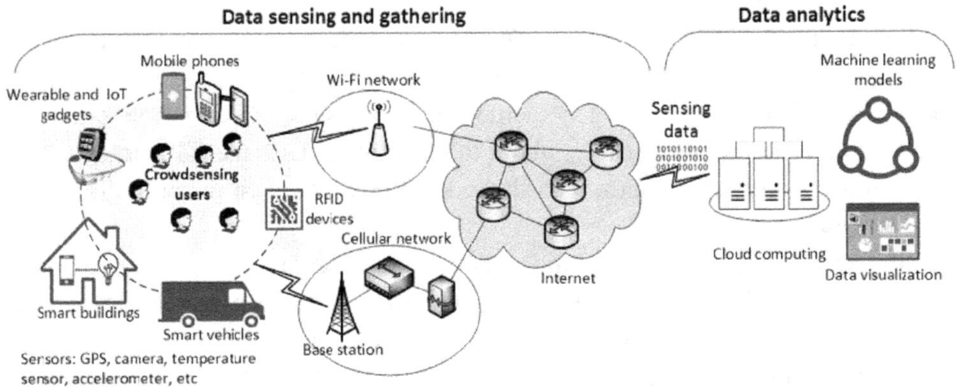

Mobile crowd-sensing is a very important concept in smart city. All the data gathered can be used for a better city planning. Crowd sensed information show urban areas where e a lot of people spend their free time and also the types of physical activities they do as well as the physical environment which can be useful to the urban planners to establish certain business and recreational centers according to the user's mobility in environment. The Urban Sense mobile application has been developed as a mobile diary service that detects and stores information about users' environment and mobile interaction by collecting following data:

- Physical environment: temperature, air pressure, air humidity, light intensity, etc.
- For the telecommunication purpose like texting someone or calling someone
- The manual work done by the phone user.
- Wireless communication devices in surrounding, e.g. Wi-Fi access points, cellular network base stations, Bluetooth devices, other mobile devices, etc.
- The number of time the phone screen goes on and off the more active the user is. This is tracked.
- Which mobile applications are most used? Also, among that which social media application is used the most
- Recording of photos/videos using phone cameras.
- The audio or the video files played
- What is the current battery level, how long it might last?

The Urban Sense demo application consists of a mobile service assigned with collection of sensing data, a mobile application used for service configuration and to get mobility information as well as a server which receives, stores, aggregates, processes and analyses crowd sensed data (Stojanovic et al, 2016).

Let us explain it through the application like WhatsApp. Every day at a particular time the backup of chats is taken at a set time. This is then stored and sent to the central server that WhatsApp has. A service provided by WhatsApp can filter which type of information to collect for the privacy preservation of the user. And this way the data is put to further possible best use. The personal or the sensitive information of the user is dropped to protect the user's privacy. This includes the contact numbers of the users and the message content. The data collected before being sent to the server is filtered by the user in mobile applications. The collected data is processed suing the spatio temporal clustering techniques and social activeness measured with the voice call history. After clustering, urban areas with intensive physical and communication activities can be visualized on a map and filtered according to the type and the period of day/week. The foremost objective of Urban Sense is to find areas with great intensity of social activities .and use this for urban planning. Urban Sense application does (Stojanovic et al, 2016).

1. Data collection
2. Filtering of collected data
3. Spatio-temporal clustering of collected data.

The mobility information of the users can also be useful in extremely complex situations like the traffic congestion, accidents, natural calamities etc.

HOW IS INTERNET OF THINGS (IOT) USED TO PROVIDE A UBIQUITOUS NETWORK OF CONNECTED DEVICES AND SMART SENSORS FOR SCC?

According to Wikipedia, the *Internet of things (IoT)* is the network of various physical devices, vehicles, home appliances and other items embedded with electronics, software, sensors, actuators, and connectivity which enable these objects to connect and exchange data. Each item is given a unique identity, and it's easy to identify by others through the embedded computing technology also, it has good capacity to connect along with the help of sensor or, actuators and different networking devices. (Devi et al, 2017) In the same line of study, A Smart city (SC) is an urban area that uses different types of electronic data collection sensors to supply information which is used to manage assets and resources efficiently. The name of smart city, as it indicates, means it has capacity to bring a new and improved life style to human beings. The automation is nice and it includes data collected from peoples, associated devices, and further assets that is deployed and help in analyzing to monitor and manage variety of services which is very important for modern trends working persons, like traffic and transportation systems, power plants, water supply networks, waste management, law enforcement, information systems, schools, libraries, hospitals, and other community services.

The emergence of smart city, and increase dependence on IoT, give the integration with information technology as well as electronics and communication engineering, even we can say it's the transformation with information and communication technology (ICT), It further catches, the numerous hardware equipment's which are connected with the IoT to optimize the efficiency of city operations and services and connect to citizens. Smart city technology allows various department members to get update the information through technology and decision with the scenario as and when it required to take, in order to avoid any long delay, and provide luxury and relax to the citizens who are availing the smart city facility.

In the past almost one decade the study, working and implementation on smart city is getting speed, because it's obvious that as the resource are getting limited, the discussion on working with smart city is taking good speed, and now even the Government of India is identifying few places in every state, where they are planning to give smart city facility to their residents.

During past two decades, the resources are getting shorter and limited, while the population is catching speed. In that scenario, it's very useful to think about the smart city and its association and implementation can be useful. The technology is the boon as well as bane, so we must ensure that we give benefits to the citizens who are facing a lot problem due to limited resources. A city which has a dynamic relationship with its citizens to provide the support is always in demand for all. That, we can call as understanding of a smart city.

Why we want to adopt Smart City is open discussion for all, but in lieu of existing system, it's going to be a buzz word "Smart city" for the coming time. It helps in saving conservation, it supports automation as well as its going to save the resources wastage, so it can be utilized for the use for others, so we are confident that it's going to give the new insight to coming generation.

Smart Lighting

The electricity is main source to give lighting, also to support number of electronic appliances. It's widely used and at large scale mostly everywhere and even it gets shortage of supply during the peak weather, in hot and cold as the appliances are required more electricity. Even in terms of back-up the generators are used to continue the use of the electric appliance to give light and another appliance supports. In the smart lighting, it utmost take care to save the energy. The lighting is controlled as the requirement and it save the excess use of lighting but making it switch off when it's not in use. Instead of manual approach, we use to take help of the sensor, who actually takes care to control the light. Even the use of light in low, medium and high is also utilised by the help of smart lighting system. With the sensors, in smart city, it possible to

Smart Parking

The places are getting reduced and the vehicles are getting increased as the year by year populations is growing, Now in the place of simple parking multi-level parking even is also a challenges, so no place for vehicle parking, It has been widely seen the lot of parking lot is get consumed or not in used by other vehicle driver due to ill-mannered parking and ill maintenance of the parking places, keeping in the same view, a sensor base parking is now a day's installed in the smart city, which has capacity to provide the smart and systematic parking scheme to the vehicle. With the help of that even its very easy to map the vehicles by driver also it gives them help to track the vehicle, which help them to avoid theft and many other vehicle related issues.

Smart Waste Management

The smart city are also adopting the smart use of waste management, as we seen generally the ratio and work for waste management is purely unequal, so with the help of smart city and IoT we can systematise the waste management system, where system will generate the alert to the main office based on the amount of waste, also it will give timely updates to the office about the cleaning issue and it will help us citizens to make fresh and happy life. The number of technique can be pulled to ensure the waste management system correction and it will help the smart city project to attract more people to give adoption from the existing system.

Smart Traffic Management System

IoT can help in smart traffic management by installing battery/solar powered wireless traffic sensors which can detect real time traffic volume, road occupancy and average travel speed of vehicles by road or lane. It can also classify them by sizes and speed categories. These sensors can be connected to the central traffic management system of the city, giving a complete picture of the present traffic situation. Based on the traffic pattern, the advanced decision-making algorithms can automatically adjust the duration of traffic lights, generate complaints against traffic rule violators and the smart cars can be connected with this system to automatically find the optimum route to a destination. Needless to say, better traffic control has a very significant positive impact on the environment with reduced emission of harmful gases. Moreover, intelligent signage's can change their display text to appropriately show the right direction to reach a specific location on the basis of real time traffic situation e.g. congestion/accident etc.

Smart Environment Monitoring

The IoT is also a useful to address about the information to collect about the environmental hazard Like the variety of sensors could be installed in the various prime location, where there is huge traffic and the machine can be trained to measure the various toxic chemicals which are very dangerous to our health like electromagnetic field, noise, temperature/humidity, CO2/toxic gases, combustion gases to detect fire. The various data can be taken in the database and a concrete decision could be taken to bring the hazardous chemical down, even it will help to give smart city, by taking preventive steps to ensure good air and hygienic facility. The less is the toxic chemicals in the environment, the lesser people will affect with related dieses symptoms.

ADDRESSING THE CHALLENGES OF SMART CITIES

Our market trials and use cases span smart metering, parking, energy, public spaces (like bus shelters), smart city private networks, crowd monitoring, and more. Our solutions help you bring the IoT to life for your city.

- **Financing:** Smart cities project is not smartly privileged, unfortunately, when it comes to funding. Financing is said to be one of the biggest challenges when it comes to the smart city challenge. The total investment approved under the smart city plans of 90 cities has gone up to Rs 1, 91,155 crores (Source: Business world). Now, that's whopping. With the presence of state-sponsored companies also the project seems to have no good start. Banks financing these projects as of now is the major reason of a considerable increase in the number of non-performing assets. The government is recently taking steps to finance these projects by making changes in the budget and we hope the problem is addressed to soon.

- **Lack of Centre-State Co-ordination:** Fruitful implementation of a project can be done only if there is a co-ordination between various government bodies. There is a need of proper regulation when it comes to planning for the development of smart cities. Both horizontal and vertical co-ordination is the requisite right now.
- **Availability of Master Plan:** Most cities in India do not have their master plans and development plans in place. This is a tragic situation if we talk about developing them into smart cities. The presence of both the requisites is the key to the implementation and encapsulation of the smart city project as that is where the changes would be monitored and there is no other way to make it simple, better and efficient. Unfortunately, most cities in India lack the presence of it.
- **No time Figure Attached to the Plan:** The entire smart city plan is a one big plan which should get all the clearances if not before time then on time. Everything should be online and timely which unfortunately is not happening in this case. The most important step to be taken in this context would be setting up a single regulatory body which monitors all the requisite approvals for the project. Doing this will address two major issues one of co-ordination and one would be the timely execution. Also, the body should be solely responsible to cater to the financial requirements.
- **Availability of Facilities:** We are very much aware of the unfortunate fact that India as of now is not that equipped when it comes to skilled manpower and advanced technology requirements for developing 100 smart cities. That is a huge number and requires lot of skilled efforts. If we talk about creating skilled labour and capacity building, not much funds have been allocated by the center and state in such initiatives. Such projects involve training, research and a hefty database for execution. This is a huge problem in our country as it is an area which has not been focused upon as of now. These programs help in many ways like time bound completion.
- **Corruption:** This point probably was meant to be from the first as this is the root cause for all above challenges. But if we talk about it solely this is also a major challenge. Both at center and state level corruption is responsible for all the co-ordination mismatch and time lag happening. The financial constraint also somehow creeps in because of this issue. Corruption in India is a challenge which has always been a reason for non-execution or ineffective execution of most big projects in the country.

THE PILLARS OF THE SMART CITY

The building blocks of smart city are in emergence, but if we closer look to that, and also we can say challenges, we say are in term of cost. In context of we seen our country, India, where government has started the smart city project in each state, with parallel multiple location. So the involvement of the PPP Model, public private model partnership is growing, because the involvement of the cost is high, and in order to reduce the costing and improving the liability, the PPP seems to be successful venture for improving the status of existing living places. In case we need to re-change, then it gets quite tougher, so keeping in view of cost and time, there are number of issues still pending for debatable and researching.

The biggest issue that 'Smart Cities' have is that there are vast amounts of money spent on relatively benign work. Remodelling of landmarks or small changes that could be classified

The design, development of Smart city and its execution cost a high volume of budget, it requires the inclusion of technologies, mobile appliances, sensors and many more technologies, which together make a platform, where we find the smart city dream true.

Further in term of Networking, as wired and wireless, wherever as require depends on situation also need to configure and must be deployed as per client –server architecture, which require a depth sketch study for automation services with the attached instruments.

In the deployment with network for protecting the network file system, prefer to adopt the clouds management system, which give services to anywhere everywhere, its connected to be part of network.

The IMPACT (Intelligent Management Platform for All Connected Things) IoT Platform, provides the necessary layers for connectivity management, application enablement and device management which is secure across all endpoints and enables you to use data and analytics to create value for your city.

The hi-tech security infrastructure is deployed to ensure the authorised user only enable to facility and get smart city services, it will enrol and permit only the authorised user to avail, also to give extended support which is step-by-step is designed for the secure smart city project.

Now number of IT firms s are also working to develops the smart city, E.g., we can say that Nokia enterprise IP communications and collaboration solution provides an open, agile, and incremental framework that lets city administrators take control of their communications services.

REFERENCES

Big Data. (n.d.). Retrieved from: wikipedia.org/wiki/Big Data

Big Data Analytics, Machine Learning and Artificial Intelligence in the Smart Grid: Introduction, Benefits, Challenges and Issues. (2017). *IEEE Smart Grid Big Data Analytics, Machine Learning and Artificial Intelligence in the Smart Grid Working Group*. Retrieved from: https://smartgrid.ieee.org/images/files/pdf/big_data_analytics_white_paper.pdf

Daki, H. (2017). *Big Data management in smart grid: Concepts, requirements and implementation*. Big Data, Open Data and Data Development. doi:10.1002/9781119285199.index

Devi, T.J.B., Subramani, A., & Solanki, V. (2017). *Smart City: IOT Based Prototype for Parking Monitoring and Management System Commanded by Mobile App*. Volume 10. Poland: ACIS. doi:10.15439/2017R23

Dhall & Solanki. (n.d.). An IoT Based Predictive Connected Car Maintenance Approach. *International Journal of Interactive Multimedia and Artificial Intelligence, 4*(3). doi: 10.9781/ijimai.2017.433

Effects of global warming. (n.d.). Retrieved from www.livescience.com

Ganti, R., Ye, F., & Lei, H. (2011). Mobile Crowd Sensing: Current State and Future Challenges. *IEEE Communications Magazine*, *49*(11), 32–39. doi:10.1109/MCOM.2011.6069707

Jha, I. S., Sen, S., & Kumar, R. (2014). Smart grid development in India — A case study. *2014 Eighteenth National Power Systems Conference (NPSC)*, 1-6. 10.1109/NPSC.2014.7103866

Mobile Crowdsensing. (n.d.). Retrieved from: wikipedia.org/wiki /Crowdsensing

Public Policy Institute. (n.d.). *Livable communities*. Retrieved from: www.aarp.org/ppi/issues/livable-communities/info-2015/

Sharma, P., & Rajput, S. (2017). *Sustainable smart cities in India: Challenges and future perspectives*. Cham: Springer International Publishing. doi:10.1007/978-3-319-47145-7

Smart City Mission. (n.d.) Retrieved from: http://smartcities.gov.in/content/

Solanki, V. K., Venkatesan, M., & Katiyar, S. (n.d.). Conceptual Model for Smart Cities for Irrigation and Highway Lamps using IoT. *International Journal of Interactive Multimedia and Artificial Intelligence, 4*(3). doi:10.9781/ijimai.2017.435

Solanki, V. K., Venkatesan, M., & Katiyar, S. (2016). Think Home: A Smart Home as Digital Ecosystem in Circuits and Systems. *Circuits and Systems, 7*(8). doi:10.4236/cs.2016.78172

Stojanovic, D., Predic, B., & Stojanovic, N. (2016). Mobile crowd sensing for smart urban mobility. In European Handbook of Crowdsourced Geographic Information (pp. 371–382). London: Ubiquity Press. doi:10.5334/bax.aa

Sun, Y., Song, H., Jara, A. J., & Bie, R. (2016). *Internet of Things and Big Data Analytics for Smart and Connected Cities*. IEEE.

Tamane, Solanki, & Joshi. (2017). The Basics of Big Data & Security Concerns. In S. Tamane, V. K. Solanki, & N. Dey (Eds.), *Privacy and Security Policies in Big Data*. Hershey: IGI Global.

The smart Grid, department of energy. (n.d.). Retrieved from: energy.gov/sites/prod/files/ oeprod/

Urbanization in India. (n.d.). Retrieved from: wikipedia.org/wiki/Urbanisation_in_India

What is IoT. (n.d.). Retrieved from: www.happiestminds.com

What is the smart grid. (n.d.). Retrieved from: www.smartgrid.gov

Chapter 2
A Collaborative Approach of IoT, Big Data, and Smart City

K. Jayashree
Rajalakshmi Engineering College, India

R. Abirami
Rajalakshmi Engineering College, India

R. Babu
Rajalakshmi Engineering College, India

ABSTRACT

During the last two decades, a number of new nations emerged and played their intense role in changing human lifestyle. The growing demand for smart city and big data stimulates innovation, and the development of new smart applications is becoming important. Internet of things comprises billions of devices, people, and services, and entitles each to connect through sensor devices. The economic development of a city leads to better life quality and improved citizen services. Thus, this chapter discusses the background of big data, IoT, and smart city. It also discusses the collaborative approach of all the above. The various related work and future research direction for implementing smart city with the concept of big data and IoT would be addressed in this chapter.

DOI: 10.4018/978-1-5225-6207-8.ch002

INTRODUCTION

Big data can be defined using five characteristics such as volume, velocity, veracity, variety and value and the major advantage of big data is, it stores numerous amount of data. Processing these large data are more complicated even when different processing tools are used. Millions and billions of data are created everyday by people be it on any social networking sites, finance, medicine, sensors, mobile applications and so on. due to the massive amount of the biological information, big data in biological processing become a common phenomenon in current industry and laboratories (Kamal et al, 2017). To manage, analyze and process all these data new processing methodologies, techniques and tools have to be used (Shukla et al., 2016). To improve the living standards of citizens government has come forward to make use of big data technology and smart cities is one of the project which is adopted by government to reduce the cost and improve the ease of living of citizens.

IoT connects all sensing, actuating and computing devices, collects the data and communicates it to various other devices using Internet (Yasir Mehmood et al., 2017). The major benefit of IoT is it connects people and things anywhere, anytime anyplace with anyone and everyone using internet. Applications like smart classrooms, smart cities, smart health are slowing emerging in software scenario. Mostly all companies are curios in developing projects which proves to improve the lifestyle and quality of citizens. In smart cities, analytics and IoT together play a vital role. These two approach helps to improve the infrastructure, transportation, parking system by reducing the traffic congestion, surveillance, electricity, health care, managing wastes by collecting the data from various devices like sensor, cameras connected in the cities. (Silva et al.,2017) Government of India has taken up new mission called smart cities, in which it aims to develop hundred cities all over the country with quality services that uplifts the quality of living and lifestyle of citizen.

Smart city applications generate huge amounts of date while big data systems utilize this data to provide information to enhance smart cities applications. The big data systems will store, process, and mine smart cities applications information in an efficient manner to produce information to enhance different smart city services. A smart city is one of concepts that argue intellectual and social capital should be considered alongside with city's tangible assets at the time of measuring the urban performance. Moreover, a city is considered to be smart when human and social capital with modern information work in conjunction to fortify the economic development of a municipality. A smart city uses digital technologies to enhance quality and performance of urban services, by carrying out the track of ill practices such as power theft and maintain security in public places (Solanki et al, 2016).

Section 2 introduces the background of IoT and big data. Section 3 discusses the related work based on big data and IoT in smart city. Section 4 explains the future research directions and Section 5 concludes the chapter.

BACKGROUND

IoT has become popular in connecting the devices with devices and with people every-second and every day with the help of low-cost wireless sensors like radio-frequency identification sensors and technologies like web technologies. These sensors and approaches are useful in collecting and analyzing the data all over the world. As the result the enormous volume of data will be collected from multiple sources. Big data handles and organizes these huge volumes, multiple forms, and rapidly changing data through various statistical and machine learning approaches (Kune et al., 2016).

To improve the economic growth and quality of life of citizens government of India has started a new initiative called Smart cities mission. In smart cities, data are generated in real time and in large volume and are collected from various sensors, devices, audio, video, networks, log files, transactions, web and through social media. To handle or to process enormous data the following terminologies are to be known: Data integration, Aggregation, Validation, Cleansing, Data anonymization and visualization, encryption, shuffling, substitution, masking and data variance.

Big data can be helpful in smart cities to extract and analyse the information from a large volume of data. In-order to extract the data from huge data the data must be explored and processed at the right time which will be achieved with proper collaboration of people,process and technology. Therefore big data which usually deals in analyzing the data can be very useful in various departments of smart cities. Big data analytic methods like edge analytics, semantic analytics, security analytics and predictive analytics can be used to analyse,process and store the data collected from different smart city applications. (Dey et al., 2018).

In smart cities data are gathered and stored in various forms, be it structured, unstructured or even semi-structured from various fields from personal to global and from various sources like websites logs, Enterprise system, GPS etc. It is necessary to understand and organize the dataset. Dataset gathered from different resources are integrated and stored using different technologies. The gathered information's are summarized and presented in the form of report this process is called data aggregation. During data validation the report is examined and data that are not relevant and data that are repeated can be removed (Sanjeev Kumar et al., 2016).

Gathered data are analyzed and transferred among different departments based on their needs. Information has to be protected and should not be misused by any of the users and at the same time the information has to reach the correct user in stipulated time, so there comes the concept of data anonymization. Data encryption, shuffling, number and date variance, truncating, masking is some of the techniques that comes along with data anonymization and visualization. In data encryption the information is encoded and transferred in unreadable form hence it achieves data confidentiality. The receiver has to decode the data to understand the information clearly. The difficulty in encryption if information or data contains different format of data, like dates, numbers then receiver has to properly decode the data first before validating it, otherwise wrong information will be transferred. Next technique is substitution, where the data will be substituted with a set of predefined symbols and letters. In shuffling certain algorithms are used and data are shuffled in random manner and substituted. But this technique cannot be used on small amount of dataset and if the algorithm is known then the information can easily be identified by intrudes.

The next technique is date and variance where the certain algorithms apply formulas to convert the data. The algorithm makes use of some random percentage on each original value in the column and changes the data so that the original data are untraceable and are protected. Nulling and truncating technique eradicated sensitive data and replaces it with null values. These techniques will not be used on vital information such as customer address, name and other contact details. In masking the important information are masked and hidden from the end users. A very good example for masking is credit and debit cards numbers, certain numbers are replaced with 'X' the important information are masked and protected.

Benefits of having a smart city

1. Effective resource utilization: Technological systems like Enterprise Resource Planning [ERP], Geographical Information System [GIS] can be used to improve the utilization of resources as they are becoming scarce and more expensive. Certain monitoring systems are used to monitor the resources repeatedly to distribute the it evenly and to control the cost. Smart cities are mainly designed for inter-connectivity and data collection to increase the communication among various application and services.
2. Better quality of life: Smart city citizens have better quality of life as all the services are provided more efficiently and quickly to everyone. This is achieved by better planning and availability of information, locations, by using better models for analyzation and to make decisions.
3. Higher level of transparency and openness: Data resource and sharing are the key aspect for achieving high level of transparency and openness. Furthermore, it helps in creating more applications and services to upgrade the quality of services offered to the citizen in smart cities. (Shukla, 2016).

APPLICATION OF BIGDATA IN SMART CITY

Big data techniques are utilized in the field of IoT, especially in smart cities to enable sufficient storage and processing of data which can improve the quality of various services in smart city. Certain big data tools and methodologies has to be used to achieve the efficient and effective analysis of data. Techniques and methodologies helps to improve the communication and collaboration among different entities and service providers in smart city which also improves the customers experience and increases the business opportunities . This helps decision makers to expand the resources and services of smart city.

The following application of big data in smart cities are listed out:

- **Smart Health Care:** There is a rapid increase in the health care data day by day (Demirkan, 2013). This is due to rapid increase in the population, which lead to different models of treatment techniques, and many decisions where taken based on previous data. Therefore proper tools has to be used to analyze, collect and maintain patients data. These data can further be used by any insurance agencies to help the patients. Good analyzes of huge health care data may result in early prediction of diseases, epidemics and cures which could improve the health condition and save the life of the people. IoT makes use of special smart equipment's which monitors, processes, senses the patients health based on physical and biomedical parameters and communicates doctors and nurses about the patients health. (Elhayatmy, G et al., 2018). This can be achieved by using special equipment's which monitors patients health behavior in home or surroundings. This can help doctors to identify the warning signs of serious illness during treatment which would save life of the patients even in a critical condition.(Roy et al., 2007).
- **Smart Transportation:** The number of accident issues can be reduced by improvising smart transportation system in cities with the help of IoT and big data. By analyzing the patterns acquired by huge amount of traffic related data from sensors can help to minimize congestion of traffic by rerouting in certain areas during peek hours, In addition to it analyzing the previous history of accidents due to factors like driver speed, helps to reduce the accident rate. Information collected from transport systems can help to optimize freight and ship movements. Smart transport system helps to increase the safety and provides more benefits to the citizens of the country. Smart transportation along with wireless communication technology will help in emergence of Intelligent Transportation System (ITS) which will take the smart city project system to the next level. (Yang et al., 2018).

- **Smart Traffic Lights:** The other aspect of transportation is having smart traffic signals and lights. In-order to identify the traffic patterns traffic signals and lights has to be interconnected with across various traffic grids placed in different areas. Sensor detects the traffic flow which includes parameters like speed of vehicles, traffic jams, traffic density waiting time at signals. The system makes decision based on these parametric data and provides suitable instructions to the signals and lights. Accurate decision will be taken only when the all data from different location are collected. Hence, to provide the best service a good intelligent decision making system which collects, stores and analyze data from all traffic signals has to be build.
- **Smart Education:** Smart education will keep people involved in active learning environment which makes their knowledge to be updated on recent technologies and information's. The data collected from various resources need to be updated clearly and correctly to provide the information relevant to the topic searched, which will increase the teaching learning process. The technology like smart boards, videos, internet and so on, need to be accessed everywhere at anytime including remote and rural areas where schooling facility will not be upto the mark or the status of the people is poor and where they could not make use of costly models. ICT provides education smart services that improves the education processes, effectiveness, productivity, and provides support for life long learning for all people. Collaboration of big data and ICT will help to create knowledge based society, which in-turn helps in the growth of the nation. These facilities may help to reduce the unemployment rate of the country.
- **Smart Grid:** Energy consumption is major factor in smart cities. In smart grid environment large amount of data are collected from different sources like smart meters, and decision are made based on power utilization by the users. This helps the decision makers to take decision in terms of supply level of electricity based on the demand of the users. The smart grids helps the researchers to analyze the power generation and power consumption data which helps them to predict the need of power supply in future.
- **Smart Governance:** Big data analytics plays an vital role in establishing smart government. The organization with common interest like NGO or other organization related to government can be quickly identified by data analysis and can collaborate and communicate with each other in-order improve the country's standard of living. In addition to it, with the help of big data analytics the government can identify the needs of the people in terms of education, health care etc., which helps them to implement policies which will be helpful to people (Nuaimi et al., 2015).

CHALLENGES

Smart city application connects millions and billions of devices, and uses various algorithms and techniques to analyze and process the data. (Yasir Mehmood et al., 2017). Combination of various IoT devices and big data lead to the emergence of smart services which enhances the opportunity for developing smart factories, smart farms and smart cities. (Ali Yavari et al., 2017). Due to the large increase in IoT devices, and the use of various big data applications to develop, deploy and to design smart cities several challenges and issues related to data sources, characteristics, information Sharing, Data Quality, Security, privacy, Cost, and smart city population has to be addressed (Nuaimi et al., 2015). However, to improve the services of the smart city, the collected data must be managed well.

1. **Data Sources and Characteristics:** In smart cities as data are collected from different devices from different locations, a large number of data will be gathered and stored. The gathered data or information can be either structured, unstructured or even semi- structured data. Hence big data has to handle all these data using complex models as traditional methods fails to store large amount of data. In addition to velocity, volume and variety (3 V's), validity, veracity, volatility are added to characteristics of data, as the result complex model also will not be able to handle the data. Challenges that occurs when considering smart city applications utilizing big data are data collection from various devices and sources, identifying the format of the data, categorizing and organizing the data.
2. **Data Information and Sharing**: Each and every agency and departments, maintain their own data center, and data are kept confidentially. Certain data cannot be shared due to some privacy conditions in such cases data sharing becomes a challenge in smart city as data has to be shared among different devices. Different approaches need to be followed to achieve proper information sharing and exchange among different devices and departments. In addition to its large amount of data are updated, and stored in different format every second on real time applications. Hence it is difficult to create a general format of data and extract information, directly from real time application.
3. **Data Quality:** Since data are gathered in different forms there is no standard way to retrieve and transform data. This may cause trustworthiness and data uncertainty. In smart cities if data collected from data are incorrect then analyzing it will not produce any quality data. This becomes a very major challenge of big data application.

4. **Security and Privacy:** As confidential data about people and government are collected and stored in the database, more security policies and procedures should be followed to keep the data protected from unauthorized user and from virus attack or bugs. In smart applications as data transfer through various networks there is high possibility of attacks, hence better policies have to be maintained. Privacy rights of organizations and individuals has to be clearly stated and protected. In certain cases, like medical records, bank and financial records are viewed by the individual which violates the privacy rights hence better techniques has to be followed to rectify this issue. Emerging research areas involves in authenticating the services with high performance security and trust among distributed clients in smart city (Sarkar et al.,2017).
5. **Cost:** The next major issue is cost. If the application is not developed and implemented properly from beginning, then it may lead to high cost of resources. For example, in traffic monitoring if the signals are tested in real time and if it is not implemented properly then it may lead to chaos as well as the cost of replacing the hardware as well as software will be high. Hence it is necessary to test the functionality of the application before launching it in public to reduce the cost.
6. **Smart City Population:** Cities population and data are proportional to each other. As the population increases, data also increases rapidly, hence variety of technical, social, economical and organizational issues arises. So it is the need for smart city applications to expand efficiently and develop quickly to manage the increasing volume and variety of data (Nuaimi et al., 2015).

RELATED WORK

Emergence of smart cities attracts researches and architects to develop a standard architectural model. Smart city applications paves way to variety of research methodologies starting from abstract concept like general design to detailed set of services on each layer. To design a standard architecture lot of study has to be on made data and testing module has to be applied on each and every layer while developing the application. To collect the data accurately and to process it, sensors has to be placed in multiple locations properly in smart home or in smart city. It could be better to design urban city architecture by using previous collected data from sensors, but it does not help in applying on real time applications. Hence Hadoop based ecosystem can be used to analyze the data for better designing of services for smart cities.

Matallah, H., et al., described a new model of storage and access to data in big data and Cloud Computing. The recent studies found out the the amount of unstructured collected and stored raised upto 90 percentage. Due to increases in size of data major problems like storage and data access arouse, even when various data processing methodologies, techniques and tools are used. Hence a need for storage space became a major thought for all researches. This new requirement led to cloud storage where Storage-as-a-Service became a new model to store and access the data.

Ali et al., developed an application using sensors in smartphone to improve the road safety and to reduce the road accidents. Sensors like accelerometer and gyroscope were used to detect and classify the behaviour of drivers and road abnormalities. To identify the behavior algorithms like K-NN and Dynamic time wrapping were used. This is used to reduce the accidents and improve the road safety in smart city transportation.

Cheng et al, developed an layered architecture called City data analytics Platform which is consists of data collection layer done by IoT broker, a repository to store data done by IoT agent, a processing module and server which provides communication facilities with external -sources. This architecture seems to achieve higher throughput in processing the data. There are several other architectures like SCOPE, FIREWARE are developed based on big data analytics. Nathalie et al. gave a different view of smart cities, in which IoT devices act as service providers like cloud based services. It achieved higher level of abstraction and helped to develop advanced applications by removing the certain drawbacks between IoT devices and logical world. Data can be analyzed by using certain optimization algorithms like genetic algorithm, artificial bees algorithm, swarm optimization, cuckoo search algorithm, bat algorithm, neural network, flower pollination. But all these computational intelligence algorithms can be effectively used on small data set, to give efficient, effective result as it involves various machine learning and soft computing approaches (Jin et al., 2014).

The next new approach is by embedding Geographical Information System [GIS] with the smart city application which helps to provide user friendly platforms which increases the touch and feel of the application. This requires combination of three dimensional and touch screen technologies with smart city application. Data will be represented based on users need and helps in fast decision making. Making use of recent technologies is an interesting area where various efficient devices and software applications are used to improve the quality of services provided to citizens in smart cities.

FUTURE RESEARCH DIRECTIONS

However, there are certain drawbacks and limitations associated with the concept such as

- Initial cost of implementing the technology is high
- Application of the concept depends upon the local development and population growth.
- The local government or local authorities must be supportive of the concept or else it may lead to failure of the work.

Hence the above drawbacks can be considered as future research direction.

CONCLUSION

The two most important concepts are Smart city and big data. Combining them together will help to achieve capability, effective governance, improved lifestyle, better flexibility and intelligent management of various resources. (Al Nuaimi et al., 2015). Big Data can be seen in all finance and business where massive amount of stock exchange, banking, online and onsite purchasing data flows through computerized system every day and are then captured and stored for inventory monitoring, customer behaviour and market behaviour. Big data applications have the potential to serve many sectors in a smart city. In addition, the big data will help decision-makers to plan for any expansion in either smart city services, resources, or areas. Thus, this has discussed about the challenges involved in using big data and IoT in smart city.

REFERENCES

Aguilera, G. (2013). *An Accelerated-Time Simulation for Traffic Flow in a Smart City*. FEMTEC.

Al Nuaimi, Al Neyadi, Mohamed, & Al-Jaroodi. (2013). Understanding the value of (big) data. In *Big Data, 2013 IEEE International Conference on*. IEEE.

Ali, A. H., Atia, A., & Mostafa, M. S. M. (2017). Recognizing Driving Behavior and Road Anomaly using Smartphone Sensors. *International Journal of Ambient Computing and Intelligence*, *8*(3), 22–37. doi:10.4018/IJACI.2017070102

Bertot, J. C., & Choi, H. (2013). Big Data and e-government: issues, policies, and recommendations. In *Proceedings of the 14th Annual International Conference on Digital Government Research*. ACM. 10.1145/2479724.2479730

Cheng. (2015). Building a big data platform for smart cities: experience and lessons from santander. *Proceedings of the 4th IEEE International Congress on Big Data*.

Dey, N., Hassanien, A. E., Bhatt, C., Ashour, A., & Satapathy, S. C. (Eds.). (2018). *Internet of Things and Big Data Analytics Toward Next-Generation Intelligence*. Springer. doi:10.1007/978-3-319-60435-0

Elhayatmy, G., Dey, N., & Ashour, A. S. (2018). *Internet of Things Based Wireless Body Area Network in Healthcare. In Internet of Things and Big Data Analytics Toward Next-Generation Intelligence* (pp. 3–20). Springer.

Fan, W. (2013). Mining big data: current status, and forecast to the future. *ACM SIGKDD Explor Newsl., 14*(2).

Gil. (2016). Internet of Things: A Review of Surveys Based on Context Aware Intelligent Services. *Sensors — Open Access Journal.*

Islam, S. M. R. (2015). The internet of things for health care: A comprehensive survey. *IEEE Access: Practical Innovations, Open Solutions, 3*, 678–708. doi:10.1109/ACCESS.2015.2437951

Jain, A. K., Murty, M. N., & Flynn, P. J. (1999). Data clustering: A review. *ACM Computing Surveys, 31*(3), 264–323. doi:10.1145/331499.331504

Jin, J. (2014). *An information framework for creating a smart city through internet of things* (Vol. 1). IEEE Internet of Things Journal.

Kamal, M. S., Dey, N., & Ashour, A. S. (2017). Large Scale Medical Data Mining for Accurate Diagnosis: A Blueprint. In *Handbook of Large-Scale Distributed Computing in Smart Healthcare* (pp. 157–176). Cham: Springer. doi:10.1007/978-3-319-58280-1_7

Khan, B. N. S., & Han, K. (2016). Internet of things based energy aware smart home control system. *IEEE Access: Practical Innovations, Open Solutions, 4*, 7556–7566. doi:10.1109/ACCESS.2016.2621752

Khan, M. (2014). Seven V's of Big Data understanding Big Data to extract value. In American Society for Engineering Education, 2014 Conference of the. IEEE.

Khan, Z. (2013). Cloud Based Big Data Analytics for Smart Future Cities. In *6th International Conference on Utility and Cloud Computing*. IEEE Computer Society. 10.1109/UCC.2013.77

Kim, G. H. (2014). *Big-data applications in the government sector*. ACM.

Kramers, A., Höjer, M., Lövehagen, N., & Wangel, J. (2014). Smart sustainable cities–Exploring ICT solutions for reduced energy use in cities. *Environmental Modelling & Software, 56*, 52–62. doi:10.1016/j.envsoft.2013.12.019

Kumar. (2016). Role of Big Data and Analytics in Smart Cities. *International Journal of Science and Research, 5*(2).

Kune, R. (2016). The anatomy of big data computing. *Software: Practice and Experience*, *46*, 79–105.

Lee, C. H. (2013). Building a generic platform for big sensor data application. *IEEE International Conference on IEEE.* 10.1109/BigData.2013.6691559

Matallah, H., Belalem, G., & Bouamrane, K. (2017). Towards a New Model of Storage and Access to Data in Big Data and Cloud Computing. *International Journal of Ambient Computing and Intelligence*, *8*(4), 31–44. doi:10.4018/IJACI.2017100103

McAuley. (2011). The Dataware manifesto. *Communication Systems and Networks (COMSNETS), 2011 Third International Conference.*

Mehmood, Y., Ahmad, F., Yaqoob, I., Adnane, A., Imran, M., & Guizani, S. (2017). Internet-of-Things-Based Smart Cities: Recent Advances and Challenges. *IEEE Communications Magazine*, *55*(9), 16–24. doi:10.1109/MCOM.2017.1600514

Michalik, P., Stofa, J., & Zolotova, I. (2014). Concept definition for Big Data architecture in the education system. *Applied Machine Intelligence and Informatics (SAMI), 2014 IEEE 12th International Symposium*, 331–334. 10.1109/SAMI.2014.6822433

Mitton, N. (2012). *Combining Cloud and Sensors in a Smart City Environment* (Vol. 1). EURASIP J. Wireless Commun. and Net.

Mohamed, N. (2014). *Real-time big data analytics: Applications and challenges. High Performance Computing & Simulation.* HPCS.

Neirotti, P., De Marco, A., Cagliano, A. C., Mangano, G., & Scorrano, F. (2014). Current trends in Smart City initiatives: Some stylised facts. *Cities (London, England)*, *38*, 25–36. doi:10.1016/j.cities.2013.12.010

Open Source PlatformF. I. W. A. R. E. (n.d.). Retrieved from https://www.fiware.org/

Pan. (2015). An Internet of Things framework for smart energy in buildings: designs, prototype, and experiments. *IEEE Internet of Things Journal, 2*(6), 527–537.

Sarkar, M., Banerjee, S., Badr, Y., & Sangaiah, A. K. (2017). Configuring a trusted cloud service model for smart city exploration using hybrid intelligence. *International Journal of Ambient Computing and Intelligence*, *8*(3), 1–21. doi:10.4018/IJACI.2017070101

Says. (n.d.). *6.4 Billion Connected Things' Will Be in Use in 2016, Up 30 Percent From 2015*. Gart. Inc.

SCOPE A Smart-city Cloud-based Open Platform and Ecosystem. (n.d.). Retrieved from https://www.bu.edu/hic/research/highlighted-sponsoredprojects/scope/

Shukla. (2016). The big data approach towards smart cities. *The Engineering Journal of Application & Scopes, 1*(2).

Silva. (2017). *Big Data Analytics Embedded Smart City Architecture for Performance Enhancement through Real-Time Data Processing and Decision-Making*. Hindawi Wireless Communications and Mobile Computing.

Solanki, V. K., Katiyar, S., BhashkarSemwal, V., Dewan, P., Venkatasen, M., & Dey, N. (2016). Advanced automated module for smart and secure city. *Procedia Computer Science, 78*, 367–374. doi:10.1016/j.procs.2016.02.076

Stankovic, J. (2014). Research Directions for the Internet of Things. *Internet of Things Journal, IEEE., 1*(1), 3–9. doi:10.1109/JIOT.2014.2312291

Su, K. (2011). Smart city and the applications. In *Electronics, Communications and Control (ICECC), 2011 International Conference on*. IEEE. 10.1109/ICECC.2011.6066743

Yang, W., Wang, X., Song, X., Yang, Y., & Patnaik, S. (2018). Design of Intelligent Transportation System Supported by New Generation Wireless Communication Technology. *International Journal of Ambient Computing and Intelligence, 9*(1), 78–94. doi:10.4018/IJACI.2018010105

Yavari, A. (2017). ConTaaS: An Approach to Internet-Scale Contextualisation for Developing Efficient Internet of Things Applications. *Proceedings of the 50th Hawaii International Conference on System Sciences*. 10.24251/HICSS.2017.715

Zanella, N., Bui, A., Castellani, L., Vangelista, & Zorzi. (2014). Internet of things for smart cities. *IEEE Internet of Things Journal, 1*(1), 22–32.

Chapter 3
Towards a Secure, Distributed, and Reliable Cloud-Based Reference Architecture for Big Data in Smart Cities

Jens Kohler
University of Applied Sciences Mannheim, Germany

Thomas Specht
University of Applied Sciences Mannheim, Germany

ABSTRACT

Current revolutions with respect to big data affect people's everyday life more and more. New ways of living, communication, and knowledge sharing have been created under the smart city umbrella. Information and communication technologies (ICT) are used to improve urban services. Examples are car sharing, energy consumption, adaptive traffic management, etc. A closer look at these services reveals that data, produced and consumed by people (or automatically by devices), are the cornerstone for working and reliable urban services. For this, large volumes of data with a great variety have to be processed at a high velocity which is commonly known as big data. This chapter designs a secure, distributed, and reliable cloud-based reference architecture that logically separates confidential data and distributes them to various databases in different clouds. It increases the reliability of the distributed data with high-availability mechanisms to protect data against various threads (e.g., hackers, terrorists, data collecting companies, etc.).

DOI: 10.4018/978-1-5225-6207-8.ch003

INTRODUCTION

Current revolutions with respect to Big Data, Internet of Things, or Industry 4.0 affect people's everyday life more and more. An example for this is the development of the World Wide Web (WWW 1.0) to the Semantic Web (WWW 3.0). Thus, new ways of how people communicate, interact, and share knowledge with each other have been created. This also holds for recent developments under the *Smart City* umbrella, which subsumes urban development strategies towards a sustainable management of resources like space, traffic, and other infrastructural assets, but also services, e.g. municipal, governmental, health care, or industrial ones (Song, Srinivasan, Sookoor, & Jeschke, 2016) (Dameri & Rosenthal-Sabroux, 2014) (Dey, Hassanien, Bhatt, Ashour, & Satapathy, 2018) (Elhayatmy, Dey, & Ashour, 2018) (Bhatt, Dey, & Ashour, 2017) (Sarwar Kamal, Dey, & Ashour, 2017) (Yang, Wang, Song, Yang, & Patnaik, 2018). Here, new Information and Communication Technologies (ICT) architectures are required to implement and improve *urban services*: concrete examples are car sharing, adaptive traffic management, energy consumption, etc. A general goal for these *urban services* is to reach a high degree of automation (without any human interaction) to increase their efficiency and to reduce costs. A closer look at these services reveals that data, produced and consumed by people (or automatically by their devices) are the cornerstones for working and reliable *urban services*. Therefore, large *volumes* of data with a great *variety* (heterogeneity) have to be processed at a high *velocity* which is commonly known as *Big Data*.

In order to address these *Big Data* challenges, NoSQL databases (in contrast to traditional relational databases) provide appealing architectures (i.e. key-value, column, document stores and graph databases) (Sadalage & Fowler, 2013) (Harrison, 2015), especially when data are processed In-Memory, i.e. not stored and analyzed on hard disk drives (HDD), but in the faster random-access memory (RAM) of a system. These non-relational architectures weaken the hard consistency criteria atomicity, consistency, isolation, and durability (ACID) of relational databases and use the basically available, soft-state, eventual consistency (BASE) model instead. This is advantageous in especially distributed systems, as the strong ACID criteria demand a tremendous synchronization overhead, whereas BASE allows several nodes to be inconsistent for a certain time. However, processing huge amounts of heterogeneous data at great pace requires adequate (and expensive) infrastructure capabilities and this is where Cloud Computing can be taken into consideration.

On the one hand, Cloud Computing promises advantages with its five key characteristics: (1) broad network access, (2) on-demand self-service, (3) rapid elasticity, (4) resource pooling, and (5) measured services (Mell & Grance, 2011). These key characteristics are also essential in a *Smart City* infrastructure, as huge data volumes require a high bandwidth to and from the cloud, especially with a great variety of different devices (1). People using urban services should be enabled to use them on-demand without any delaying registration and approval processes (2). The infrastructure should be able to handle dynamic changing loads automatically (3), e.g. when big social events with many people take place). The concrete hardware infrastructure must be abstracted from with the usage of virtualization technologies. This enables *resource pooling*, where different hardware capabilities are jointly used to provide multiple services (4). Last but not least, the provided services must be monitored transparently for the *cloud providers* and *consumers* (5) to create reliable and trustworthy services and to establish adequate billing models.

On the other hand, data processed in a cloud infrastructure are highly sensitive and confidential and thus a very attractive goal for hackers, terrorists, data collecting companies, and the like. Questions regarding data security, privacy, and trust are not trivial to solve, especially in a globalized world where people from different countries move, travel, and visit various *Smart Cities* with different national laws. Basically, security goals with respect to confidential and sensitive data can be formulated as the so-called CIA (confidentiality, integrity, and availability) principle:

- *Confidentiality* states that only authorized entities (e.g. people or devices) are able to access data,
- I*ntegrity* means that data is accurate and complete, and
- *Availability* defines that data must be accessible when they are requested.

Furthermore, *privacy* describes data in a sense that no conclusions about a certain entity can be drawn. Lastly, *trust* in an architecture cannot be technologically established or created; *trust* has to be earned over a certain period of time. There are various factors that contribute to build *trust*: e.g. the transparent and comprehensible implementation of the CIA principle, or data privacy with corresponding technologies like encryption, etc., but eventually, *trust* can be regarded as an individual level of confidence of an entity into a 3rd party system, service, architecture, etc.

With respect to these facts, this work designs a *secure, distributed, and reliable cloud-based reference architecture* for the management of *Big Data* that logically separates confidential data and distributes them to various databases in different clouds. It increases the *reliability* of the distributed data with high-availability mechanisms. The foundation of the proposed *reference architecture* is built on relational and NoSQL databases. A *vertical partitioning* strategy is used to logically separate confidential data into small chunks, where ideally each chunk contains data that are not usable without the other chunks. These chunks are then distributed across different cloud infrastructures to achieve a higher degree of data *security*, *privacy*, and eventually *trust*. This approach was first proposed in (Kohler & Specht, 2012) for relational data and is also known as SeDiCo (a framework for a secure and distributed cloud data store). This work now adapts SeDiCo and transfers it to the above-mentioned NoSQL architectures. Furthermore, the reliability of the databases that store the respective chunks is improved with the usage of *Docker Swarm*, that implements a high-availability mechanism if one or more databases are inoperable. Finally, the proposed reference architecture is discussed and analyzed to clearly identify its strengths but also potential weaknesses with respect to its usage in a *Smart City*. Hence, this work contributes to the design and creation of working *secure, distributed, reliable,* and *trustworthy urban services.*

The main contributions of this chapter can be summarized as:

- A definition of a *secure*, *distributed*, and *reliable cloud-based reference architecture* for the management of *Big Data* scenarios.
- A proof of the technological feasibility of the *cloud-based reference architecture* for *Big Data.*
- An analysis of the strengths and weaknesses of the proposed *cloud-based reference architecture.*
- A demonstration how the proposed *cloud-based reference architecture* can be applied in a *Smart City* scenario.
- The integration of *Big Data* and *Cloud Computing* without losing data security and privacy.

The remainder of this chapter is organized as follows. First, in the background section, basic notions required for a common understanding are outlined. This is followed by a concrete problem formulation and requirements for a *secure, distributed, and reliable cloud-based reference architecture* for the management of *Big Data.* Moreover, the *Design Science Research* method as the used research method is sketched. After this, the authors' approach to develop such an architecture is presented. Then, the results are discussed and concluded. Finally, further research directions that are addressed by the authors are mentioned and explained.

BACKGROUND

In order to create a common understanding, this section defines central notions for the entire chapter.

Smart City

Firstly, it can be stated that there is actually no concrete, exact and precise definition of a *Smart City*. In literature, it is often understood as a consistent strategy for the management of infrastructural capabilities and *urban services* of a city, e.g. (Schaffers et al., 2011) (Hollands, 2008) (Galdon-Clavell, 2013). The main goal of the strategy is to improve the quality of the people's lives. Recent developments in the field of ICT lead to the fact that *Smart City* strategies are heavily based (and dependent) on current Information Technologies (IT). Examples for this include the development of the Internet in combination with increased network bandwidth and also the trend towards smart mobile devices. This enables a faster information flow with respect to traffic management (real-time information about traffic jams, arrival and departure times of trains, etc.).

Based on recent developments in the field of machine learning, autonomous driving cars, and Artificial Intelligence (AI) in general, it can be expected that the trend of integrating ICT infrastructures in various *Smart City* aspects will further increase. This is also what (Dustdar, Nastić, & Šćekić, 2017) (Bassi, 2017) recognized when they argue that ICT infrastructures for storing, processing and analyzing data to improve the quality of *urban services* are essential in future *Smart City* considerations. Moreover, this stresses a data-centric focus, as *urban services* rely on data that are produced and consumed by Smart City inhabitants or their devices. Finally, one aspect to improve the people's lives refers to the optimization of these prevailing ICT-based *urban services*.

Urban Services

Similar to the definition of a *Smart City*, there is currently no concise and consistent definition of the term *urban service*. There are approaches in relevant literature (e.g. see above) that classify *urban services* according to their functionality, i.e. municipal, governmental, medical, or infrastructural services, etc. (Zheng, Capra, Wolfson, & Yang, 2014) (Ali, Atia, & Mostafa, 2017) (Jain & Bhatnagar, 2017), but what they have in common is an implicit assumption that *urban services* are provided and consumed by the entities (i.e. inhabitants, companies, local councils, etc.) of a *Smart City*. Hence, *urban services* can be seen as a conglomerate of all provided and consumed services in a *Smart City*. Finally, providing and consuming services require a kind of *marketplace* or *service broker* in order to bring service providers and service consumers together.

Transferred to nowadays ICT architectures, this is analogous to well-known service-oriented architectures (SOA) that include *service providers* that offer one or more services, *service consumers* that are able to use them, and a *service registry* (or directory) that acts as a broker where services are listed and can be found (Erl, 2005). Figure 1 illustrates this in the famous SOA triangle.

Adapted to the *Smart City* context, the *service registry* in Figure 1 can be regarded as nowadays well-known App Stores (analogous to e.g. Google's Google Play, or

Figure 1. SOA triangle

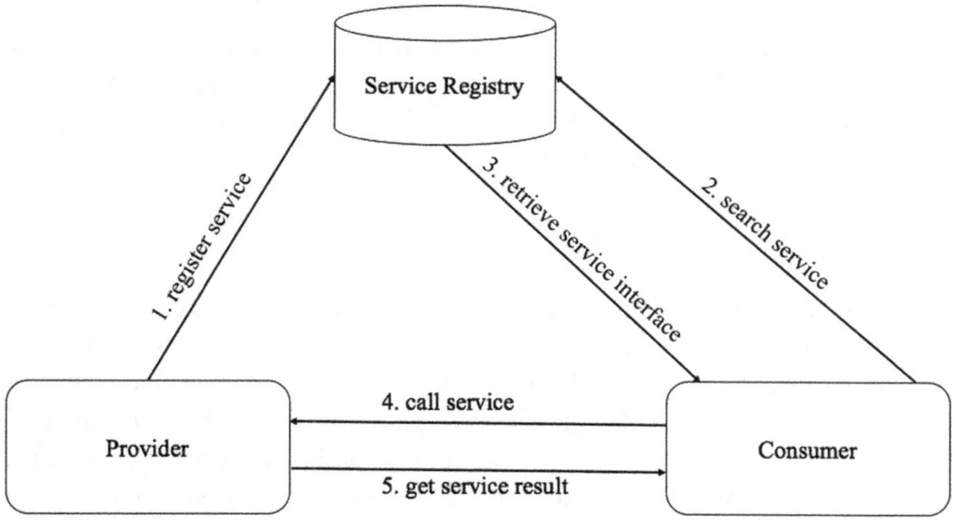

Apple's App Store). An example for an intelligent transportation system based on a SOA can be found in (W. Zhang, Qi, & Deng, 2017). Such architectures lead to interesting new challenges, especially when a significant amount of people is involved as either *service consumers* or *providers*. Regarding such a service-oriented architecture in more detail, there are two challenges that illustrate the necessity of a *secure, distributed, and reliable cloud-based reference architecture*:

1. **Big Data:** How to efficiently, securely, and reliably store and manage data produced by a large number of *service consumers* and *providers* (e.g. *Smart City* inhabitants, visitors, tourists, etc.) with a plethora of different devices (mobile and smartphones, tablets, laptops, etc.)?
2. **Cloud Computing:** How to design a *Smart City* ICT infrastructure from an architectural point of view, to handle the previously described *Big Data* challenges?

These two essential challenges are defined in greater detail in the following paragraphs, which follows the before-mentioned goals of bringing together the two current trend topics: *Big Data* and *Cloud Computing*, without losing data security and privacy. This also represents the novelty of the depicted *cloud-based reference architecture*.

Cloud-Based Architecture for Big Data

Literature regarding *Big Data* stresses three main components (3Vs): *volume, velocity*, and *variety* (Baron, 2013). There are also works that count even more, like e.g. *veracity* (Claverie-Berge, 2012) or *value* (Swoyer, 2012), but for the scope of this chapter, the basic 3Vs are enough.

Volume

Global consumer IP traffic grows continuously about 23% per year, therefore the volume of this data will reach 98,9 petabytes/month in 2017 (Cisco, 2014). This tremendous volume of data must be dealt with adequate database technologies that include the storage, manipulation, and analyzation of these heterogeneous data. As a promising technology, NoSQL with its four basic architectures (key-value, document, column stores, and graph databases) can be mentioned here (Sadalage & Fowler, 2013) (Harrison, 2015).

Velocity

Huge unstructured data not only have to be stored but also analyzed. Moreover, this has to be done efficiently in a short amount of time (i.e. in milliseconds). Here also, *In-Memory Databases* can be used to store, manage, and analyze relevant data in near real-time (Plattner & Zeier, 2012).

Variety

This large amount of unstructured data requires new storage techniques and technologies. Applying relational database schemas and transferring these data into the third normal form is not possible in a reasonable timeframe, i.e. in near real-time (Plattner, 2013). The same holds for conventional Business Intelligence (BI) approaches, like aggregating data in an extract, transform, and load (ETL) process, or the separation of transactional data (OLTP) and analytical data (OLAP). At this point, In-Memory database systems are a promising way to handle such huge amounts of heterogeneous data (Plattner, 2013).

In connection with a *Smart City* scenario, these 3Vs are considered inevitable. It becomes even more clear, if not only data are considered that are constantly created by *service providers* and *consumers*, but also data that are automatically created, stored and analyzed by (smart) devices, e.g. GPS location data of smartphones, traffic data acquisition, accident statistics, weather data, car sharing usage, etc. Then, additionally, not only current data but also the long-lasting storage, aggregation, and analyzation of all these data to create historical views that are the foundation for future analytics are challenging aspects.

However, managing data in a cloud infrastructure requires an enormous amount of trust into the cloud provider, as the cloud provider is also able to access the data. Hence, countermeasures like encryption or logical data distribution can be used. However, these techniques have a negative impact on the overall performance, as proved in (Kohler & Specht, 2014).

The current above-cited literature reveals that there is no best-fitting *reference architecture* realizable with current technologies for the management of *Big Data*. Quite the opposite is the case: there is a plethora of different technologies which all are designed for dedicated tasks or issues related to *Big Data*. The Apache Foundation (Apache Foundation, 2017) is a good starting point to get an impression of current state-of-the-art technologies and tools. However, since there is no prevalent generalized reference architecture for various Big Data scenarios, and with the considerations above, *Cloud Computing* offers architectural and technological capabilities that are promising to address current *Big Data* challenges.

Cloud Computing

The most common definition for Cloud Computing was created by the National Institute of Standardization and Technology (NIST) which describes the cloud with 3 *service models*: *Software as a Service* (SaaS), *Platform as a Service* (PaaS), and *Infrastructure as a Service* (IaaS), and 4 *deployment models*: the *public*, *private*, *hybrid*, and *community* model (Mell & Grance, 2011). The *service models* define *how* services are operated, whereas the deployment models express *where* services are operated. Above that, NIST defines 5 essential key characteristics of *Cloud Computing*: broad network access, on-demand self-service, rapid elasticity, resource pooling, and measured services. A closer look at these characteristics can be found in the introduction above. Therefore, this paragraph outlines the *service* and *deployment models* in more detail.

Service Models

For *consumers,* the crucial service model is the SaaS model (Figure 2). Here, ready-to-use *urban services* are provided and these services are based on the lower layers: PaaS and IaaS, respectively.

These lower layers are of greater importance for *service providers*. The PaaS model is primarily used by service developers to implement *urban services*, which are then operated on virtualized ICT infrastructure provided by the IaaS model. This virtualization enables the creation of clearly defined interfaces and so the

Figure 2. Exemplified cloud service models in a Smart City environment

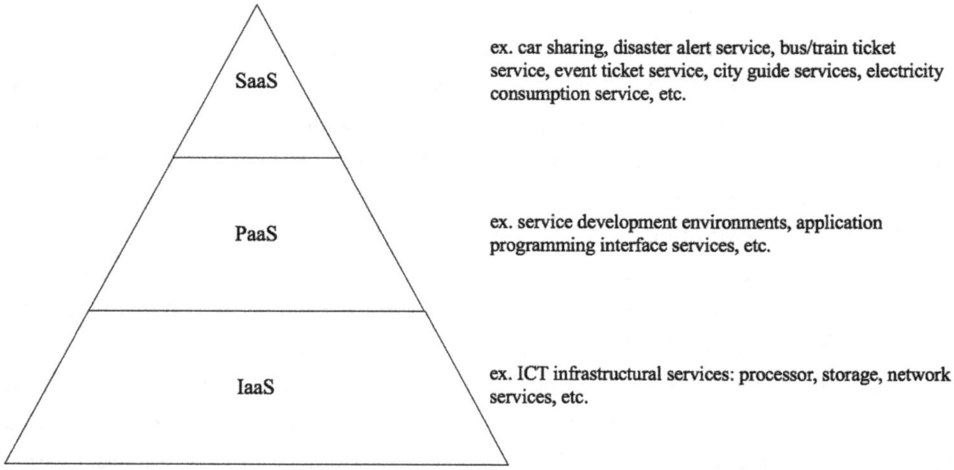

concrete underlying hardware is abstracted from, which basically is the foundation for the *resource pooling* and the *rapid elasticity* for the cloud. Moreover, complete transparency for the layers above is provided.

Deployment Models

As Figure 3 illustrates, the 4 *deployment models* define where a cloud infrastructure is operated: a *private* cloud is hosted by the owner of the underlying ICT infrastructure (e.g. town council, service providers, etc.).

A *hybrid* cloud has a *private* and a *public* part. This can be useful if the infrastructure resources of a *private* cloud are not sufficient and additional resources from another provider are cheaper than investing in further hardware capabilities.

A *public* cloud is completely operated by a 3rd party cloud provider and its resources are rentable on-demand. This is interesting if a *service provider* has no own infrastructure or no know-how to operate, manage and maintain one.

Finally, a *community* cloud is the combination of different *private* clouds where cloud providers connect their infrastructures to use the respective hardware resources more efficiently and to share administration, management and maintainability efforts.

To sum all up, it can be stated that *Cloud Computing* offers various opportunities but also challenges for *urban services*, i.e. a dynamic scalability (rapid elasticity) because of the resource pooling capabilities based on virtualized infrastructures. Defining an adequate *cloud infrastructure* demands a careful analysis of the concrete architecture with respect to the above-mentioned *service* and *deployment* models. On the one hand, a public cloud does not require any administration or maintainability

Figure 3. Cloud deployment models in a Smart City environment

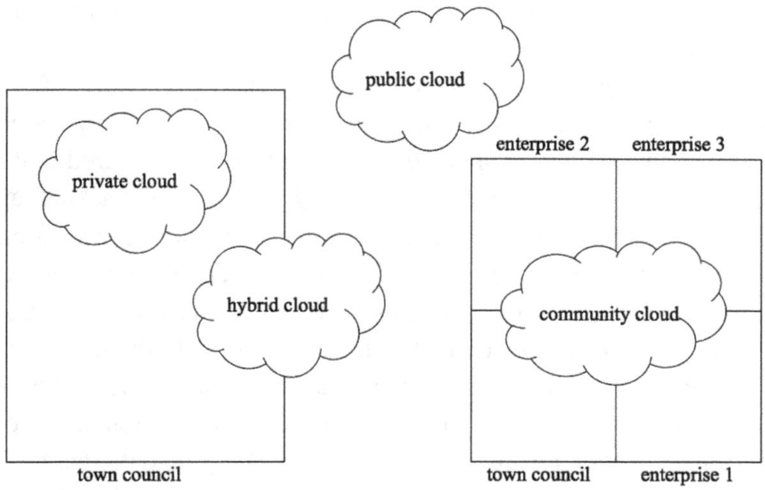

effort or know-how concerning the ICT infrastructure. On the other hand, all *urban services* and the corresponding data are stored and operated at a 3rd party cloud provider. Hence, a great amount of *trust* into the cloud provider, closely aligned with adequate and transparent measures to preserve *security* and *privacy*, is required. Approaches towards a trustworthy cloud infrastructure for a Smart City environment based on Fuzzy Logic can be found in (Sarkar, Banerjee, Badr, & Sangaiah, 2017) and an improved access to cloud capabilities based on the Hadoop File System can be found in (Matallah, Belalem, & Bouamrane, 2017).

The following paragraph takes a closer look at *security*, *privacy*, and *trust* and how the terms relate to each other.

Security, Privacy, and Trust

A basic definition of *security* can be seen as the preservation of *confidentiality*, *integrity*, and *availability* of data (DIN ISO 27000, 2011). This is also known as the CIA-Principle (see examples above). *Security* relates to all kinds of data (general data) that is produced or consumed by either *service consumers* or *providers*. Thus, it is a broader definition than *privacy*, as *privacy* only considers personal data that are closely related to an individual or entity (Pearson, 2013).

In contrast to *security*, *privacy* focuses only on personal data that belong to, describe or identify a concrete entity uniquely, i.e. an urban inhabitant as a *service consumer*, a visitor, a tourist, etc. (Pearson, 2013). Prime examples of such identifying data are social identification numbers, credit cards, or bank account numbers, email addresses, and the like. With such data, an entity can be identified directly or indirectly. Because of that, there are *anonymity* and *pseudonymity* that prevent others to draw conclusions about an entity.

Hence, *security*, *privacy*, *anonymity*, and *pseudonymity* are of particular importance in *Big Data* and in *Smart City* scenarios, where large amounts of general and personal data are collected. Firstly, it has to be ensured that data are anonymized and secondly, 3rd parties (e.g. *service providers*, hackers, and other attackers) should not be able to deduce information that belongs to a certain entity. This however also reveals a dilemma, because it is a challenging task to preserve a certain degree of anonymity but at the same time provide useful information for the *service consumers* or *providers*. Examples for this are manifold: *service consumers* want to receive specific information about traffic or train arrival and departure times, but then *service providers* need to know the current location of the respective *service consumer*. Another example is *consumer-identifying* data that are used for accounting services where *service consumers* have to uniquely identify themselves for an accurate and proper invoicing.

The considerations above finally lead to the notion of *trust* that can be built with a proper implementation of *security* and *privacy* features. As already pointed out, *trust* is regarded as an individual level of confidence of an entity into another one (e.g. a 3rd party). *Trust* is a very sensitive topic, as it takes a certain amount of time until it is built, but can be destroyed with just a single incident (e.g. a data breach, unauthorized access or sharing of data, etc.).

These factors influenced the development of a framework focused on database data that vertically partitions data and distributes the respective partitions across different clouds: *SeDiCo*. Because of the logical separation of data, each cloud only stores an independent part of the data which is not usable without the other parts.

SeDiCo

As already pointed out, the basic idea is a vertical partitioning method of database data. This is possible in relational as well as in NoSQL database architectures, but in case of NoSQL with certain limitations that can be found in (Kohler, Lorenz, Gumbel, Specht, & Simov, 2017). The vertical partitioning and distribution approach can best be illustrated and motivated with a simplified example of *service consumer* data. *Service consumers* have to register themselves to use a certain *urban service* and for accounting purposes, they have to provide their credit card number and their birthdates. The *service consumer* data are stored by a *service provider* in a NoSQL *document store* (other database architectures work analogously), illustrated in Figure 4.

Obviously, the *service consumer data* can be distinguished in *general*, i.e. name, and *personal* (sensitive) data, i.e. credit card number or birthdate. In order to

Figure 4. Consumer data in a NoSQL document store

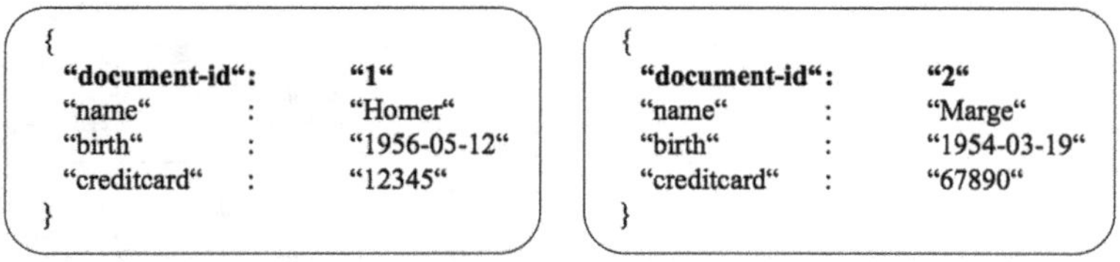

increase the level of *security* and *privacy,* these data are now vertically partitioned according to Figure 5.

Due to this vertical distribution (Figure 5) the respective data are independent of each other (from a *creditcard* in *partition2* its corresponding *customer* in *partition1* cannot be deduced). Thus, it becomes possible to store *partition1* in a certain (e.g. external) cloud and *partition2* in another one. Finally, only the *service provider* is able to join the respective partitions via the replicated *document-id* (Figure 5). Figure 6 outlines the entire *SeDiCo* approach.

Other NoSQL architectures, i.e. key-value, graph, or column stores, as well as relational database architectures work analogously. A more detailed and technological view of the framework can be found in e.g. (Kohler, Simov, Fiech, & Specht, 2015).

Formally, the example above can be described as a query π against a database relation (i.e. table) R with attributes A, denoted as $R(A)$. This query is then

Figure 5. Vertically partitioned document store

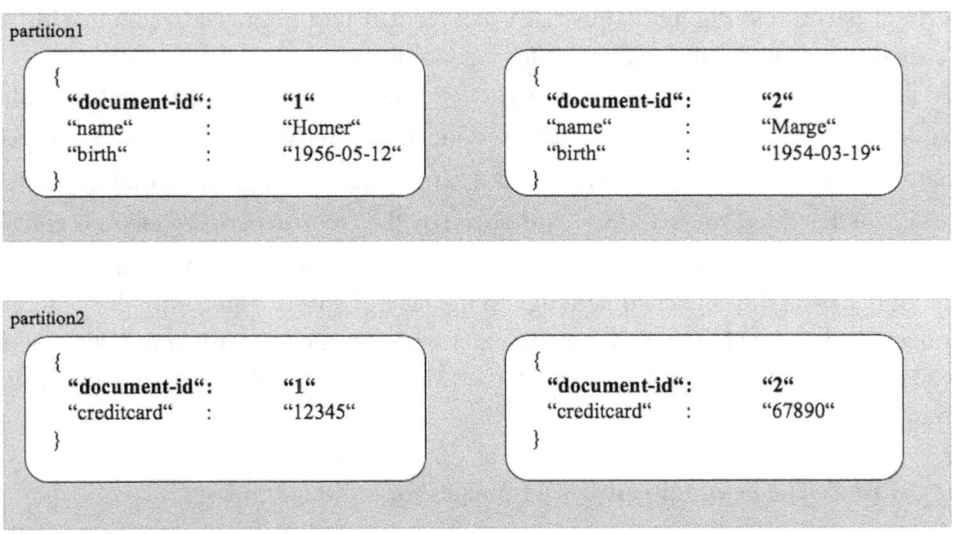

Figure 6. SeDiCo architectural overview

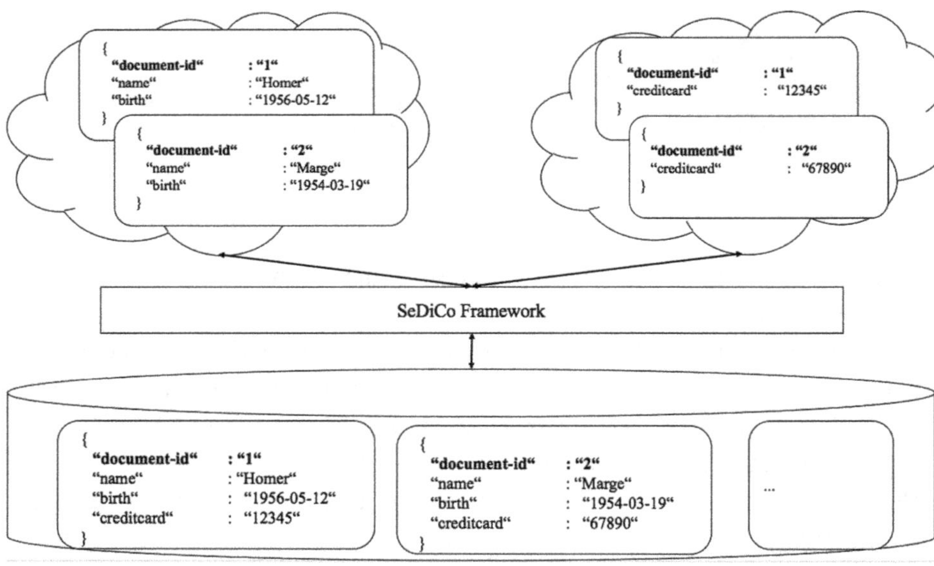

vertically partitioned into so-called *reconstruction queries* $\pi_{v1\omega1}, \ldots, \pi_{vn\omega n}$ according the number of vertical partitions n, which are also represented as relations with corresponding attributes, i.e. $S\left(B\right), T\left(C\right), etc.$ The set of attributes $\{ a_1, \ldots, a_j \}$ that are contained in the original query π are also vertically partitioned and added to the respective *reconstruction query* $\pi_{v1\omega1}$ and $\pi_{v2\omega2}$ (Figure 7). This results in the construction of the *reconstruction queries* that are issued against the respective vertical partitions. Figure 7 illustrates the vertical partitioning algorithm with 2 vertical partitions for the sake of better readability.

Then, the result sets of the *reconstruction queries* are joined into a final result set RS_{final}. The proof of this algorithm is done in two steps. First, it has to be proved that the algorithm always terminates for arbitrary queries π, and second, it has to be shown that the result set RS_{final} produced by the *reconstruction queries* is equal to the result set that would be produced by the original query π. Note that the proof can only be roughly sketched here due to the lack of space. Thus, only the general concept is presented here. It is assumed that with the general outline the interested reader is able to comprehend the basic idea and that the concrete proof is straightforward.

1. **Proof:** The algorithm always terminates for arbitrary queries π.

Figure 7. Vertical partitioning algorithm

```
Data:
B;                  /* the set of attributes for partition S_v(B) */
C;                  /* the set of attributes for partition T_v(C) */
ω = {a_i, ..., a_j};          /* the set of query restrictions */
Result:
Π_{v1ω1} and Π_{v2ω2};          /* the reconstruction queries */
begin
    ω1 ← {a1};       /* the set of restriction for query Π_{v1ω1} */
    ω2 ← {a1};       /* the set of restriction for query Π_{v2ω2} */
    foreach a ∈ ω \ {a1} do
        if a ∈ B then
            ω1 ← ω1 ∪ {a};
        else
            ω2 ← ω2 ∪ {a};
        end
    end
    return Π_{v1ω1}, Π_{v2ω2}
end
```

Due to the finite set of relations $R,S,T,etc.$ and attributes A,B,C it is guaranteed that the algorithm always terminates in a finite number of steps. As the algorithm always terminates, it is guaranteed that for each query π and for each vertical partitioning schema, corresponding *reconstruction queries* can be built.

2. **Proof:** The result set of the *reconstruction queries* RS_{final} is equal to the result set of the original query RS.

As all query attributes $\{a_1,\ldots,a_j\}$ of the original query π are considered and vertically partitioned (i.e. disjoint) into the attributes of the *reconstruction queries* (ω_1) and (ω_2), it is guaranteed that no attribute is missed or left out. Thus, each tuple that is collected by the original query π is also collected by one of the reconstruction queries $\pi_{v1\omega1}$ and $\pi_{v2\omega2}$. After this, the result sets of the *reconstruction queries* are joined into the final result set RS_{final}. To formally close the proof, it has to be shown that $RS_{final} \subset RS$ and that $RS \subset RS_{final}$, which is assumed to be straightforward here.

PROBLEM FORMULATION

In order to concretize the above-mentioned research question concerning *Big Data* and *Cloud Computing*, possible attack vectors with respect to the CIA principle are outlined here. The attack vectors depicted in Table 1 are a summarization and are inspired by (Wang, Ali, & Kelly, 2015) (Ijaz, Shah, Khan, & Ahmed, 2016) (K. Zhang et al., 2017) (Bartoli, Hern, Hernández-Serrano, & Soriano, 2011) (Elmaghraby & Losavio, 2014). In order to clearly identify potential bottlenecks, points of attacks, and other weaknesses, these attack vectors (without any claims of completeness) have a key impact on the preservation of the CIA principle.

With respect to Table 1, it must be noted that a complete list of attack vectors is rather unrealistic and impossible because of the individual character, different requirements and conditions of each *Smart City* strategy. Hence, each *Smart City* strategy demands its own individual *security* and *privacy* analysis. For this, the following literature can be recommended: (DIN ISO 27000, 2011) (OWASP, 2017).

The challenges with respect to *Big Data* and *Cloud Computing* and the attack vectors show the ICT-demanding character of future *Smart Cities*. They further show the need for a *secure, distributed, and reliable* ICT architecture. Thus, all in all, the research problem:

How to design a secure, distributed, and reliable infrastructure from an architectural point of view, such that it preserves the CIA principle?

is investigated in this chapter with the following research methodology.

Table 1. CIA principle and attack vectors

CIA Principle	Possible Attack Vectors

Confidentiality	• Intruders to physical and virtual systems
	• Data theft, caused by:
	  • unintentional publication of data
	  • intruders
	  • incorrect usage of data and/or systems

Integrity	• Unwanted data modification, caused by:
	  • hard and software failures
	  • intruders
	  • incorrect usage of data and/or systems

Availability	• Data loss, caused by:
	  • hard and software failures
	  • infrastructure overload
	  • natural hazards
	• Denial of Service (DoS) attacks
	• Distributed Denial of Service attacks (DDoS)

RESEARCH METHODOLOGY

The definition and implementation of a *secure, distributed, and reliable cloud-based reference architecture* for the management of *Big Data* in a *Smart City* scenario is conducted with the Design Science Research methodology presented by (Hevner & Chatterjee, 2010). Following this methodology means to create so-called *artifacts*; in the case of this chapter: software components that result in an overall *reference architecture*. Figure 8 illustrates the entire methodology with the interaction between the *Environment* (in the case of this chapter: the *Smart City*), the used methodology (in the middle of Figure 8) and the *Knowledge Base* (in this case this chapter: the chapter itself with all the mentioned references).

Hence, the proposed *cloud-based reference architecture* can be regarded as an *artifact* to enable the management of *Big Data* in a *Smart City* context.

APPROACH

A key issue for the approach of a *secure, distributed, and reliable* infrastructure is *failure safety*, i.e. the operation of an architecture, system, etc. in case of one or more failures. The notion of a *failure* is manifold and can be categorized with respect to a distributed system according to (Tanenbaum & Van Steen, 2007) as follows:

- **Availability:** As the probability of a distributed system to work correctly
- **Reliability:** As a time-interval in which a distributed system runs without an error
- **Safety:** As the fact that in case of one or more failures no catastrophe occurs, and

Figure 8. Design science research methodology

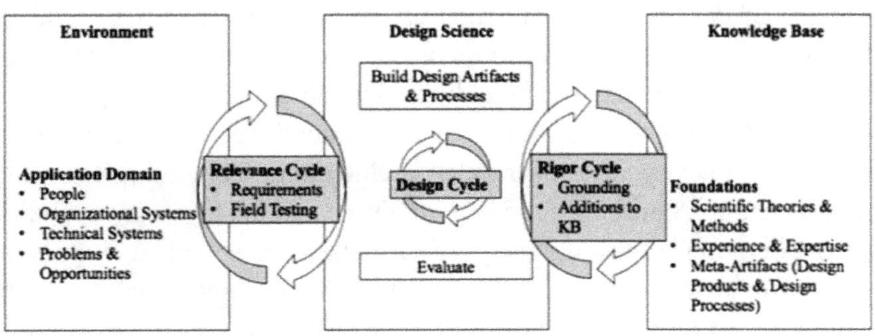

- **Maintainability:** As the degree of easiness to repair a failing distributed system.

Availability and Reliability

Achieving a highly-available architecture can be achieved using replication of services or data. On the one hand, the more replicas exist, the higher is the general availability of a distributed system or architecture, as in case of a failing service or corrupt data, a corresponding replica (copy) can take over. On the other hand, the communication overhead for the synchronization of the replicas increases with a larger number of existing replicas, and that might decrease the overall performance. In a distributed architecture, the same holds for reliability. As soon as a replica fails, its functionalities can be offered by another one. Thus, the probability of a failure that affects the entire distributed architecture can be minimized.

Safety

The notion of *safety* is more difficult. Especially in a *Smart City* environment, in which critical data and services are managed and processed (e.g. traffic light control, life-sustaining measures in hospitals, etc.). Currently, distributed architectures also take advantage of multiple replicas of the data and services. Here also, the probability of a failure that refers to *safety* is minimized with the usage of replication. Moreover, with the usage of replication a distributed architecture might even become able to repair itself: in case of a failing component another one can take over automatically as long as the former one is restored or recovered (also automatically) to its last working state.

Maintainability

Although distributed architectures with replicated services and data are useful with respect to *availability*, *reliability,* and *safety*, *maintainability* has also to be taken into consideration. Not only is the higher communication overhead for the synchronization of the replicas a major concern, but additionally, the increasing complexity of the entire architecture. It is obvious that such above-mentioned fail-over mechanisms increase the complexity of a distributed architecture. This complexity can be reduced to a certain degree with the help of automated deployment capabilities in e.g. Cloud Computing. With the usage of virtualization, so-called *snapshots* (i.e. saving the current status) of a virtual machine (VM) become a viable possibility. Hence, in case of a failure, the component can be reverted to the last correct working *snapshot*. Another possibility is to deploy a VM redundantly (again with the help of virtualization), i.e. to not only provision one VM but several copies of it. However, what still remains is a failover mechanism that defines which VM is in charge in case of a failing VM. Thus, it is possible to manage the increased complexity with a higher degree of automated deployments in a cloud infrastructure, but the eventual challenge of a higher complexity remains.

Nevertheless, the above-mentioned discussion shows that the usage of replication for *high-availability*, *reliability*, and *safety* outweighs the disadvantages (i.e. *maintainability*) by far. Hence, replication is considered a feasible mechanism to create a *secure*, *distributed*, and *reliable Smart City* ICT-focused architecture.

As the vertical partitioning and distribution approach of SeDiCo uses different clouds, *Cloud Computing* is another paradigm that has to be considered in more detail. As already outlined above, a basic technological principle of *Cloud Computing* is *virtualization*. Virtualized infrastructures, completely abstracted from their underlying hardware offer resource pooling and rapid elasticity, 2 of the 5 cloud key characteristics. Above that, virtualization can be used to replicate components (services, data, etc.) of a distributed architecture. Especially in a cloud with (theoretically) unlimited computing resources, it is a viable approach to provision more than one virtual machine (VM) to process a given task or to offer a certain service. Thus, in case of a failing VM, another one is able to take over until the original VM is restored. Despite the fact that virtualization enables effective high-available and reliable system architectures, the overhead for the actual virtualization (Grund, Schaffner, Krueger, Brunnert, & Zeier, 2010) and the increased complexity are becoming major drawbacks. In particular, the increased overhead is in the focus of emerging container-based technologies, like Docker (Docker, 2017a), rkt (CoreOS, 2017), Docker Swarm (Docker, 2017b) and Kubernetes (Kubernetes, 2017).

With respect to the development of a secure, *distributed, and reliable* infrastructure for a *Smart City*, both technologies, *virtualization* and *container-based* ones are promising, as they do not mutually exclude each other. Quite the opposite: they are can be used complementarily. In order to use advantages of both technologies, virtualization can be considered as the fundamental technology (as it is in *Cloud Computing*) for containers. Thus, even containers become independent from the underlying physical operating system. Figure 9 illustrates this in an architectural overview.

With respect to the above-mentioned discussion, Docker (Docker, 2017a) as the most prominent and most widely used container technology implementation has to be mentioned. Its concrete architecture is outlined in Figure 10.

This is the first step towards a high-available and reliable architecture, however, such architectures still contain a single-point-of-failure: either the underlying physical hardware (hosting *n* VMs) or the Docker instance (hosting *n* containers).

Figure 9. Virtualized and container-based architecture

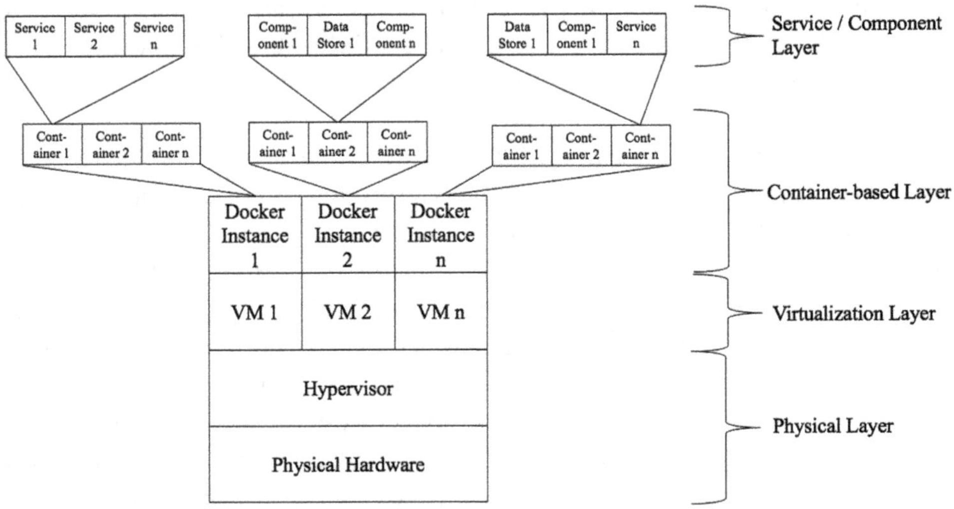

Figure 10. Docker architecture

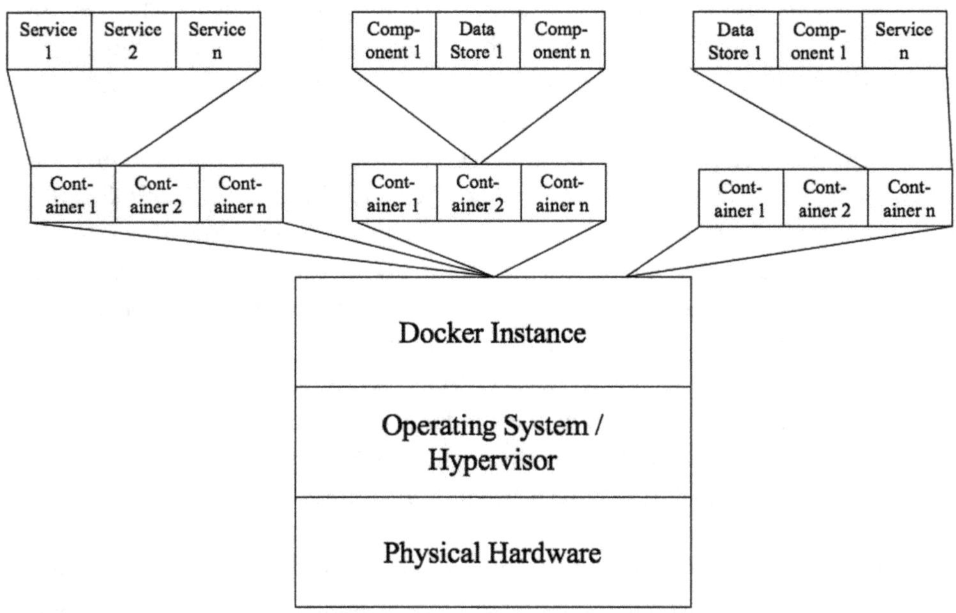

So far, *virtualization* and *containers* are considered from a fundamental technological view, but *high-availability* and *reliability* features have been neglected. From a virtualization point of view, this leads to *Cloud Computing*, where several VMs can be hosted in different locations and in which they are connected to each in a network, such that in case of failures other VMs can take over. Thus, the VMs are synchronized and with the distribution across different (physical) locations (i.e. cloud data centers), the probability of an entire system failure is minimized.

What still remains is the *high-availability* and *reliability* of *containers*. Here also, a synchronization and distribution logic is required. This is actually offered by approaches like *Docker Swarm* (Docker, 2017b) or *Kubernetes* (Kubernetes, 2017) to mention the two most prominent representatives. In order to remain consistent with Figure 10, Docker Swarm is considered in more detail here (Figure 11). Generally, a Docker Swarm consists of *n* connected Docker containers, managed by so-called Docker managers.

Figure 11 shows that *1* Docker *manager* is responsible for *n* swarm nodes, and if a swarm node fails, the *manager* transfers the requests for this swarm node to another one. Furthermore, if a *manager* fails, another *manager* is capable of taking over and to keep the swarm working. Note that one manager is operated on a single (either physical or virtual) machine. Accordingly, the distribution and the connectedness of the containers enable a *high-available* and *reliable* infrastructure. Regarding the *synchronization* of the containers (via the managers), and also with respect to the synchronization of different VMs in a cloud (via load balancers, e.g. HAProxy (HAProxy, 2017)), a best practice concerning the number of load balancers and

Figure 11. Docker Swarm architecture

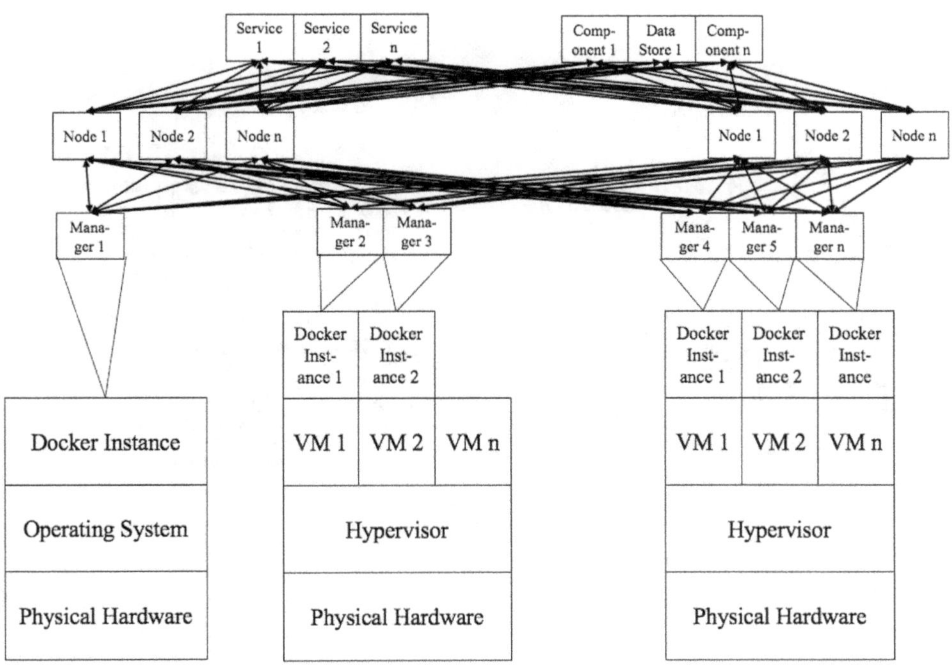

managers has to be mentioned. Every synchronization requires a certain kind of control instance that is – in case of a failure - in charge of deciding which other VM or manager should take over. Therefore, a *leader* must be elected between all managers or load balancers respectively. This leads to the fact that a certain manager must always be selected as the *leader*, which is responsible for the consistent and effective synchronization of all containers (i.e. *nodes* and *services*). Accordingly, the question of how such a leader is elected comes up. For this, Docker uses the *Raft Consensus* (Ongaro & Ousterhout, 2014) algorithm. In summary, the leader is elected based on timeouts and the one manager that responds at first usually becomes the most votes from the other participating managers. Then, all operations between the containers and the clients have to go through the leader, which realizes distributed consistency over all containers. In order to implement an efficient and quick voting phase for a distributed architecture with *n* managers, an uneven number of managers is recommended to find a majority. Above that, it can be stated that a distributed architecture with *n* managers is able to work correctly with at most *(n-1)/2* failing managers.

The summary of the above-mentioned discussion leads to the architectural overview of a secure, reliable, and trustworthy infrastructure for a Smart City, illustrated in Figure 12.

Figure 12. Approach for a secure, reliable, and trustworthy Smart City ICT infrastructure

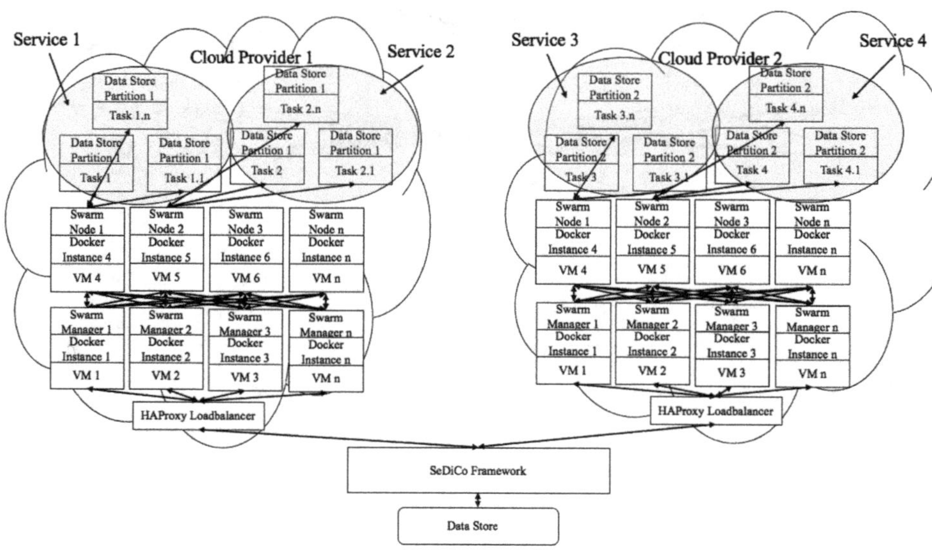

Here, data are stored in a relational or NoSQL database, denoted as *Data Store*. Then, the *SeDiCo framework* enables a vertical partitioning and distribution of the data into several clouds. The VMs as the basic key technology of the clouds are monitored and managed by *HAProxy*, a *load balancer* implementation. This load balance not only distributes the incoming requests evenly among the VMs but also ensures that failing or erroneous VMs are shut down and correct working ones are instantiated. In the described architecture (Figure 12), every VM hosts a single *Docker* instance, either a *Swarm Manager* or a *Swarm Node*, as the fundamental *container* technology. Hence, restoring failing or erroneous VMs becomes more effective and comprehensible. In summary, these containers build the respective *Docker Swarm*, in which the *services* (i.e. *urban services*), are operated. It has to be noted, that *1…n tasks* define *1 service*. Finally, a failing VM is restored by the *load balancer* (*HAProxy*) and a failing container is restored by a respective *manager* of the *Docker Swarm*. This enables a *high-available*, *reliable* and *secure* infrastructure for *urban services*, at the cost of a high infrastructure complexity. The maintainability has to be analyzed and discussed in further real-world experiments and scenarios, as a complex infrastructure usually implies a high maintainability effort. However, in the architecture depicted in Figure 12, most of the provisioning (i.e. installation or deployment) of VMs or containers is automated by scripts and the failover mechanisms of the *load balancer* and the *container swarm* enable a so-called self-healing system without any user interaction.

DISCUSSION OF RESULTS

This discussion identifies advantages, but also bottlenecks and drawbacks of the proposed *cloud-based reference architecture*, and finally points to further research directions that address these challenges. A more detailed analysis of the architecture depicted in Figure 12 with respect to the CIA principle is outlined in Table 2.

Unfortunately, a more detailed analysis would exceed the boundaries and would lose the general character of this work. Yet, more concrete lists of attack vectors, vulnerabilities and threads can be found in (OWASP, 2017). Furthermore, (DIN ISO 27000, 2011) defines security measures and processes to manage and to behave in case of an attack.

Another issue concerning the above-mentioned architecture is the usage of container-based technologies (in the concrete case *Docker* with its *Docker Swarm* implementation). Compared to virtualization, container technologies are considered agiler and more flexible and enable individual system architectures and environments due to their efficient and automatable provisioning capabilities. Further examples of such individual and flexible architectures can be found in e.g. (Yu & Huang, 2015)

Table 2. Proposed architecture mapped to CIA principles

CIA Principle	Attack Vectors	Techniques	Implementations
Confidentiality	• Intruders to physical and virtual systems • Data theft, caused by: • unintentional publication of data • intruders • incorrect usage of data and/or systems	Partitioning	Vertical partitioning (SeDiCo)
		Authentication	DB authentication
		Authorization	DB authorization
		Encryption	To be done
Integrity	• Unwanted data modification, caused by: • hard and software failures • intruders • incorrect usage of data and/or systems	Authentication	DB authentication
		Authorization	DB authorization
		Isolation levels	DB isolation levels
		Replication	Docker Swarm, Cloud Virtualization (e.g. Snapshot functionality)
Availability	• Data loss, caused by: • hard and software failures • infrastructure overload • natural hazards • Denial of Service (DoS) attacks • Distributed Denial of Service attacks (DDoS)	Backup/Recovery	Docker Swarm, Cloud Virtualization (e.g. Snapshot functionality)
		Replication	Docker Swarm, Cloud Virtualization (e.g. Snapshot functionality)

(Kecskemeti, Marosi, & Kertesz, 2016) (Fazio et al., 2016) (Affetti, Bresciani, & Guinea, 2015) (Viennot, Lécuyer, Bell, Geambasu, & Nieh, 2015). Yet, the developed architecture in Figure 10 shows that the two technologies do not mutually exclude each other; in contrast: they can be used complementarily. Hence, it is not a decision between, but among them. Comparing container-based technologies with virtualization in the context of this work results in two main considerations: *security* and *performance*.

Security

From a security-based perspective, it can be stated that container technologies are an emerging field and there are current research works that deal with security-related aspects, e.g. (Bui, 2015) (Combe, Martin, & Di Pietro, 2016). Here, virtualization benefits from a longer history and utilization in the academic world but also in the industry. However, a closer look into both technologies is necessary to stress the security-related differences. Here, it is worth mentioning that both, containers and virtualization, use operating system fundamentals stemming from the 1960s (*chroot*, *zones*, *namespaces*, etc.). On the one hand, with container-technologies, there are *cgroups* that handle the OS processes and manage the access to them for the containers, *namespaces* which manage the physical resources shared by the OS for the containers, and the *union file system*, a file system organized in layers for the creation and operation of the containers. On the other hand, there is virtualization, that uses emulation, abstraction, sharing, and optimization of operating system functions and resources (Li & Kanso, 2015) (Morabito, Kjaellman, & Komu, 2015) (Pahl & Lee, 2015). Hence, both technologies rely on a long history and well-investigated technologies, but virtualization is currently still prevalent, especially in *Cloud Computing* environments, where it demonstrates its strengths with e.g. (in contrast to container technologies) the complete isolation of VMs from each other. Docker is a lightweight container technology compared to VMs. Because of the resource sharing mechanisms, the overhead is considerably smaller and the provisioning is faster. However, this comes at cost of flexibility, as Docker or other container technologies are tightly bound to the underlying operating system (OS), whereas a Hypervisor completely abstracts from the fundamental OS functionalities and system calls. Also, security features have to be taken into account, as a Hypervisor hides the OS, whereas a container uses dedicated OS capabilities directly in a resource sharing manner.

Performance

Considering the performance, containers also known as a lightweight technology, benefit from the resource sharing capabilities, whereas in virtualization, the complete isolation from the physical resources, i.e. the emulation is responsible for a significant performance loss.

In order to compare the performance loss in concrete numbers with respect to the architecture in Figure 12, (Julian, Shuey, & Cook, 2016) determined that with Docker, the performance loss in a web application operated on top of a high-performance cluster (HPC) system environment is neglectable (0.4%). However, it has to be noted that physical limitations also have to be considered. Experiments with the above-depicted architecture proved that if the (virtual) resources (i.e. the *managers*) are operated at their full (physical) capacity the system performance collapses until a complete system failure due to a system overload. This also has to be considered in case of attacks like denial of service (DoS) or distributed denial of service (DDoS) attacks, in which a tremendous amount of network and system load is generated.

Comparing these numbers to (Grund et al., 2010), where a virtualized in-memory database workload was evaluated shows that the performance degrades by 7% when a virtual machine instead of a physical machine is used. The performance degrades even more (up to ~60%), if more virtual machines (4 VMs) on one physical host are used, although enough physical resources are provided by the host.

Even though the two studies are not directly comparable (due to different physical hardware setups and different benchmarks), these numbers indicate that from a performance point of view, containers are advantageous. Moreover, transferring these numbers to the architecture denoted in Figure 12, it has to be noted that the major performance loss (factor ~20 to ~500, depending on the implementation) is determined by the vertical data distribution of the SeDiCo framework as investigated in (Kohler & Specht, 2014). Further optimization approaches that use caching or parallel data access via query rewriting can be found in (Kohler, Simov, & Specht, 2015) (Kohler, Simov, Fiech, et al., 2015) (Kohler & Specht, 2015).

Persistent Data Store

Additional components that realize a storage engine for all containers would reduce the synchronization overhead (depicted with arrows in Figure 12) between them. Also, the provisioning of new tasks and nodes in case of a failure would be more efficient with a decentralized, but persistent data storage that provides a uniform access to all containers, as persistent data would not have to be replicated to all containers.

However, on the other hand, with such a cluster-based storage, the synchronization would just be transferred to a lower level in the architecture, namely to the underlying storage cluster. Considering the advantages in the context of this work, they are considered feasible but come at cost of a much higher complexity (i.e. for the setup, installation, and maintainability of an extra storage cluster). Therefore, this opportunity is not considered in closer detail here. Finally, it can be stated that such additional storage plug-ins are able to improve the performance (failover, response time, provisioning of new components, etc.) of the architecture, but also introduced additional complexity. Therefore, an improved performance has to be carefully weighed against an increased complexity.

CONCLUSION

In order to conclude the results of this chapter, it can be stated that the authors developed a *secure, distributed, and reliable cloud-based reference architecture* for the management of *Big Data*. The entire architecture is illustrated in Figure 12. Although the proposed *reference architecture* is a generic one, it contains concrete, ready-to-use and current technologies. Compared to other *Big Data reference architectures*, e.g. (Chen, Mao, & Liu, 2014) (Pääkkönen & Pakkala, 2015) (Demchenko, De Laat, & Membrey, 2014) or (Philip Chen & Zhang, 2014) the presented *reference architecture* in this chapter uses concrete technologies and preserves data security and privacy although operable in cloud infrastructures.

Moreover, it has a strong focus on reliability in case of system failures and therefore, the authors developed a self-healing architecture, which uses replication and is able to repair itself automatically without any user interaction in case of a failure. Hence, the combination of *Big Data* and *Cloud Computing* (which also emphasizes the novelty of the chapter) can be regarded as successful with the proposed *secure, distributed, and reliable cloud-based reference architecture*.

FUTURE RESEARCH DIRECTIONS

This chapter outlined the development towards a *secure, distributed, and reliable cloud-based reference architecture* for the management of *Big Data* in a *Smart City*. The focus of this chapter was the conceptualization and implementation of such an infrastructure to prove its feasibility. Nonetheless, the analysis and discussion of the above-mentioned approaches and technologies show that the performance has not been analyzed so far. The chapter mentioned some implications on the performance due to the usage of virtualization, container-based technologies or for the usage of a cluster-based storage engine for all services. Yet, concrete and comparable performance measurements that clearly demonstrate advantages and bottlenecks of the respective technologies, ideally in real-world scenarios, would lead to improved analyses and investigations.

One step towards such performance experiments is the development of cloud benchmarks that consider container-based solutions, e.g. (Varghese, Subba, Thai, & Barker, 2016) (Morabito et al., 2015) (Kozhirbayev & Sinnott, 2017) (Anderson et al., 2016) (Amaral et al., 2015) (Preeth, Mulerickal, Paul, & Sastri, 2016). However, a standard benchmark such as TPC (TPC, 2016) for databases or YCSB (Patil et al., 2011) for cloud infrastructures has not been able to assert itself.

Another aspect of these new container-based technologies is *security*. This is of particular importance in *Cloud Computing* environments with their multi-tenancy, where several (competitive) cloud consumers share the same physical resources. Currently, this is realized with virtualization as the basic technology and containers are placed on top of virtualized infrastructures. Here, a thinkable approach (also with respect to performance considerations) would be to remove the above-mentioned virtualization overhead and just use containers for the isolation of the cloud consumers. This would foster more advanced investigations of container-based security issues. Currently, there are analyses that take container-based security on the level of the OS functions: *cgroups*, namespaces and the union file system into account (Bui, 2015) (Combe et al., 2016). Here (Bui, 2015) determined that the core OS functionalities provide a good isolation for the containers via *cgroups* and *namespaces*, but the networking model used for the containers suffer from vulnerabilities such as e.g. MAC (media access control address) flooding and ARP (address resolution) spoofing. Transferring these vulnerabilities to a cloud environment, additional investigations how to avoid them (such as e.g. network filters) would be necessary.

REFERENCES

Affetti, L., Bresciani, G., & Guinea, S. (2015). ADock: A Cloud Infrastructure Experimentation Environment Based on Open Stack and Docker. *Proceedings - 2015 IEEE 8th International Conference on Cloud Computing, CLOUD 2015*. 10.1109/CLOUD.2015.36

Ali, A. H., Atia, A., & Mostafa, M.-S. M. (2017). Recognizing Driving Behavior and Road Anomaly using Smartphone Sensors. *International Journal of Ambient Computing and Intelligence*, *8*(3), 22–37. doi:10.4018/IJACI.2017070102

Amaral, M., Polo, J., Carrera, D., Mohomed, I., Unuvar, M., & Steinder, M. (2015). Performance evaluation of microservices architectures using containers. *Proceedings - 2015 IEEE 14th International Symposium on Network Computing and Applications, NCA 2015*. 10.1109/NCA.2015.49

Anderson, J., Hu, H., Agarwal, U., Lowery, C., Li, H., & Apon, A. (2016). Performance considerations of network functions virtualization using containers. *2016 International Conference on Computing, Networking and Communications, ICNC 2016*. 10.1109/ICCNC.2016.7440668

Apache Foundation. (2017). *Welcome to The Apache Software Foundation!* Retrieved January 2, 2018, from https://www.apache.org/

Baron, P. (2013). *Big Data für IT-Entscheider*. München: Hanser Verlag. doi:10.3139/9783446433922

Bartoli, A., Hern, J., Hernández-Serrano, J., & Soriano, M. (2011). Security and Privacy in your Smart City. *Smart City Counc*, 1–6. 10.1109/MCOM.2013.6525606

Bassi, A. (2017). Looking at Smart Cities with an Historical Perspective. In V. Angelakis, E. Tragos, H. C. Pöhls, A. Kapovits, & A. Bassi (Eds.), *Designing, Developing, and Facilitating Smart Cities: Urban Design to IoT Solutions* (pp. 3–15). Cham: Springer International Publishing; doi:10.1007/978-3-319-44924-1_1

Bhatt, C. M., Dey, N., & Ashour, A. (2017). *Internet of things and big data technologies for next generation healthcare*. Springer. doi:10.1007/978-3-319-49736-5

Bui, T. (2015). Analysis of Docker Security. *Computing Research Repository*. Retrieved from http://arxiv.org/abs/1501.02967

Chen, M., Mao, S., & Liu, Y. (2014). Big data: A survey. In Mobile Networks and Applications (Vol. 19, pp. 171–209). Academic Press. doi:10.100711036-013-0489-0

Cisco. (2014). *Cisco Visual Networking Index: Global Mobile Data Traffic Forecast Update, 2013–2018*. Retrieved July 7, 2014, from http://www.cisco.com/c/en/us/solutions/collateral/service-provider/visual-networking-index-vni/white_paper_c11-520862.html

Claverie-Berge, I. (2012). *Solutions Big Data IBM*. Retrieved August 11, 2014, from http://www-05.ibm.com/fr/events/netezzaDM_2012/Solutions_Big_Data.pdf

Combe, T., Martin, A., & Di Pietro, R. (2016). To Docker or Not to Docker: A Security Perspective. *IEEE Cloud Computing*, *3*(5), 54–62. doi:10.1109/MCC.2016.100

Core, O. S. (2017). *rkt | rkt Container Engine with CoreOS*. Retrieved October 4, 2017, from https://coreos.com/rkt/

Dameri, R. P., & Rosenthal-Sabroux, C. (2014). *Smart City and Value Creation*. Academic Press. 10.1007/978-3-319-06160-3_1

Demchenko, Y., De Laat, C., & Membrey, P. (2014). Defining architecture components of the Big Data Ecosystem. In *2014 International Conference on Collaboration Technologies and Systems, CTS 2014* (pp. 104–112). Academic Press. 10.1109/CTS.2014.6867550

Dey, N., Hassanien, A. E., Bhatt, C., Ashour, A., & Satapathy, S. C. (2018). *Internet of things and big data analytics toward next-generation intelligence*. Academic Press.

DIN ISO 27000. (2011). Retrieved February 1, 2016, from https://www.beuth.de/de/norm-entwurf/din-iso-iec-27000/243433889

Docker. (2017a). *Docker - Build, Ship, and Run Any App, Anywhere*. Retrieved October 4, 2017, from https://www.docker.com/

Docker. (2017b). *Swarm mode overview | Docker Documentation*. Retrieved October 4, 2017, from https://docs.docker.com/engine/swarm/

Dustdar, S., Nastić, S., & Šćekić, O. (2017). *Smart Cities - The Internet of Things, People and Systems*. Cham: Springer International Publishing; doi:10.1007/978-3-319-60030-7

Elhayatmy, G., Dey, N., & Ashour, A. S. (2018). *Internet of Things Based Wireless Body Area Network in Healthcare*. Academic Press. 10.1007/978-3-319-60435-0_1

Elmaghraby, A. S., & Losavio, M. M. (2014). Cyber security challenges in smart cities: Safety, security and privacy. *Journal of Advanced Research, 5*(4), 491–497. doi:10.1016/j.jare.2014.02.006 PMID:25685517

Erl, T. (2005). *Service-Oriented Architecture: Concepts, Technology, and Design*. City. doi:10.1109/SAINTW.2003.1210138

Fazio, M., Celesti, A., Ranjan, R., Liu, C., Chen, L., & Villari, M. (2016). Open Issues in Scheduling Microservices in the Cloud. *IEEE Cloud Computing, 3*(5), 81–88. doi:10.1109/MCC.2016.112

Galdon-Clavell, G. (2013). (Not so) smart cities?: The drivers, impact and risks of surveillance- enabled smart environments. *Science & Public Policy, 40*(6), 717–723. doi:10.1093cipolct070

Grund, M., Schaffner, J., Krueger, J., Brunnert, J., & Zeier, A. (2010). The Effects of Virtualization on Main Memory Systems. In *Proceedings of the Sixth International Workshop on Data Management on New Hardware* (pp. 41–46). New York, NY: ACM. 10.1145/1869389.1869395

HAProxy. (2017). *HAProxy - The Reliable, High Performance TCP/HTTP Load Balancer*. Retrieved October 9, 2017, from http://www.haproxy.org/

Harrison, G. (2015). *Next generation databases: NoSQL, NewSQL, and Big Data*. Academic Press.

Hevner, A., & Chatterjee, S. (2010). *Design Research in Information Systems. Design Research in Information Systems* (Vol. 22). Boston, MA: Springer US; doi:10.1007/978-1-4419-5653-8

Hollands, R. G. (2008). Will the real smart city please stand up? *City, 12*(3), 303–320. doi:10.1080/13604810802479126

Ijaz, S., Shah, M. A., Khan, A., & Ahmed, M. (2016). Smart Cities: A Survey on Security Concerns. *International Journal of Advanced Computer Science and Applications, 7*(2), 612–625. doi:10.14569/IJACSA.2016.070277

Jain, A., & Bhatnagar, V. (2017). Concoction of Ambient Intelligence and Big Data for Better Patient Ministration Services. *International Journal of Ambient Computing and Intelligence, 8*(4), 19–30. doi:10.4018/IJACI.2017100102

Julian, S., Shuey, M., & Cook, S. (2016). Containers in Research: Initial Experiences with Lightweight Infrastructure. *Proceedings of the XSEDE16 Conference on Diversity, Big Data, and Science at Scale*. 10.1145/2949550.2949562

Kecskemeti, G., Marosi, A. C., & Kertesz, A. (2016). The ENTICE approach to decompose monolithic services into microservices. *2016 International Conference on High Performance Computing and Simulation, HPCS 2016*. 10.1109/HPCSim.2016.7568389

Kohler, J., Lorenz, R., Gumbel, M., Specht, T., & Simov, K. (2017). A Security by Distribution Approach to Manage Big Data in a Federation of Untrustworthy Clouds. In N. D. S. Tamane & V. K. Solanki (Eds.), *Privacy and Security Policies in Big Data*. IGI Global. doi:10.4018/978-1-5225-2486-1.ch005

Kohler, J., Simov, K., Fiech, A., & Specht, T. (2015). On The Performance Of Query Rewriting In Vertically Distributed Cloud Databases. *Proceedings of The International Conference Advanced Computing for Innovation ACOMIN 2015*.

Kohler, J., Simov, K., & Specht, T. (2015). Analysis of the Join Performance in Vertically Distributed Cloud Databases. *International Journal of Adaptive, Resilient and Autonomic Systems, 6*(2), 65–87. doi:10.4018/IJARAS.2015070104

Kohler, J., & Specht, T. (2012). SeDiCo - Towards a Framework for a Secure and Distributed Datastore in the Cloud. *Proceedings of Chip-to-Cloud Security Forum 2012*.

Kohler, J., & Specht, T. (2014). Vertical Query-Join Benchmark in a Cloud Database Environment. In *Proceedings of 2nd World Conference on Complex Systems 2014* (pp. 143–150). Agadir, Morocco: IEEE. 10.1109/ICoCS.2014.7060950

Kohler, J., & Specht, T. (2015). A Performance Comparison Between Parallel And Lazy Fetching in Vertically Distributed Cloud Databases. In *International Conference on Cloud Computing Technologies and Applications - CloudTech 2015*. Marrakesh, Morocco: IEEE Computer Society. 10.1109/CloudTech.2015.7337005

Kozhirbayev, Z., & Sinnott, R. O. (2017). A performance comparison of container-based technologies for the Cloud. *Future Generation Computer Systems, 68*, 175–182. doi:10.1016/j.future.2016.08.025

Kubernetes. (2017). *Kubernetes - Production-Grade Container Orchestration*. Retrieved October 4, 2017, from https://kubernetes.io/

Li, W., & Kanso, A. (2015). Comparing containers versus virtual machines for achieving high availability. *Proceedings - 2015 IEEE International Conference on Cloud Engineering, IC2E 2015*. 10.1109/IC2E.2015.79

Matallah, H., Belalem, G., & Bouamrane, K. (2017). Towards a New Model of Storage and Access to Data in Big Data and Cloud Computing. *International Journal of Ambient Computing and Intelligence*, *8*(4), 31–44. doi:10.4018/IJACI.2017100103

Mell, P., & Grance, T. (2011, September). *The NIST Definition of Cloud Computing*. Retrieved February 1, 2016, from http://csrc.nist.gov/publications/nistpubs/800-145/SP800-145.pdf

Morabito, R., Kjaellman, J., & Komu, M. (2015). Hypervisors vs. lightweight virtualization: A performance comparison. *Proceedings - 2015 IEEE International Conference on Cloud Engineering, IC2E 2015*. 10.1109/IC2E.2015.74

Ongaro, D., & Ousterhout, J. (2014). In Search of an Understandable Consensus Algorithm. Atc '14, 22(2), 305–320. doi:10.1145/1529974.1529978

OWASP. (2017). *Open Web Application Security Project (OWASP) Web Site*. Retrieved September 29, 2017, from https://www.owasp.org/index.php/Main_Page

Pääkkönen, P., & Pakkala, D. (2015). *Reference Architecture and Classification of Technologies, Products and Services for Big Data Systems*. Big Data Research. doi:10.1016/j.bdr.2015.01.001

Pahl, C., & Lee, B. (2015). Containers and clusters for edge cloud architectures-A technology review. *Proceedings - 2015 International Conference on Future Internet of Things and Cloud, FiCloud 2015 and 2015 International Conference on Open and Big Data, OBD 2015*. 10.1109/FiCloud.2015.35

Patil, S., Polte, M., Ren, K., Tantisiriroj, W., Xiao, L., López, J., … Rinaldi, B. (2011). YCSB++. In *Proceedings of the 2nd ACM Symposium on Cloud Computing - SOCC '11* (pp. 1–14). New York: ACM Press. 10.1145/2038916.2038925

Pearson, S. (2013). Privacy, Security and Trust in Cloud Computing. In *Privacy and Security for Cloud Computing SE - 1* (pp. 3–42). Springer London. doi:10.1007/978-1-4471-4189-1_1

Philip Chen, C. L., & Zhang, C.-Y. (2014). Data-intensive applications, challenges, techniques and technologies: A survey on Big Data. *Information Sciences*, *275*, 314–347. doi:10.1016/j.ins.2014.01.015

Plattner, H. (2013). *A Course in In-Memory Data Management: The Inner Mechanics of In-Memory Databases*. Berlin, Germany: Springer. doi:10.1007/978-3-642-36524-9

Plattner, H., & Zeier, A. (2012). *In-Memory Data Management. Springer*. Berlin, Germany: Springer; doi:10.1007/978-3-642-29575-1

Preeth, E. N., Mulerickal, J. P., Paul, B., & Sastri, Y. (2016). Evaluation of Docker containers based on hardware utilization. *2015 International Conference on Control, Communication and Computing India, ICCC 2015*. 10.1109/ICCC.2015.7432984

Red Hat. (2017). *Ceph Ceph storage - Ceph*. Retrieved October 4, 2017, from http://ceph.com/ceph-storage/

Sadalage, P. J., & Fowler, M. (2013). *NoSQL distilled : a brief guide to the emerging world of polyglot persistence*. Addison-Wesley.

Sarkar, M., Banerjee, S., Badr, Y., & Sangaiah, A. K. (2017). Configuring a trusted cloud service model for smart city exploration using hybrid intelligence. *International Journal of Ambient Computing and Intelligence*, *8*(3), 1–21. doi:10.4018/IJACI.2017070101

Sarwar Kamal, M., Dey, N., & Ashour, A. S. (2017). *Large Scale Medical Data Mining for Accurate Diagnosis: A Blueprint*. Cham: Springer; doi:10.1007/978-3-319-58280-1_7

Schaffers, H., Sallstrom, A., Pallot, M., Hernandez-Munoz, J. M., Santoro, R., & Trousse, B. (2011). Smart Cities and the Future Internet: Towards Cooperation Frameworks for Open Innovation. *2011 17th International Conference on Concurrent Enterprising*, 1–11. 10.1007/978-3-642-20898-0

Song, H., Srinivasan, R., Sookoor, T., & Jeschke, S. (2016). *Smart cities: Foundations, principles, and applications*. Academic Press.

Swoyer, S. (2012). *Big Data - Why the 3Vs Just Don't Make Sense*. Retrieved August 11, 2014, from http://tdwi.org/articles/2012/07/24/big-data-4th-v.aspx

Tanenbaum, A. S., & Van Steen, M. (2007). *Distributed Systems: Principles and Paradigms, 2/E. Communication*. Retrieved from http://www.pearsonhighered.com/academic/product/0,0132392275,00+en-USS_01DBC.html

TPC. (2016). *TCP Website*. Retrieved February 1, 2016, from http://www.tpc.org/

Varghese, B., Subba, L. T., Thai, L., & Barker, A. (2016). Container-based cloud virtual machine benchmarking. In *Proceedings - 2016 IEEE International Conference on Cloud Engineering, IC2E 2016: Co-located with the 1st IEEE International Conference on Internet-of-Things Design and Implementation, IoTDI 2016*. IEEE. 10.1109/IC2E.2016.28

Viennot, N., Lécuyer, M., Bell, J., Geambasu, R., & Nieh, J. (2015). Synapse: A Microservices Architecture for Heterogeneous-Database Web Applications. *Proceedings of the Tenth European Conference on Computer Systems - EuroSys '15*. 10.1145/2741948.2741975

Wang, P., Ali, A., & Kelly, W. (2015). Data security and threat modeling for smart city infrastructure. *2015 International Conference on Cyber Security of Smart Cities, Industrial Control System and Communications, SSIC 2015 - Proceedings*. 10.1109/SSIC.2015.7245322

Yang, W., Wang, X., Song, X., Yang, Y., & Patnaik, S. (2018). Design of intelligent transportation system supported by new generation wireless communication technology. *International Journal of Ambient Computing and Intelligence*, *9*(1), 78–94. doi:10.4018/IJACI.2018010105

Yu, H., & Huang, W. (2015). Building a Virtual HPC Cluster with Auto Scaling by the Docker. *Computing Research Repository*. Retrieved from http://arxiv.org/abs/1509.08231

Zhang, K., Ni, J., Yang, K., Liang, X., Ren, J., & Shen, X. S. (2017). Security and Privacy in Smart City Applications: Challenges and Solutions. *IEEE Communications Magazine*, *55*(1), 122–129. doi:10.1109/MCOM.2017.1600267CM

Zhang, W., Qi, Q., & Deng, J. (2017). Building Intelligent Transportation Cloud Data Center Based on SOA. *International Journal of Ambient Computing and Intelligence*, *8*(2), 1–11. doi:10.4018/IJACI.2017040101

Zheng, Y., Capra, L., Wolfson, O., & Yang, H. (2014). Urban Computing: Concepts, Methodologies, and Applications. *ACM Trans. Intell. Syst. Technol.,* *5*(3), 38:1-38:55. 10.1145/2629592

Chapter 4
Cloud- and IoT-Powered Smart Connected Cities

Pankaj Kumar
National Institute of Technology Hamirpur, India

Lokesh Chouhan
National Institute of Technology Hamirpur, India

Ankit Songara
National Institute of Technology Hamirpur, India

ABSTRACT

The internet of things (IoT) is rising at a stunning rate. Consistently, every day, the presence of new gadgets, cameras, cell phones, and sensors these are associated with the internet. It is envisioned IoT will find, coordinate, and make utilization of such gadgets and their information in the advance of new service and items that can change and emphatically blow our lives. Presently, IoT and cloud computing are the hottest area used by the researchers. In any case, to the best of the authors' insight, these works do not have a point-by-point investigation of the new Cloud-IoT worldview, which includes totally new applications, challenges, and research issues. To connect this hole, this chapter gives a review on the reconciliation of Cloud and IoT. Beginning by describing basics of both IoT and cloud computing, the authors examine their complementarity, itemizing what is at present heading to their integration. At the end this chapter, the authors describe the smart roadside assistance for smart communities.

DOI: 10.4018/978-1-5225-6207-8.ch004

INTRODUCTION

Internet of things (IoT) is a new revolution that is speedily growing in the field of the technology world. The main motivation of IoT concept to provide advanced connectivity and pervasive presence of things such as actuators, RFID tag, and sensing device (Bitam, 2015). Unique addressing schemes are available to connect and communicate with the things or object over the internet, e.g., smart homes, car sensors, appliances in smart cities, robots, etc. In IoT, connectivity and near-field communications create enormous new interactive features which are presuming to provide a promising area to upgrade the role of industrial systems such as transport and industrial systems. Systems authority will be able to track vehicles location, predict its future and possible road traffic. IoT can be acknowledged in three ideal models: 1) Internet situated (middleware), 2) Things arranged (sensors), 3) Semantic arranged (information). This result in the generation of tremendous measures of data which must be stored processed and presented in a consistent, proficient and effectively interpretative frame (Gurav, 2010.). These models will represent of an organization that are wares and address in a way like customary wares. cloud computing can give the virtual foundation to such utility processing which coordinates checking devices, storage, analytics, perception stages and customer conveyance. Internet of things (IoT) is a fundamental part of the today's development of the smart city. Individuals could remotely get to and interface with an extensive variety of gadgets coordinated with sensors, from home apparatuses, wearable gadgets to ecological screens. With such huge scope potential in our day by day life, IoT with lessened vitality utilization (the 'green' quality) has attracted more and more attention. In later a long time, vitality effective systems administration and processing have been broadly considered from numerous points of view, for example, the structure plan, the calculation outline, and the asset reusing

Although a few studies on IoT have been led, none of them is specifically focused on building parts of IoT. With the point of investigating IoT architecture designs, this study is conducted. Accordingly, this work is propelled by the requirement for arranging models, as will be required later on to oblige trillions of devices.

The commitments of the article are various: First, this proposal investigate, highlight, and report premier research propels made in IoT design recently.

- Then this chapter is categorize and classify IoT structures and devise a scientific categorization
- Then identify and plot the key prerequisites for future cloud based IoT architecture and smart city concept.
- A few prominent case studies on smart city based cloud of things are discovered and presented

IoT EVOLUTION

Norman Joseph started the concept of "web of things" in 1949 when he invented standardized tag known as bar code after then got the patent for the standardized tag, then Morton heilig developed "Sesorama" in the year 1950, the primary head mounted to show that gave a client the experience of riding a motorcycle. After that internet act as the primary segment in the year 1962, which started as the important component in Defense Advanced Research Projects Agency (DARPA) then enhanced the technology and it changed into ARPANET (Gurav, 2010.) (He W Yan, 2014.).

Similar to traditional post system in the computer network, the ability to communicate between two end points to understand the sender and receiver location, find how to relay the message to the destination. Standardization of message and results taking ability, where deciding factor that makes today internet a ubiquitous network of global connecting. This standardization attributed to internet protocol suite. The global addressing system in internet protocol (IP) suite is called IPv4 that provide approximately 4 billion unique addressing for the end point of communication.

Today world population has already surpassed 7 billion people. As consequences, the unique addressing has to be share for security perspective and communication between the end device and user. The need of IoT indicates that the user wants to connect with the device to interact with each other. It is an informational network that used for monitoring of information about real-world. The interfacing between the Internet and machine to machine (M2M) is done by communication that considered as most important part of IoT network (Jiang L. D., 2014).

Later in section, will discuss the key aspects of IoT that are used to setting up of secured communication, accessing the resources, user identification and authentication. In addition, with the growing technology the IoT need a fast wireless Internet access such as Wi-Fi and 4G-LTE to maintain the devices and network connectivity 4G technology is good for current scenarios, but after five or ten years 4G technology will not meet the future requirement, then it will shift to 5G technology to meet the requirement, to provide end to end latency, improve coverage and enhance the data rate.

Internet of Things Vital Research and Innovation Motivation

IoT will make a tremendous collection of billions of object imparting each other. IoT isn't subversive unrest over the current advances, it is the complete usage of existing advances, and it is the production of the newly similarity modes.

In IoT, applications, administrations, middleware parts, systems, and end hubs will be basically composed and utilized as a part of completely new ways. IoT offers a way to investigate complex procedures and connections.

The IoT implies a cooperative communication between the genuine/physical and the advanced/virtual universes as shown in Figure1: physical elements have advanced partners and virtual portrayal; things move toward becoming setting mindful and they can detect, convey, connect, and trade information, data, and learning. New open doors will meet business necessities, and new administrations will be made in light of continuous physical world information. Everything from the physical or virtual world will perhaps be connected by the IoT (A. Munir, 2017). Connectivity between the things might be accessible to all with minimal effort and may not be claimed by private elements. For IoT, smart adapting, quick sending, best data comprehension and deciphering, against misrepresentation what's more, pernicious assault and security insurance are fundamental necessities.

Internet of Things Standards and Infrastructure

The IoT standard framework contains the architecture standards, the application necessities measures, the communication protocols principles, the distinguishing proof measures, the security principles, the application measures, the information benchmarks, the data handling measures, and the general population benefit stage principles. The structures of these IoT frameworks are divided and can't correspond and coordinate the information from various storehouses; these segregated IoT

Figure 1. Internet of things basic scenario

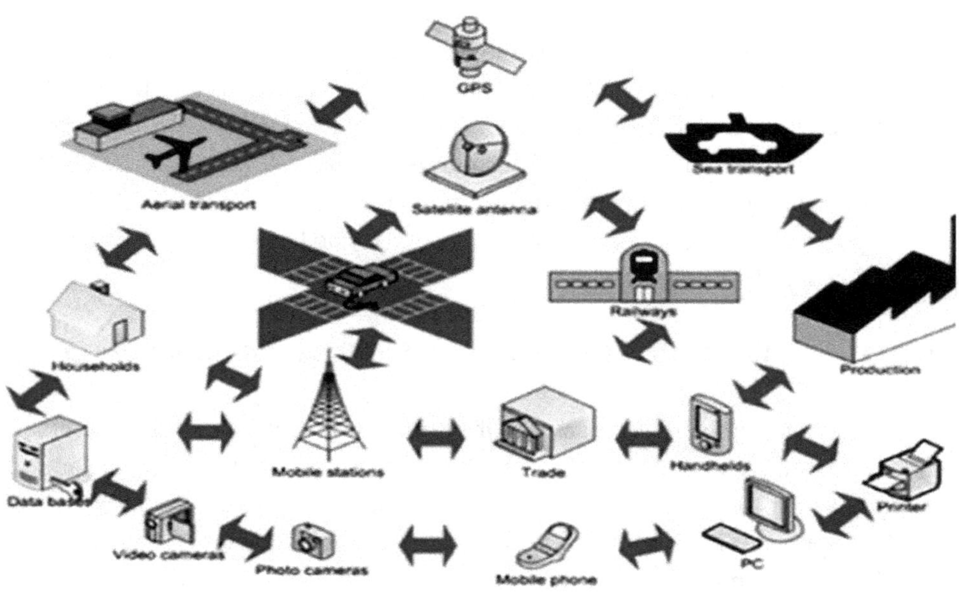

arrangements utilize private conventions and cause numerous issues in data sharing, innovation multiplexing, arrange administration, and updating every one of these issues are obstructing the advancement of IoT. In a request to reduce the aggregate IoT cost and offer data, we have to incorporate various capacities and assets into a bigger framework. IoT in this way should be planned with an open and IoT design with open interfaces and assets, considering distinctive business situations, application-based necessities, and current innovations (T.H. Luan, 2016) (A. Botta, 2014).

The advancement of the present web into the system of the connected object that not just reaps information from the environment and interface with the physical environment actuation. In addition, it utilizes existing internet gauges to provide services to data exchange, analytics, applications, and interchanges. In IoT, biological systems are formed known as smart things, for example, small physical gadgets as far as memory limit, calculation ability, vitality independence, and communication abilities, by using the communication interface such as Bluetooth, RFID, sensor, Wi-Fi, and actuator.

Internet revolution prompted the association between individuals at an uncommon level and pace. Future revolution will be the association between items to make smart objects. In 2011, a number of IoT devices on the planet surpassed the total number of world population. At present, there are 9 billion IoT devices and it is estimated that it will achieve 25 billion gadgets by 2020. As indicated by the global system for mobile communications (GSMA), this adds up to 2.3 trillion (dollar) income open doors for the versatile system alone spreading over vertical fragments, for example, health sectors, smart transport, utilities and custom hardware (A.V. Dastjerdi, 2016.). The clients traverse from a person to national level associations tending ranging issues.

An open and generic IoT design is an incorporated solution with interoperability. It will have the following attributes:

- **Standard Interface and Convention:** By comparing different private IoT frameworks, a generic IoT foundation has a similar equipment and programming interfaces, and conventions.
- **Open and Working:** A general IoT architecture framework is deployed to assume control open IoT applications with open working capacity. An open IoT framework would thus be able to incorporate various IoT applications into one design.
- **Open, Versatile, and Adaptable:** An open IoT architecture with open assets, open norms, and open interfaces can without much of a stretch broaden its usefulness and the size of execution. It would thus be able to adjust to various necessities including specialized improvements adaptably.

Rapid advances in communication and technology abilities of these small gadgets in the course of the most recent decade have made correspondence over longer separations with forced continuous imperatives, feasible

The association of these IoT gadgets with inheritance frameworks requires the effective and secure use of the internet communication foundation. Therefore, not only IoT devices set up and utilize remote correspondence to pass on a status report of a specific event or action, yet can likewise give end-clients a more elevated amount of trust in the security and realness of all administrations gave in that. The level and method to give protection and security to an IoT foundation will change in assorted variety, depiction, and in addition many-sided quality, contingent on the IoT gadgets being used. For example, a cell phone associated with a 3G correspondence channel to the internet will execute an alternate arrangement of conventions for securing the channel, instead of an asset obliged RFID tag, that will perpetually have a tweaked form of an asset requesting mystery key verifier. The design of internet of Things stage is outlined in Figure 2. It also highlights the standard and protocols used. The platform has three primary parts: the sensor and actuator systems, the internet of Things (IoT) server and the client interfaces for representation and management (S. Chen, 2015).

- **Sensor and Actuator:** It convey in a reliable bidirectional path with the IoT server: they can utilize remote conventions (e.g. ZigBee, wi-fi, Bluetooth

Figure 2. The IoT network infrastructure with few protocols and standards

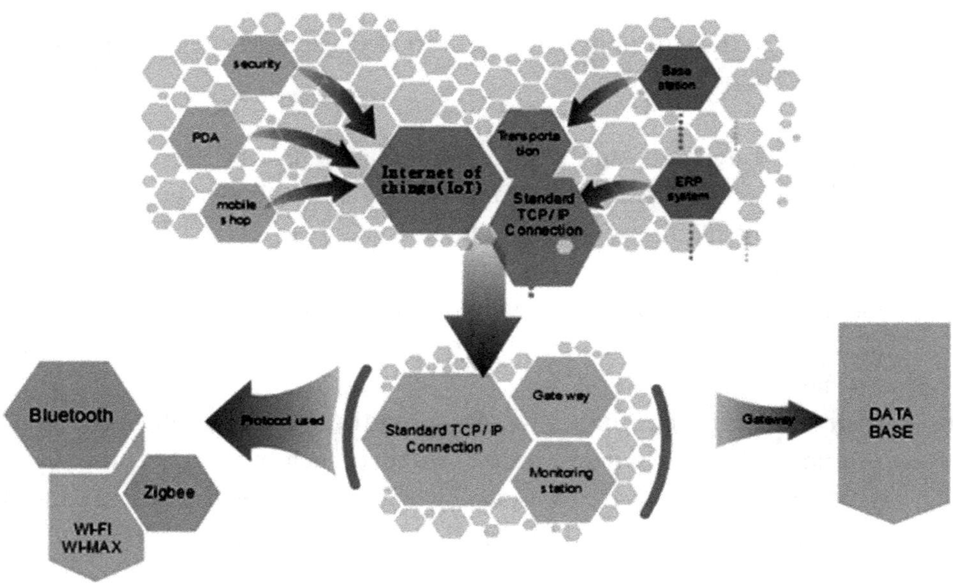

Low Vitality, and so on.) or wired conventions. The correspondence amongst nodes and the IoT server takes after the TCP/IP customer server demonstrate, straight forwardly or through an entryway. Sensors send messages in their local arrangement to the IoT server, where the information administration unit extricates data and enters it in an "all inclusive" organization in the sensor database. At the point when sensors should be arranged or cross-examined, the configurator unit readies an order as indicated by the objective sensor convention.

- **IoT Server:** The IoT server changes over the crude payload from heterogeneous hubs into a "universal" configuration, containing object identifier, question sort, an estimation unit, information field, geological position, and timestamp. At that point, it makes information accessible to applications and clients. Along these lines, information representation and handling is isolated from estimation and information gathering and does not have to consider the correspondence convention of the beginning source. In addition, the IoT server gets information from clients with a specific end goal to arrange and deal with the Sensor and actuator.
- **User Interfaces:** It enables clients and outsider programming to get to live and historical sensor information. Similar interfaces additionally permit clients with organization benefits to oversee systems and single nodes.

This chapter introduces a taxonomy that will help in characterizing the segments required for the IoT segment perspective. Three type of components of IoT that empower consistent ubiquitous computing are: a) sensors, actuators correspondence equipment. b) Middleware - on request capacity and processing instruments for information investigation. c) Presentation - novel straightforward perception and interpretation tools which can be generally gotten to on distinctive stages and which can be intended for various applications.

This section discusses a few enabling technologies of IoT.

Radio Frequency Identification (RFID)

RFID technology is a leading achievement in the inserted correspondence worldview which empowers plan of microchips for remote information correspondence. It also offers assistance in program ID of the device that is going too connected about as an electronic scanner tag. RFID and Electronic Product Code (EPC) are the essential technologies in the internet of things. RFID tag made of the silicon chip and an antenna, RFID tags uses electromagnetic fields to automatically identify and track tags attached to objects. EPC global Network (EPC Network) provides support for development and implementation for RFID technologies (H. Gupta, 2016). The module of RFID used in the vast variety of commercial applications such as aces control, automatic identification, and Inventory management etc. RFID consists of three components: A Smart Label, RFID reader, and an antenna. Power consumption and operating time are major factors in battery powered RFID readers.

Wireless Sensor Networks (WsN)

Recent technological advances in low power coordinated circuits and remote correspondences have made accessible effective, minimal effort, low power small devices for use in remote detecting applications. The combination of these factors has enhanced the practicality of using a sensor arranged to comprise a substantial number of intelligent sensors, empowering the accumulation, preparing, examination and dissemination of important data, accumulated in an assortment of conditions. Active RFID is almost the same as the brought down end wireless nodes with constrained handling capacity and storage. The challenges are to understand the huge capacity and analysis of network are considerable. Sensor information is shared among sensor hubs and sent to a dispersed or concentrated framework for examination.

WsN Hardware and Communication Protocol Stack

In wireless network contains system interfaces, system handler, handset units and power supply. Quite often, they contain different A/D converters for sensor interfacing and more present-day sensor hubs can impart utilizing one recurrence band making them more adaptable. The nodes are relied upon to be conveyed in a specially appointed way for general applications (S. Gilbert and N. Lynch, 2002). Planning a suitable topology, steering and media access control (MAC) layer is basic for versatility and lifespan of the conveyed arrange. Nodes in a WsN need to convey among themselves to transmit information in one or multi-distance to a sink. Node dropouts and subsequently corrupted system lifetimes are visited. The correspondence stack at the sink node ought to have the capacity to connect with the outside world through the internet to go about as an entryway to the WsN subnet and the internet (J. Han, 2011,).

Organization of electrical and electronics designers (IEEE) and the internet engineering task Force (IETF) are outlining new interchanges and security conventions that will assume a central part in empowering future IoT applications.

These techniques are outlined in accordance with the requirements and attributes of devices and low rate remote correspondences, such qualities have additionally impacted past outlines the designs of applications employing remote sensor systems

(RSN) detached from the internet and various research recommendations on security mechanisms, the new institutionalized arrangements are being intended to ensure interoperability with actual internet devices and ensure that detecting devices can communicate with other internet substances with regards to future IoT circulated applications (E. Brewer, 2012) (J. Han, 2011,). The communication protocols accessible or being planned at the IEEE and IETF presently empower an institutionalized protocol stack talked about in and delineated in Figure 3. The mechanism framing this stack should accordingly empower communications protocols including obliged detecting devices while adapting to the Figure3.

Communication protocols in the IoT, prerequisites of low-energy communications situations and the objectives and the lifetime of IoT applications. From a base of the approach, the accompanying is the principle characteristics of the different protocols in this stack (Brewer, 2012) (E. Brewer, 2012):

- IEEE 802.15.4 sustained the low power communications at the physical and medium access control layers. In this manner, sets the rules for interchanges at the lower layers of the stack furthermore, lays the ground for IoT interconnection conventions at upper layers.
- Low-energy communication situations utilizing IEEE 802.15.4 extra at most 102 bytes for the transmission of information at higher layers of the stack, esteem substantially less than the greatest transmission unit (MTU) of 1280 bytes required for IPv6. The 6LoWPAN adjustment layer tends to this viewpoint by empowering the transmission of IPv6 parcels over IEEE 802.15.4. 6loWPAN moreover executes instruments for bundle fracture and reassembly, among different functionalities.

Figure 3. Communication protocols in the IoT

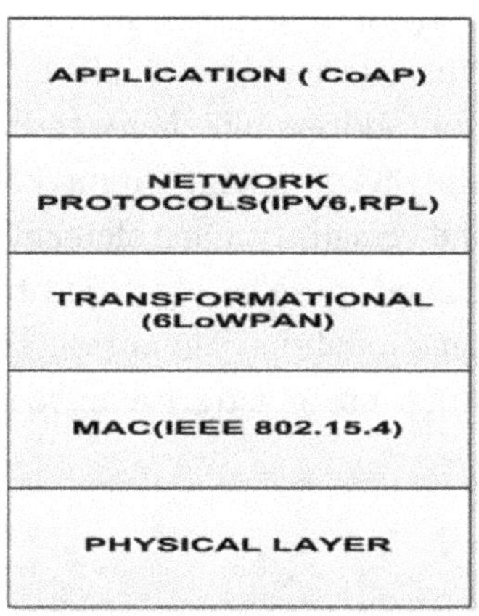

- The Constrained Application Protocol (CoAP) supports transmission at the application layer. CoAP protocol is right now being planned at the IETF to give interoperability in conformance with the illustrative state exchange design of the internet.

Secure Information Aggregation

An efficient and secure communication solution is required to enhance the lifetime of the network. It also guaranteeing actual information gathered from sensors. Node failure being a typical characteristic of WSNs, network topology ought to have the ability to recover itself.

Middleware

A system to combine digital foundation with an administration situated design (ASD) and sensor systems to give path to heterogeneous sensor assets in a sending freeway. This depends on the idea to isolate assets that can be utilized by a few applications. A platform with having the free middleware for creating sensor applications is required, for example, an Open Sensor Web Engineering (OSWA).

Addressing Schemes

The capacity to exceptionally recognize 'Things' is basic for the achievement of IoT. This won't just enable us to interestingly recognize billions of devices yet additionally to control remote devices through the internet. On the internet, the addressing schemes are used in a hierarchical manner, the reason being that in the router are design to forward packet is taken by looking destination address. Since the receiving the packet from a host and it may have to forward to any host in the network. So it is necessary that the router must have the forwarding table for the entire possible host which is connected to the internet, this would make the forwarding table to very large network and therefore addressing schemes are designed to be hierarchical.

The few of most basic highlights of making a remarkable address are uniqueness, dependability, persistence, and versatility. Each element that is as of now associated and those that will be associated must be identified by their novel recognizable identification, location, and functionalities. The current IPv4 may support to an extent where a gathering of cohabiting sensor gadgets can be recognized geologically, be that as it may, not exclusively.

Data Storage and Analytics

Cloud and big data are in demand and need these days. IoT devices connect such abundance amount of data and it cannot store in the form of the table. So it needs to depict them in real time so that intelligent decision can be made correctly. Subsequently, server farms which keep running on the gathered vitality and which are brought together will guarantee vitality productivity and in addition unwavering quality (EU). IoT devices collect data and big data system analyzes them. Data analytics can churn out all kind of analytics using the different techniques (P. Domingos and G. Hulten, 2000). The information must be put away and utilized cleverly for shrewd checking and activation. It is essential to create artificial intelligence algorithms which could be incorporated or circulated based on the need.

Novel combination algorithms should be designed to comprehend the data gathered. These frameworks indicate attributes, for example, interoperability, mix, and versatile interchanges. They additionally have a modular engineering both as far as equipment framework configuration and also programming advancement and are generally extremely appropriate for IoT applications. Cutting edge nonlinear, worldly machine learning strategies in view of a transformative calculation, hereditary calculations, neural systems, and other computerized reasoning methods are important to accomplish mechanized basic leadership.

Figure 4. Roadmap of key technological with regards to IoT application areas

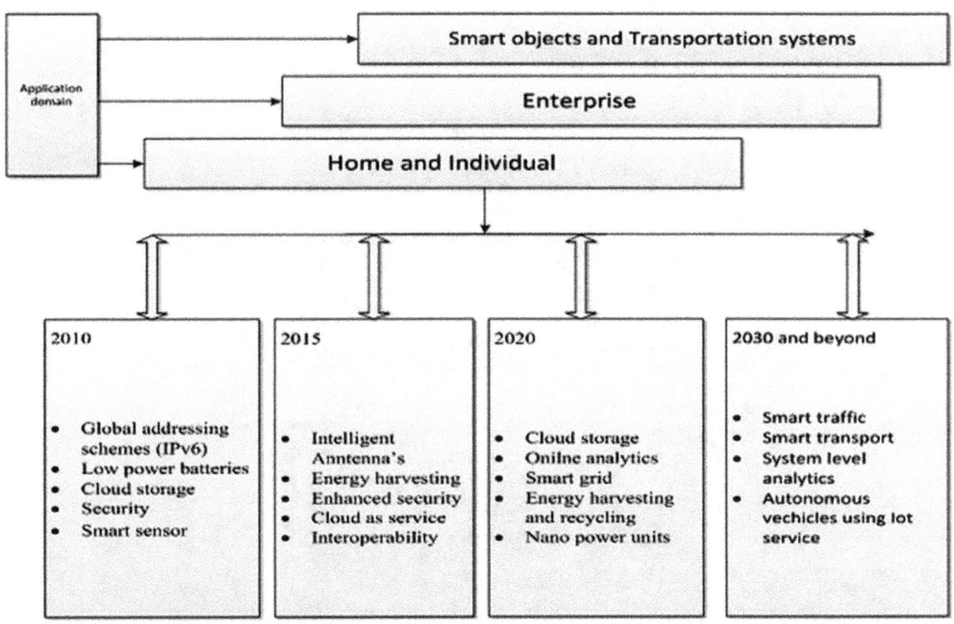

Some of the open challenges are discussed on the IoT components displayed before. The challenges incorporate IoT particular difficulties, for example, security, participatory detecting, data analytics and geographic information system (GIS)-based perception separated from the standard WSN challenges counting engineering, vitality proficiency, security, conventions, and Quality of Service. Figure.4 describe the roadmap of technologies that how technologies enhanced with the time from the year 2010 to the upcoming year 2020.

APPLICATIONS OF IOT

The most majority of the day by day life applications that we typically observe area of now smart environment, yet they can't speak with each other what's more, empowering them to speak with each other and offer valuable data with each other will make an extensive variety of creative applications. These developing applications with some self-sufficient abilities would positively enhance the nature of our lives. A couple of such applications are as of now in the market, we should take the case of the Google Auto which is an activity to give a self-driving auto involvement with continuous movement, street conditions, climate and other data trades, all because of the idea of IoT.

There are various conceivable future applications that can be of awesome preferred standpoint. In this area, we exhibit few of these applications as shown in figure 5 (A. Botta, 2014).

Figure 5. Application areas of IoT

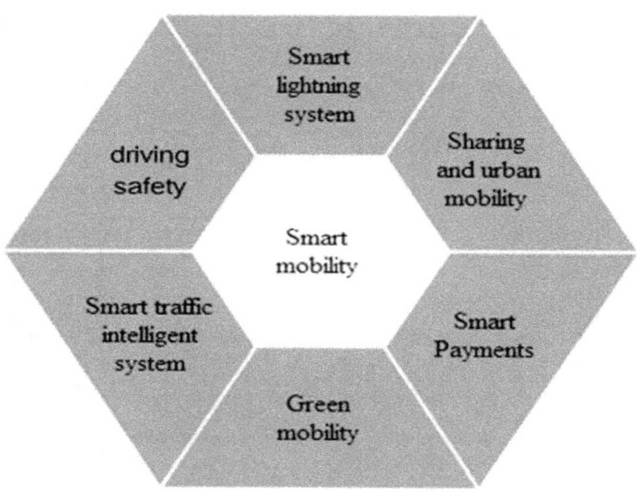

Smart Homes

Smart homes application bring revolution step of success in the technology world. It is likely the most popular feature of IoT application which is used on large scale, nowadays most cheap and readily available applications are used by the consumer such as heaters, fan, lights, Air conditioner (AC) etc. but the setup cost of smart homes are very expensive. Smart homes make the object to more demanding, more helpful, and more comfortable.

Smart Parking

Smart parking used for optimizing vehicles parking spaces and find out the arrival and departure of time of vehicles, which saves the time and fuels of vehicles. Everybody who has ever been disappointed driving around urban regions looking for stopping has longed for an answer that could rapidly lead them to that slippery spot. This worry pulled in vital ventures from devoted industry segments to support stopping incomes through technology-enabled solutions. Parking industry is being upset by new advancements that empower urban areas to decrease activity clog and carbon discharge. IoT penetrates with the world of parking to streamline forms that convey smart parking solution, which broadens and oversee parking inventories. In this specific context, IoT utilizes embedded sensor's systems to associate physical parking spot frameworks with data and correspondence technologies, where cloud-based intelligent services are given. This bury availability move is additionally driving socio-economic changes, where information released from physical frameworks is prompting profitability increases through new applications and new plans of models (A. K. Dey, 2001.).

Smart City

A Smart city is another powerful application of IoT which includes smart surveillance, smart energy management, water distribution, smart security and environment monitoring. Smart City is the result of the quickened advancement of the new age data innovation and information-based economy, in view of the system combination of the internet and media communications organize, communicate arrange, remote broadband system and different sensors systems where IoT innovation as its center. The principle highlights of a smart city incorporate a high level of data innovation joining and a far-reaching use of data assets. The basic parts of urban advancement for a smart city ought to incorporate smart innovation, brilliant industry, smart administrations and brilliant life.

Agriculture Purposes

IoT devices are design to check the soil moisture and level of nutrients, and investigate the requirement of fertilizer. As we all know World population grew at very high rate, which increases supply of food demand. Smart farming helps the farmer to increase their productivity.

E-Health Care

IoT Research in healthcare sector shows empowering people to live healthier. Sensor, actuator gather information and relay to other end for analysis of an individual's health. It will reduce the risk of sudden death cases by pre-analysis of patient/ individual. IoT frameworks connected to healthcare enhance existing innovation and the general routine with regards to medicine. They grow the span of experts inside an office and a long ways past it. They increment both the exactness and size of medical data through different information gathering from extensive arrangements of true cases. They likewise enhance the exactness of medical care conveyance through more complex joining of the social healthcare framework.

The existing internet protocol will be unable to deal with billions of smart device. Hence, how to interface a few objects to the web and how to distinguish extraordinarily the information created from those items is the principal worry in web arranged vision. Things-situated vision essentially alludes to the following of each question with specific innovation, for example, RFID tag. RFID tag is most well-known innovation in setting up an IoT domain. Cloud computing is a computing paradigm, where all figuring assets, for example, memory, storage, preparing abilities are leased from outsider supplier through the internet (P. Barnaghi, 2015).

This paradigm is dissimilar to the conventional method for calculation, where the clients have finish set of assets in an independent PC to play out any calculation and processing undertaking. In cloud computing, every one of the assets is accessed to through internet application and it is a potential stage for various organizations, associations, analysts for its capacity to scale up and downsize the assets according to the prerequisite.

INTEGRATION OF CLOUD COMPUTING WITH INTERNET OF THINGS

IoT depends on smart and self-healing devices that are interconnected with dynamic system framework. It represents to a standout amongst the most troublesome advancements, empowering omnipresent and unavoidable figuring situations. IoT characterized and largely portrayed by certifiable smart things, generally circulated, with constrained capacity and handling limit, which includes concerns with respect to execution, dependability, security, and privacy (P.R, " Jul – Aug. 2012.).

Need of Integration of Cloud With IoT

The Internet made a revolution in the history of networking, gives methods for asset sharing to worldwide clients. At present, the internet worldview is marginally altered with the development of cloud processing, which made assets accessible to everybody with an immense vault of a repository of data storage. Cloud computing with enormous information applications gives unmistakable advantages including expanded system get to, adaptability, use viability, and asset sharing. Because of this, the distributed computing worldview appropriation is broadly expanding and actualized in different areas like social communication, government, social health applications, scientific selections, etc.

In this specific circumstance, ability to uniquely identify things is basic for the achievement of IoT.

Integration of cloud-Iot provides different services such as configure connections, verify data, view data provide security and storage in very efficient manner. Cloud and IoT have seen a free development. Be that as it may, a lot of basic favorable position is the consequence of their reconciliation has been distinguished in the literature; anticipate what's to come in future. From one perspective, the internet of things can profit by cloud practically boundless limit and assets to compensate for the specialized constraints. Specifically, cloud computing can give a successful idea for management of IoT and piece and utilization of things or information applications. Cloud computing is sub-area of the IoT and it can take benefit from IoT, by extending out its degree to manage things, in reality, more disseminated and dynamic route and give new benefits on countless life situations (R. Buyya, 2008.) (A. Botta, 2014). The correlative qualities of CC and IoT are alluring a direct result of the distinctive recommendations detailed in the survey and empowering CC-IoT worldview. CC-IoT gives a successful and effective arrangement, for example, following and overseeing anything whenever from wherever to utilize a custom entryway and inherent applications. The integration of both technologies take care of the vast majority of these issues likewise giving an additional functionality, for example, simple entry, convenience, and diminished sending costs. Utilizing Cloud-IoT worldview to make smart administrations and utilization of new scene in light of the extension of the cloud by objects such as Video Observation, detecting and actuation, sensor, and database as a service.

Design Issues of Cloud –IoT

Cloud-IoT will make more advancement in the field of technology world by creating the business and research scope for the future generation. Privacy and security are the main design issues, security turns out to be vital in the cloud, where there is a substance of private and open clouds, utilized by organizations. It won't be that easy to permit everything turns out to be a part of IoT and afterward having every one of the assets accessible through the cloud. Some of the issue that must be dealt with to permit cloud-IoT for the advancement of the technology. Heterogeneous systems will be included, with the unique set of information and services. The system must have the adaptability to help a wide range of information, as indicated by their necessities, with the quality of service. Some of the design issues are discussed below (al, 2009.) (A. Botta, 2014).

Managing User Access and Devices

If there arise an occurrence of cell phones, similar to portable sensor hubs on vehicles and different articles, need mapping in the new system they have quite recently entered, then the nodes are verified by using unique identification over the internet in IoT environment. In large scale network, IPv6 is used to help sort of pervasive systems by providing IP addressing to the things.

Identity Relationship Management

The user access management has evolved identity relationship management; the ability to access data "anytime anywhere" has created the business to change the way to secure data. Mostly user access transforms the traditional security approach to identity security model, where security is not defined based location of user weather inside the firewall or outside, nowadays the average person carry 3 devices, then the traditional model does not apply for this every device has to be identified uniquely or individual. Adding connectivity to the devices leads to poor authentication in typical IP-connected IoT devices such as smart light bulbs, IP cameras, NAS and wireless printers. When the traditional model shift to new Identity Relationship model where the people and devices are connected, people have access to system facility to connect with another device. The concept of a user has changed to identity and the access to the relation. Need certain ACLs and RBAC policies can be implemented with adequate technological support, token granting services for things belong to a group.

Energy Efficiency

Energy efficiency is one of the important issues; in which sensor organizes all over the place and network with the cloud will prompt a considerable measure of information correspondence, which devours a lot amount of energy. A normal wireless node is made out of four parts: a detecting unit, handling unit, handset, and power unit. In the event of video detecting, encoding, and translating, strategic maneuvers an imperative part. Regularly, video encoding is more mind-boggling, as thought about to disentangling. Having proficient use of energy resources, efficient services of resources are required. There ought to be implied for sensors to create control from the surrounding. Additionally, viable rest mode can be convenient for the sensor for enhancing the life of the sensor.

Protocol Support

Protocols are used for various things to be associated with the web. Regardless of the possibility that there are homogenous elements, but in case of different IoT devices the sensors have to deal with the different platform. A portion of the protocols will be upheld at the gateway, due to the independent platform of devices, make the protocols do not have a support, everything relies on the gateway and the sensor being utilized. Mapping of a standardized protocol for the protocol supports at the entrance gateway.

Security and Privacy

There is no overall framework for how to recognize and manage identities across different solutions. There is no single standard for connecting devices to the internet of things. Lack of standardization, devices, and sensors speak a lot of different protocols and most of them are not HTTP. That is why application development in the IoT is hard to be implemented. There is a lack of decent application integration layers (A. K. Dey, 2001.). If the rate of IoT device growth continues, there will be billions of IoT devices that collect, store, and transfer data through networks. This information could easily be used to create detailed accounts of private lives of millions of users. These devices can even be triggered to send information autonomously or by default, without user awareness.

ARCHITECTURE MODEL FOR CLOUD-IoT

Cloud computing and IoT are two altogether different innovations that are both as of now part of our life. Their selection and utilize are required to be increasingly inescapable, making them critical segments of the future internet. A novel paradigm where Cloud and IoT are combined is anticipated as problematic and as an empowering agent of countless situations (Daniel J. Abadi, " 2009) (Bitam, 2015).

Then again, Cloud computing has for all intents and purposes boundless abilities as far as capacity and processing power is a significantly more develop innovation and have the majority of the IoT issues at any rate in part settled. Along these lines, a novel IT worldview in which Cloud and IoT are two integral innovations consolidated is relied upon to upset both present and Future internet. Integration of cloud and IoT, generally call this new paradigm as Cloud – Internet of things. In this segment it describes cloud-IoT Structure to deal with the real-time information created from different smart devices and describe how the communication process appeared in figure 6, the proposed cloud -IoT structure can be isolated into cloud layers, the interface layer and IoT layer with different important segments (Solanki, 2016).

This framework also motivated to plan the cloud stage that can deal with the real-time information. In spite of the fact that, IoT is the response for many issues while making human life more agreeable and extravagant, this additionally presents many difficulties. Capacity and registering ability are the significant difficulties of IoT gadgets, for example, sensors (MafruzZamanAshrafi, 2007.).

Figure 6. Three layer architecture of cloud-IoT

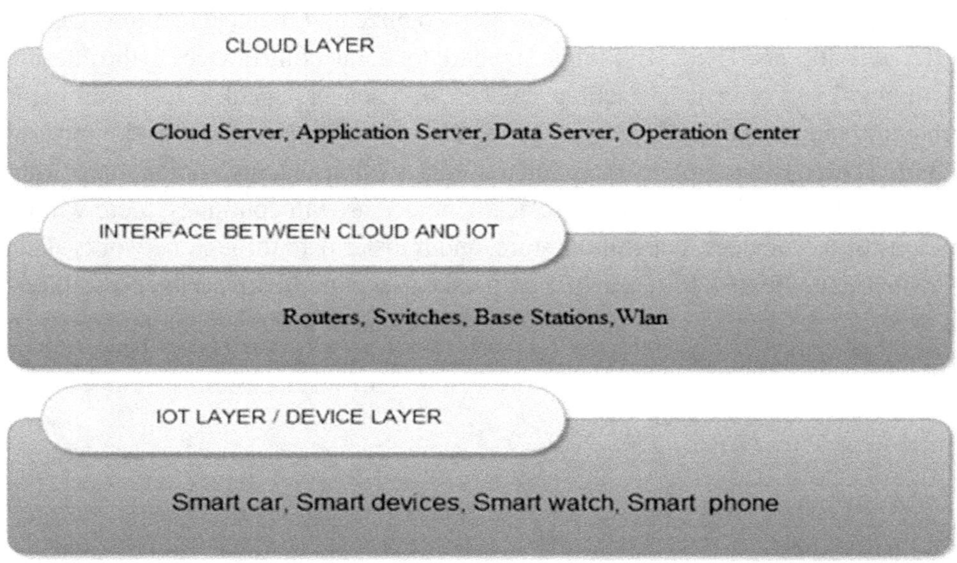

Cloud plays a very vital role in this architecture for cloud representative, evaluate, purchaser, and transporter. Out of sight, Cloud services for career purposes act as the spine of the cloud condition as all interchanges are made through the cloud transporter. IoT application will be used for various usages in daily life and it will change the traditional lifestyle of people too smart lifestyle. The proposed architecture of cloud –IoT gives a powerful designing, where services are enabled by the Cloud-IoT service provider in the form of services such as software as a service (SaaS), platform as a service (PaaS), and infrastructure as a service (IaaS).

Cloud-IoT also provides the functionality to smart things such as smart healthcare, cities, smart transport and logistics etc. As shown in figure 8 there are three layers (Bitam, 2015) (Gurav, 2010.):

1. IoT sub-layer includes all the wireless smart devices such as the smart watch, smartphone, and smart traffic, these devices can produce real-time information and continuously sense and monitoring the environment with the help of the wireless sensor. The information is produced so often that the total measure of information progresses toward becoming in a huge number of information and later on collected at cloud storage.
2. Middle/Interface layer gives the interface between the IoT and cloud layer. Interface layer relays the beacons from one place to another with the help of the above hardware devices for further computation and storage segments, for example, base stations, LAN, Bluetooth, RFID, and NFC. Cloud computing is the center of the internet of Things, which implies joining figuring and availability in every one of the things around us.
3. Cloud layer intended to give essentially storage and analytical capabilities; cloud services can make the computational task very efficiently. Integration of cloud-IoT makes the framework more effective; real-time information is being handled by cloud handler layer (CHL) (D. Giusto, 2010.). Diverse sorts of solicitations, for example, demand to investigate satellite information, therapeutic picture and so forth can be created by clients in IoT layer. On the other hand, real-time and group information additionally created by IoT layer. On the basis of information, cloud layer expects that all the broadened information and solicitations require the particular arrangement of projects to store and to process. Cloud receives the request with the help of virtualization segment in the CHL through any sort of sensors, for example, temperature and pressure sensors. The received request got demand can be for the computation and storage purposes.

Cloud layer included these types of sub layers as shown in figure 7 (D. Giusto, 2010.) (A. Botta, 2014):

Figure 7. Layer-wise design of the Cloud Structure

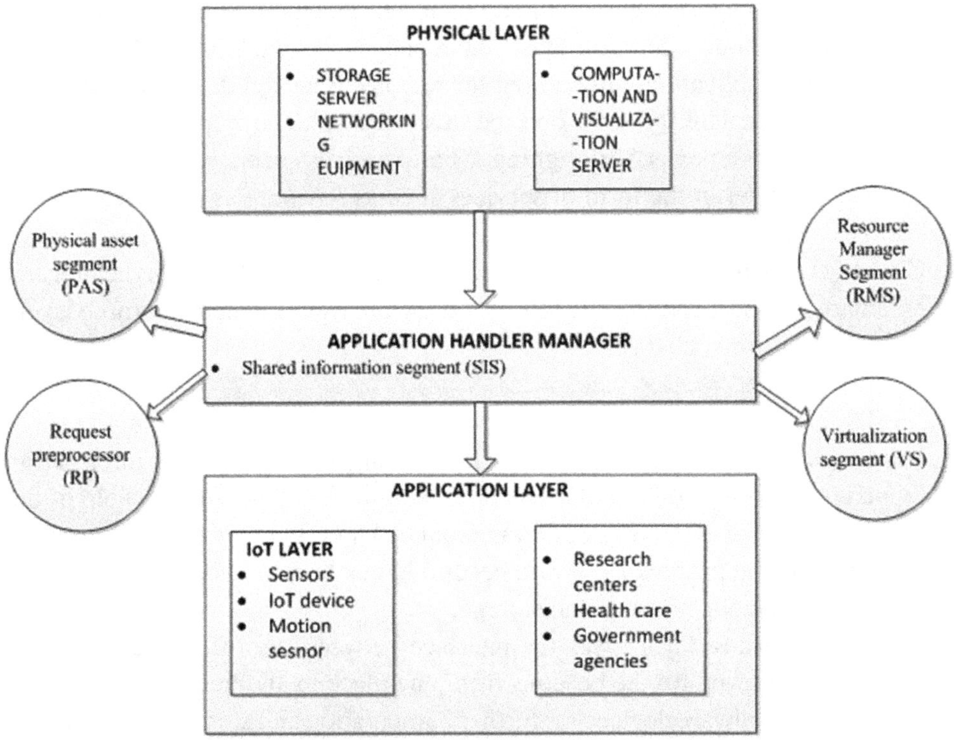

- **Resource Manager Segment (RMS):** Provide service to application handler with the help of virtualization, it also disperses the accessible physical assets among various application handlers.
- **Shared Information Segment (SIS):** Shared information segment is intended to monitor the information, which is required by the various applications. For example, multiple-information created from various smart devices such as smart traffic; smart camera and e-health can be utilized by various association and research centers. Diverse arrangements of information can be computed by various applications that depend on the type of information or datasets.
- **Physical Asset Segment (PAS):** The resources that are appropriated to application segment handlers and the acknowledgment choice of the prerequisites are overseen by the PAS. On the off chance that total request from all application segment handlers is more than the accessible physical assets at the cloud supplier, the appropriations are made in view of the need and the executed technique will be based on priority strategy.
- **Virtualization Segment (VS):** Virtualization can keep running in numerous Operating systems in the meantime in single PC framework, by utilizing

virtualization software. Most of the time business trends and marketing campaign come in & out; with no time you need to react on the business solution within a time frame. It also tries to adopt new technology like cloud integration and Hadoop technology most of the trends other vender or solutions, you don't need to write code to implement the solution, you just plug & plug the data virtualization, and get the data right there in less time. Working framework distinctive sorts of virtualization could be capacity virtualization, arrange virtualization and server virtualization.

- **Request Preprocessor (RP):** Application segment handler sends the request to cloud with the assistance of virtualization segment. Cloud segment process the received request send to other segments for further storage and visualize the data. The RP segment removes the ambiguity from the data cloud storage and sends the data in the specific format for further operation in the cloud layer.

Enabling Smart Communities Through Big Data Analytics and Cloud of Things

As technology becomes an active component in everything that individual feel and touch. In human history with the deployment of satellite, sensor, RFID tag and GPS to enable to analyze and visualize vast amount of data and make sense of physical environment around us. This presents enormous opportunity to tackle system complexity challenges such as climate change and infrastructure. The cities are changing with the evolution of technology and its integration into the urban fabric. Process decision for smart communities as shown in figure 8 highlight all the steps required for decision in smart communities that is generating new concept of smart city, for some time as smart city having promising area that is more efficient structurally and functionality, but the unavailability of computing power mean it now need to both faster and cheaper to process data, emergency data analysis and visualization technique, we are now able to understand the flaws and relationship of assets like energy, water transport within our cities (H. Sundmaeker, 2010).

According to Cisco by 2010, 12.5 billion devices were connected to the internet, by 2015 it will 25 billion devices connected and by 2020 it is predicted that 50 billion devices connected to internet. According to research conducted in U.K (United Kingdom), department business innovation and skills, the smart cities market will be worth 400 billion dollars by the year 2020 that represent big business. The fundamental quality of the big analytics is to collect the immense measure of data from various part of smart communities (D. Giusto, 2010.) (A. Munir, 2017). With the rise in business and industrial application of cloud-based IoT, researcher more focus on data storage communities to provide the data handling, management, and

Figure 8. Process decision for smart communities

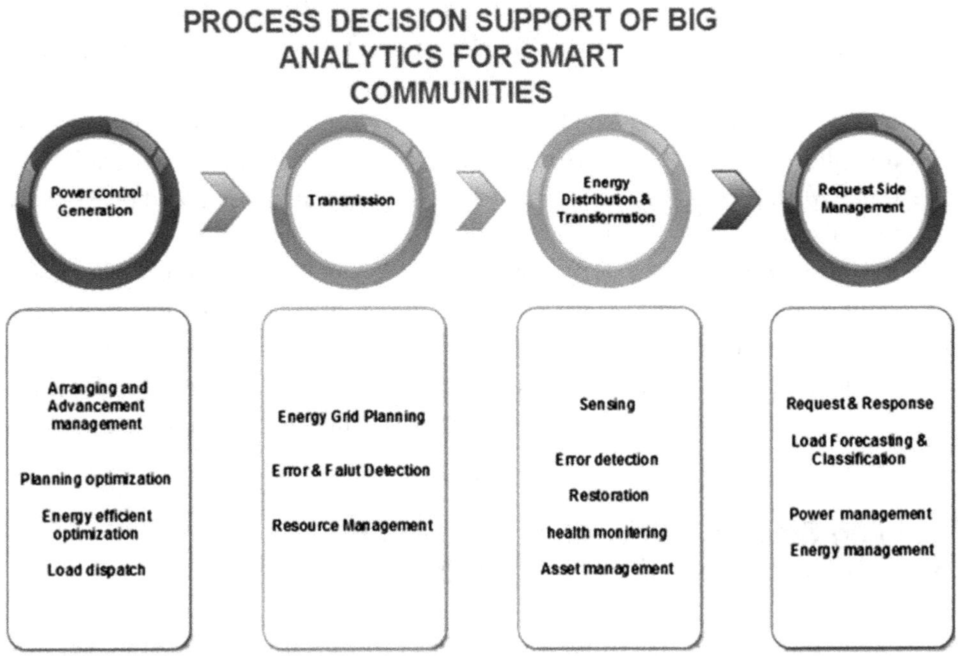

data visualization. For this, big data analytics the perfect solution is cloud-based IoT. Enormous information developing abruptly, right now at an anticipated rate of 30% development in the measure of worldwide information produced every year versus just 4% development in worldwide IT spending (Elhayatmy, (2018).).

OVERVIEW OF SMART CITY COMMUNITIES

The concept of the smart city begins in 2008, based on a theory that with the increasing population the migration from the rural area to urban areas is increasing. By 2040, it is estimated that 65% of the population will live in cities; migration of people to the urban area in hope for better education, medical and jobs purpose. Handling such a large population is very difficult task. However, there are numerous challenges in growing realization of a city that providing efficient services such as smart transport management, accommodation, road safety, intelligent traffic management, security and privacy of data. We need highly advanced infrastructure, communication technology, and sustainable real estate (Dey N. H., (2018).).

Expert from various areas define smart city, such as (Yang, (2018)):

- A smart city is characterized as the idea that utilizations data and correspondence innovation to give better administrations to the general public.
- A concept of smart city, that it incorporates correspondence innovation with IoT with a point of view to give high caliber of life, reasonable monetary improvement, condition, versatility, individuals, and government.
- Shrewd Urban areas as a connected innovation term regularly alludes to oversee assets savvy meters and other foundation for power, water supply and waste administration.

The main aim of smart city concept is to merge communication technology and Internet of things (IoT) technology in a secure fashion to improve the efficiency of services (Padwalkar, February 2014.). In addition, the smart city makes efficient use of the public resources, reducing the operational and high-quality services offered to the city (Bhatt, (2017)).

As indicated by the IEEE Savvy Urban communities Specialized People groups (Dr. Ovidiu Vermesan, 2013):

- "A smart city unites innovation, government, and society to empower the accompanying qualities: a savvy economy, keen versatility, a savvy situation, brilliant individuals, keen living, and shrewd administration".
- A few creators depict a smart city as "a city well performing in a forward-looking manner in economy, individuals, administration, versatility, condition, and living, based on the savvy blend of gifts and mind full residents".

Smart mobility is an extremely expansive vision, and the examination into the smart mobility is still in its early stages. Subsequently, there are no standard definitions for mobility. Definitions concur that to accomplish smart versatility, data and correspondence advancement ought to be utilized. Savvy portability frameworks generally incorporate the fields of smart transportation frameworks, car innovation, data interchanges innovation and installed frameworks. Smart versatility frameworks can gather information from various sources, such as movement information from activity administration frameworks, transport timetables from transport frameworks, swarm information from residents, sensor information from vehicles, activity lights, parking garages, and streets (A. Alamri, 2013.)

As appeared in Figure5, there are more ranges of enthusiasm inside smart mobility. As indicated by smart versatility world, these are the principle zones of a research area. Every one of these definitions concurs that to accomplish smart mobility, information and communication. Advancements ought to be utilized, smart mobility frameworks generally incorporate the fields of intelligent transportation frameworks, automotive technology, information and interchanges technology and embedded systems. Smart mobility is a key idea to change urban transport and change the way our urban communities move. A few activities have been produced the world over to give greater versatility for subjects (M. M. Rathore, 2015.).

There are some research areas in smart mobility (Sarkar, (2017)):

- **Driving Safety:** Gives innovation to protected and secure portability, enabling autos to connect with different vehicles and the foundation around them.
- Smart lightning system- utilizes light-emitting diode (LED) with associated controls to give better lightning, cut vitality utilization, lessen movement clog and progress movement stream (Nataša Gospić, 2015).
- **Sharing and Urban Mobility:** Shared transport frameworks corporate auto sharing, bike sharing, carpools, and vanpools. In Addition, multimodal frameworks can utilize unique and ideally joined transport modes inside the outing chain flawlessly to approach more prominent manageability in urban transport
- **Green Mobility:** Minimizes the natural effect caused by the transportation part without affecting the development energy.
- **Smart Payment Systems:** Actualized to overcome the constraint of the regular installment strategies by redoing the installment technique by means of stopping meter and different innovations.

CLOUD-IOT BASED ROADSIDE SYSTEM IN SMART COMMUNTIES

IoT keeps on growing its materialness to numerous buyer markets. The open doors incorporate a fundamental situation like interfacing things to a cloud stage to a mind-boggling situation like associated roadside foundation, path location and following of vehicles from self-driving autos. More prominent accessibility of the IoT items, administrations, and economical network are making new business cases in different spaces. In intelligent transportation system (ITS) market, roadside assistance is a critical shopper portion which aids a few operations counting wellbeing and crisis cautions. But the joined cost of sending, support and different assets is very costly (A. Cammarano, Oct. 2012,).

The roadside help scene ought to be moved up to fit into a standard, shopper-driven and open APIs based IoT structure as shown in figure 9. This paper recognized research, building challenges identified with the proposed IoT structure and examine commitments in this paper are:

- Making a conjunction of conveyed information investigation (edge registering) and distributed computing.
- Buyer-driven and even IoT application advancement,
- Consistent help of interoperability among building squares,
- IoT and Web of Things (WoT) measures based execution to break information storehouses
- Open interfaces and APIs permitting outsider designers to make reasonable roadside help application.

The smart roadside framework is characterized by the utilization of smart sensor and analytics hardware in a coordinated way to enhance well-being and productivity of the transport system IoT, to avoid congestion and life threaten accidents on roads (A. Cammarano, Oct. 2012,) (A. Botta, 2014). The architecture for smart roadside system is shown in Figure9. Architecture is divided into 2 portions, namely an application layer, processing/storage layer and perception layer. The system considers the vehicular assets from which benefit ask for starts at the discernment layer. The smart roadside segment (SRS) of a vehicle contain a module to start particular user demands. The handler management and capacity layer parts are sent in a street side segment which is linked with the static cloud management unit for getting and storing information, Static cloud management also provide the computation and processing assistance to the vehicle queries from anything such as traffic purposes, registration queries, and smart parking (H. N. M. Quoc and D. Le Phuoc, 2015).

The correspondence interface between the vehicle and smart roadside segment is access control management. SRS removes the demand from the payload and advances it the proper web benefit giving the help. This enables the structure to give open and secure APIs to outsider application improvement. The application improvement is stretched out from which enables any designer to center on the application rationale of novel roadside application benefit (A. A. Cardenas, Jun. 2008,) (Ali, (2017)).

The SRS having the following functionalities are:

- Monitor the vehicle by using wireless devices such as smart phone and laptop.
- Handle the traffic management.
- It also provide energy to the wireless equipment installed in vehicles.
- Provide the assistance to any type of queries.
- Emergency calling.

It is likewise wanted to convey an arrangement of association with the crisis administrations to give speedier discovery and restriction of street mischances, to give quick help to casualties in instances of crises. The layered framework also provide security and privacy to avoid the threats in network. To enhance movement

Figure 9. IoT layered framework for smart roadside system

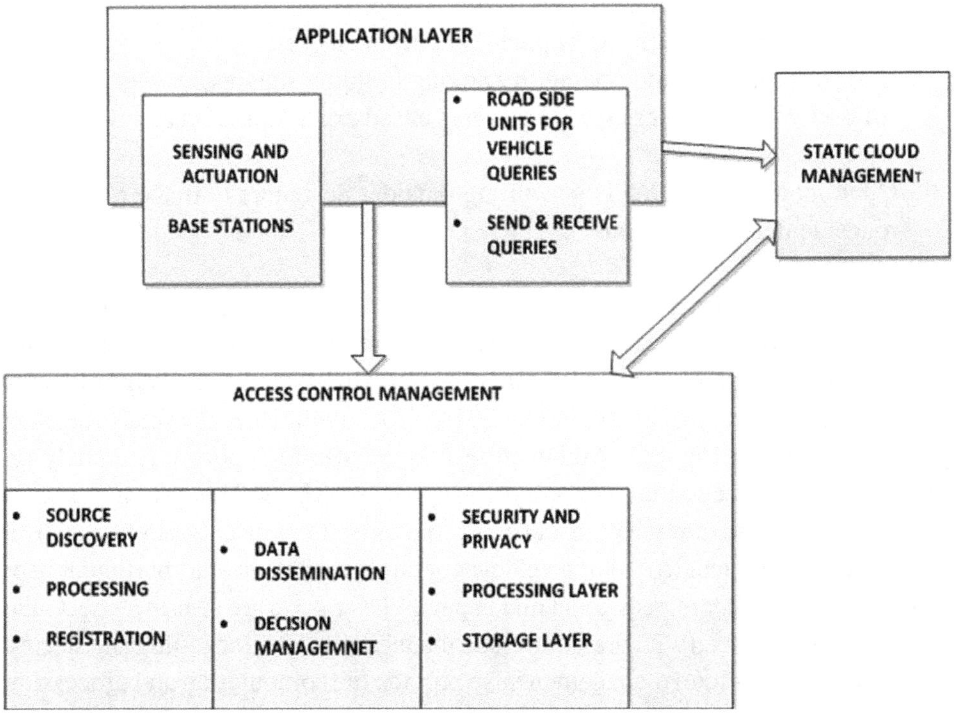

control gadgets it is proposed an inventive gadget, for example, the screen with an intuitive interface (S. Capkun, Jan./Mar. 2003.). There is the capacity to make a foundation to scale, contingent upon the changing urban communities conditions and the quantities of vehicles, and furthermore the capacity to the regional augmentation and expansion of new "hubs" of cloud framework. The smart roads side framework will be implemented in future so that it allows the researcher to concentrate more on consumer-driven applications.

PRAGMATIC ENCOUNTERS OVER THE WORLD

In view of the US National Insight Chamber report, IoT is a standout amongst the best devices on the US interests on the approach to 2025. The quantity of interconnected gadgets outperformed the populace in 2011. This slant was altogether developing as the quantity of interconnected gadgets was around 9 billion out of 2012 and they will achieve 24 billion until 2020 (S. Distefano, 2012,) (Ricardo Faria, 2017).

Table 1. Approaches and systems of urban areas identified with smart cities

City	Experience
Barcelona	Execution of sensor innovation, utilizing the information examination of movement streams to plan a novel transport arrange and the usage of brilliant activity
Amsterdam	Diminishing the activity, vitality sparing, and expanding the wellbeing level
Singapore	Actualized a framework intended to upgrade movement stream and keep street activity running wellbeing. A portion of the key activities in the transportation segment incorporate one observing, a far-reaching gateway serving all drivers and vehicle proprietors in the nation
Santa Cruz	Break down the information of violations to estimate the requirements of police what's more, amplify the nearness of police in the required spots
New York City	The 'One New York Design' reported in the year 2015 is an exhaustive arrangement for a practical and strong city
Anyang (South Korea)	The city has conveyed a wise transport framework to give activity administration and movement data administration. Sensors send ongoing activity data to the control focus, which utilizes the knowledge to give data to general society by means of the web and portable application
Yinchuan(china)	Stipulated the organization of a security cloud stage including 30,000 coordinated cameras. A segment of this comprises of cutting edge HD cameras with savvy examination capacities to guarantee the security of the city and its inhabitants.

CONCLUSION

The integration of cloud-IoT represents to the following enormous future's leap head on the internet. The new applications arising from this incorporation known as Cloud-IoT open new directions for business and research. This chapter surveyed the literature keeping in mind the end goal to distinguish the corresponding parts of Cloud and IoT and the fundamental drivers for coordinating them into a special situation. Since the reception of the Cloud-IoT worldview empowered a few new applications. In additionally, analyzed such difficulties with a specific end goal to distinguish research directions. This chapter also reviewed accessible stages and activities by contrasting their main aspects and recognized open issues and future research bearings in this field. Thanks to the Cloud-IoT worldview regular day to day existence and exercises will be possibly enhanced for everybody: smart cities communities will empower more productive benefits and advance new business openings, pervasive, healthcare applications will enhance the personal satisfaction for some patients, and so on. These new application situations act critical research difficulties such like the heterogeneity of included gadgets and advancements; the required execution, unwavering quality, adaptability, and security; protection safeguarding; legitimate and social perspectives. The open issues of Cloud-IoT worldview relate principally power and vitality productivity, SLA implementation, estimating and charging, security and protection. envisioned future directions incorporate the distinguishing proof of the authoritative answer for naming and tending to things, the vast scale support for multi-organizing, and the meeting toward a typical open administration stage condition.

SUMMARY

- In this chapter, Smart communities through big data analytics and cloud-IoT is highlighted. In the first section brief introduction about the IoT, its behavior and requirement of technology and cloud-IoT are discussed. Moreover, IoT layered architecture is also illustrated.
- Integration mean is the ability to use the functionality of previous model's technology and integrate it, with the new one technology. Cloud-IoT covers the requirement of cloud computing, its elements, design issues and architecture model. In addition, various design issues are challenges of cloud-IoT are discussed in third section.
- In the fifth section, overview of smart city communities are described. Moreover, this section also discussed smart communities through big data analytics based cloud-IoT.
- Priority based MAC protocols and modeling, i.e., POMDP and queuing theory are explained in the fifth section.
- Pragmatic encounters of smart city are described as case study in the sixth section.
- In the last section, conclusion is summarized, which is evidently demanding of specific cloud and IoT requirement and new framework architecture for successful deployment of smart city framework in the future.

REFERENCES

Alamri, A., Ansari, W. S., Hassan, M. M., Hossain, M. S., Alelaiwi, A., & Hossain, M. A. (2013). A survey on sensor-Cloud: Architecture, applications and approaches. *International Journal of Distributed Sensor Networks*, *2013*, 1–18.

Ali, A. H., Atia, A., & Mostafa, M. S. M. (2017). Recognizing Driving Behavior and Road Anomaly using Smartphone Sensors. *International Journal of Ambient Computing and Intelligence*, *8*(3), 22–37. doi:10.4018/IJACI.2017070102

Anand & Padwalkar. (2014). Internet of Things –A Future of Internet: A Survey. *International Journal of Advance Research in Computer Science and Management Studies, 2*(2).

Anisha, Kumar Reddy, Reddy, & Reddy. (2012). *Third Party Data Protection Applied To Cloud and Xacml Implementation in the Hadoop Environment With Sprawl*. Academic Press.

Armburst, M. (2009). Above the Clouds: A Berkeley View of Cloud Computing. Berkeley, CA: Univ. of California.

Ashrafi. (2007). *A Data Mining Architecture for Distributed Environments*. London, UK: Springer.

Bakmaz. (2015). Machine – To – Machine Communications Toward Smart Logistics Systems. *2nd Logistics International Conference*.

Barnaghi, Bermudez-Edo, & Tönjes. (2015). Challenges for Quality of Data in Smart Cities. *J. Data and Information Quality, 6*(2-3), 6:1–6:4.

Bhatt, C., Dey, N., & Ashour, A. S. (Eds.). (2017). *Internet of things and big data technologies for next generation healthcare*. doi:10.1007/978-3-319-49736-5

Bitam, S., Mellouk, A., & Zeadally, S. (2015). Vanet-cloud: A generic cloud computing model for vehicular ad hoc networks. *IEEE Wireless Communications, 22*(1), 96–102. doi:10.1109/MWC.2015.7054724

Botta, A., Donato, W. d., Persico, V., & Pescapé, A. (2014). On the Integration of Cloud Computing and Internet of Things. *2014 International Conference on Future Internet of Things and Cloud (FiCloud)*, 23–30. 10.1109/FiCloud.2014.14

Brewer, E. (2012). CAP twelve years later: How the "rules" have changed. *Computer, 45*(2), 23–29. doi:10.1109/MC.2012.37

Buyya, R., Chee Shin, Y., & Venugopal, S. (2008). Market-Oriented Cloud Computing: Vision, Hype, and Reality for Delivering IT Services as Computing Utilities. *10th IEEE Conference on High Performance Computing and Communications*. 10.1109/HPCC.2008.172

Cammarano, A., Petrioli, C., & Spenza, D. Pro-energy: A novel energy prediction model for solar and wind energy-harvesting wireless sensor networks. *Proc. IEEE 9th Int. Conf. Mobile Ad-Hoc Sensor Syst. (MASS)*, 75–83. 10.1109/MASS.2012.6502504

Capkun, S., Buttyan, L., & Hubaux, J.-P. (2003, January/March). Self-organized publickey management for mobile ad hoc networks. *IEEE Transactions on Mobile Computing*, *2*(1), 52–64. doi:10.1109/TMC.2003.1195151

Cardenas, Amin, & Sastry. (2008). Secure control: Towards survivable cyber-physical systems. *Proc. 28th Int. Conf. Distrib. Comput. Syst. Workshops*, 495–500.

Chen, S., Li, M., Ren, K., Fu, X., & Qiao, C. (2015). Rise of the Indoor Crowd: Reconstruction of Building Interior View via Mobile Crowdsourcing. *Proc. of ACM SenSys*, 5971. 10.1145/2809695.2809702

Daniel, J. (2009). Data Management in the Cloud: Limitations and Opportunities. *A Quarterly Bulletin of the Computer Society of the IEEE Technical Committee on Data Engineering*.

Dastjerdi, A. V., & Buyya, R. (2016). Fog Computing: Helping the Internet of Things Realize its Potential. *Computers*, *49*(8), 112–116. doi:10.1109/MC.2016.245

Dey, A. K. (2001). Understanding and using context. *Personal and Ubiquitous Computing*, *5*(1), 4–7. doi:10.1007007790170019

Dey, N., Ashour, A. S., & Bhatt, C. (2017). Internet of things driven connected healthcare. In *Internet of things and big data technologies for next generation healthcare* (pp. 3–12). Cham: Springer. doi:10.1007/978-3-319-49736-5_1

Dey, N., Hassanien, A. E., Bhatt, C., Ashour, A., & Satapathy, S. C. (Eds.). (2018). *Internet of Things and Big Data Analytics Toward Next-Generation Intelligence*. Springer. doi:10.1007/978-3-319-60435-0

Distefano, S., Merlino, G., & Puliafito, A. (2012). Enabling the Cloud of Things. In *Innovative Mobile and Internet Services in Ubiquitous Computing (IMIS), 2012 Sixth International Conference on*. IEEE.

Domingos, P., & Hulten, G. (2000). Mining high-speed data streams. In *Proceedings of the sixth ACM SIGKDD international conference on Knowledge discovery and data mining*. ACM. 10.1145/347090.347107

Elhayatmy, G., Dey, N., & Ashour, A. S. (2018). Internet of Things Based Wireless Body Area Network in Healthcare. In *Internet of Things and Big Data Analytics Toward Next-Generation Intelligence* (pp. 3–20). Cham: Springer. doi:10.1007/978-3-319-60435-0_1

Faria, R. (2017). *Smart Mobility: A Survey*. IEEE.

Gilbert, S., & Lynch, N. (2002). Brewer's conjecture and the feasibility of consistent, available, partition-tolerant web services. *ACM SIGACT News, 33*(2), 51–59. doi:10.1145/564585.564601

Giusto, D., Iera, A., Morabito, G., & Atzori, L. (Eds.). (2010). The Internet of Things. Springer.

Gondchawar & Kawitkar. (2016). IoT based Smart Agriculture. *International Journal of Advanced Research in Computer and Communication Engineering, 5*(6).

Gupta, Nath, Chakraborty, & Ghosh. (2016). *SDFog: A Software Defined Computing Architecture for QoS Aware Service Orchestration over Edge Devices*. arXiv.org

Gurav, U., & Shaikh, R. (2010). Virtualization: a key feature of cloud computing. In *Proceedings of the International Conference and Workshop on Emerging Trends in Technology*. ACM. 10.1145/1741906.1741957

Han, Haihong, Le, & Du. (2011). Survey on NoSQL database. *Pervasive computing and applications (ICPCA), 2011 6th international conference on*, 363–366.

He, W., Yan, G., & Xu, L. (2014). Developing vehicular data cloud services in the iot environment. *IEEE Transactions on Industrial Informatics, 10*(2), 1587–1595. doi:10.1109/TII.2014.2299233

Jiang, L., Da Xu, L., Cai, H., Jiang, Z., Bu, F., & Xu, B. (2014). An iot-oriented data storage framework in cloud computing platform. *IEEE Trans. Ind. Inform., 10*(2), 1443–1451.

Kamal, S., Ripon, S. H., Dey, N., Ashour, A. S., & Santhi, V. (2016). A MapReduce approach to diminish imbalance parameters for big deoxyribonucleic acid dataset. *Computer Methods and Programs in Biomedicine, 131*, 191–206. doi:10.1016/j.cmpb.2016.04.005 PMID:27265059

Luan, Gao, Li, Xiang, Wei, & Sun. (2016). *Fog Computing: Focusing on Mobile Users at the Edge*. arXiv.org

Munir, A., Kansakar, P., & Khan, S. U. (2017). *IFCIoT: Integrated Fog Cloud IoT Architectural Paradigm for Future Internet of Things*. arXiv preprint arXiv:1701.08474

Peng, M., Yan, S., Zhang, K., & Wang, C. (2016). Fog Computing Based Radio Access Networks: Issues and Challenges. *IEEE Network, 30*(4), 46–53. doi:10.1109/MNET.2016.7513863

Quoc, H. N. M., & Le Phuoc, D. (2015). An Elastic and Scalable Spatiotemporal Query Processing for Linked Sensor Data. In *Proceedings of the11th International Conference on Semantic Systems, ser. SEMANTICS '15*. New York, NY: ACM. 10.1145/2814864.2814869

Rathore, Ahmad, Paul, & Rho. (n.d.). Urban planning and building smart cities based and internet of things. *Computer Network*. DOI:10.1016/jcomnet.2015.12.023

Sarkar, M., Banerjee, S., Badr, Y., & Sangaiah, A. K. (2017). Configuring a trusted cloud service model for smart city exploration using hybrid intelligence. *International Journal of Ambient Computing and Intelligence*, *8*(3), 1–21. doi:10.4018/IJACI.2017070101

Solanki, V. K., Katiyar, S., BhashkarSemwal, V., Dewan, P., Venkatasen, M., & Dey, N. (2016). Advanced automated module for smart and secure city. *Procedia Computer Science*, *78*, 367–374. doi:10.1016/j.procs.2016.02.076

Sundmaeker, Guillemin, Friess, & Woelfflé. (2010). Vision and challenges for realizing the Internet of Things. *Cluster of European Research Projects on the Internet of Things - CERP IoT*.

Vermesan & Friess. (2013). Internet of Things: Converging Technologies for Smart Environments and Integrated Ecosystems. IERC.

Yang, W., Wang, X., Song, X., Yang, Y., & Patnaik, S. (2018). Design of Intelligent Transportation System Supported by New Generation Wireless Communication Technology. *International Journal of Ambient Computing and Intelligence*, *9*(1), 78–94. doi:10.4018/IJACI.2018010105

Chapter 5
Internet of Things and Its Impacts in Computing Intelligence:
A Comprehensive Review - IoT Application for Big Data

Md. Golam Sarowar
East West University, Bangladesh

Md. Sarwar Kamal
East West University, Bangladesh

Nilanjan Dey
Techno India College of Technology, India

ABSTRACT

In the age of information superhighway, internet of things (IoT) is a hotly debated subject. The difference between ancient era and current time is the very lager digital communications among the people in whole world. The entire universe is moving towards digitalization and the key framework is the internet and connecting devices. The devices that are connected with internet are called internet of things. So, the impacts and influences of IoT are very significant in current computational analysis and dealings. Realizing the worth of big data in internet of things (IoT), this contribution gives instances of M2M messaging mechanism, cloud-based embedded computing, mobile computing, and cellular networking technologies in alliance with IoT devices, sensors, gadgets, etc., location spoofing detection for IoT, clustering tools for next generation, and details observation of methods, prospects, processing of big data which has a massive growth after the evolution of IoT. Finally, the concluding announcement interprets principal scheme of this work and future research directions in this field.

DOI: 10.4018/978-1-5225-6207-8.ch005

INTRODUCTION

The terms "Internet of Things" or simply known as IoT delegates the interconnection of various physical devices, smart phones, vehicles etc. by means of network connectivity. For making the process dynamic, various kinds of electronic devices, sensors, actuators and software are mounted with those physical devices to gather transactional data as well as exchange it. Specifically IoT assist the objects which are used to be perceived the actual situation and respond according that. Moreover, IoT being significantly valuable because of its remotely controlled system through a basic network infrastructure (Marotta et al, 2014). Moreover, all the significant opportunities of this new invention of information technology paves our physical world towards a new computer based, efficient, accurate and dominant infrastructure reducing human manipulation. Additionally, in this mint of condition IoT is declared as "the infrastructure of the information society" by Global standard initiative. Since IoT is combined with enormous types of sensors as well as actuators, it encloses high technological context likes intelligent transportation, Artificial intelligence applications, robotics, smart homes, smart society etc. as well. All of this technologies are being manipulated uniquely using interconnected computing system and software around its existing network ranges. Identically, The internet of Things are treated as the improved use of modern connectivity through advanced devices, systems which is fully machine based compressing a wide range of protocols, domains, services as well as applications (Holler et al, 2014). Therefore, IoT is basically refers to interconnection between machine to machine, device to device. However, the internet of things (IoT) is currently indicating the biggest upcoming technological evolution than ever (Raggett, 2015; Want et al, 2015). The whole universe is stepping towards the always connected situation from person to devices and machine to machine. Because of expansion of internet of things (IoT), the size of regular transferred data is going beyond the limit of human manipulation. In addition, for achieving some significant information analyzing this billions of data badly need more flexible, efficient, improvised processing approach. Since next generation is thought to be appeared with global computing system instead of desktop computing, therefore IoT will perform main role to paves present situation towards tomorrow's advanced and improved world. The universal architecture of IoT can be illustrated as follows.

Currently, all the information are being generated with high velocity, dimension, structured, semi-structured and sometimes unstructured. Thus, this upcoming data have to be cleaned for further manipulation. Recent year explosion of data is due to use of billions of sensors, actuators, smart phones, social media services. Moreover, thousands of devices gathering information simultaneously. Storage capacity maintenance and machine learning approaches are quite easy to manage than before which results in involvement of big data concept for present generation as well as

Figure 1. Basic infrastructure of IoT

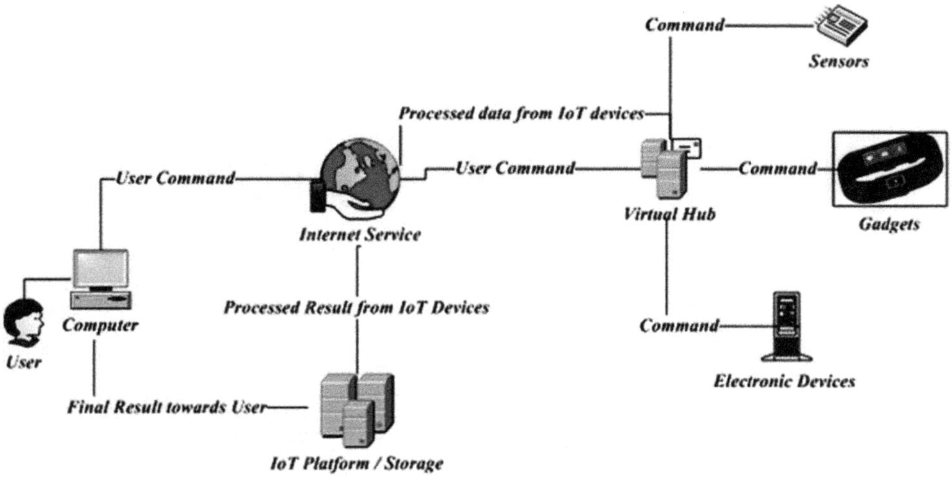

next generation (Gupta et al, 2012). Big data analytics along with the process cloud computing is seemed to be adequate for solution of this types of vast and complex data problems. A lots of research has been executed on big data analytics and various approaches are implemented though currently lots of approaches associated with cloud computing like systems for large-scale analytics based on the MapReduce parallel programming model, Hadoop etc (Dean et al, 2008; Lam, 2011). being the most relevant implementation ever for big data manipulation paradigms. Initially data management system along with big data analytics proposed for basic solution of this problems but thereafter cloud computing approaches have been introduced which outperforms all the previous proposed methods to handle big data. From the last very recent years the prediction statistics of rising till 2020 is likely,

Since big data is indicated by high-elasticity, high velocity, huge number of data, high volume, therefore for this phenomenon initially map reduce approach becomes fruitful to maintain or summarize this kinds of data (Dean et al, 2004; Chaturvedi et al, 2013).In general, the map reduce process manipulate the huge number of datasets to process and identify some information in automatic and distributed way. Basically this is the combination of two primitive processes called map and reduce. Because of this two general properties anyone can reduce the dimensional problem without worrying about other phenomenon. Thereafter hadoop (White, 2012; “Apache Hadoop Project”, 2015). becomes more familiar and effective to the researchers to cope with big data problems. Hadoop is more flexible because of its features like online processing, ability of performing faster processing than the other distributed computing algorithms.

Figure 2. Probable growth of technological data till 2020

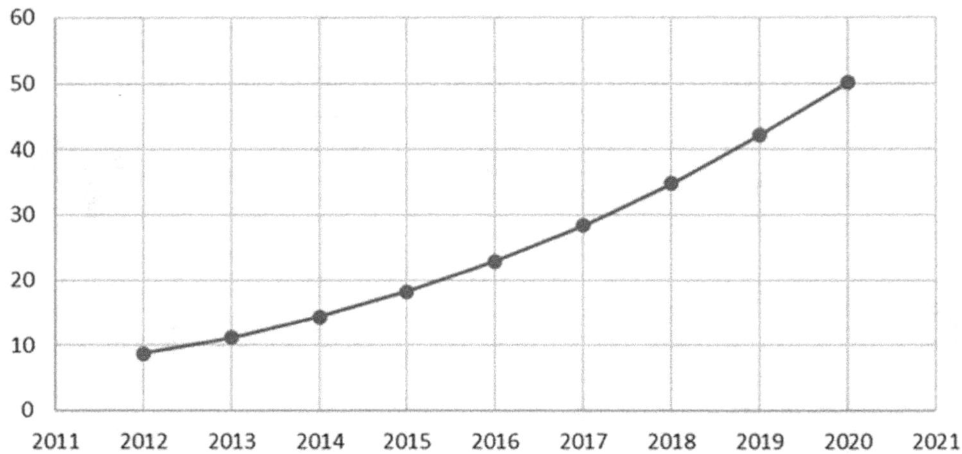

Additionally there are three types of existing IoT architectures present currently are service based solution, Cloud based solution and the rest one refers to actor based framework (Altawy et al, 2016; YUvaraj, 2016; Dinh et al, 2016; Ayagi et al, 2016; Kholod et al, 2016) . Service based solution usually illustrate service oriented features. The second one denoted by cloud based solution of IoT is related with the total number of devices or sensors a user can control. Moreover it limits the range as well as connected IoT devices associated with specific user. Sequentially the actor based framework acts in similar way except the feature of reusability of IoT devices or sensors connected with a network or domain. The overview of services based IoT architecture is illustrated using following figure 3.

Actually, this work delegates with some recent trends in information technology. It is basically representation of some basic ideas along with some evolutionary approaches to analyze big data trend embedded with internet of things. Combining lots of work conducted on current literature this work comprehensively interpret the effectiveness, efficiency, memory complexity of some mechanisms. Since, Internet of things (IOT) is recent invention thus some basic level knowledge including middleware, internet protocols, communication scheme associated with it has been provided here for better clarification. Moreover, mathematical derivation of various mechanism as well as theoretical has been emphasized here. Additionally, lots of evolutionary theorems in the area of big data analytics have been illustrated technically and presented graphically as well. Therefore, this work is able to provide a clear vision of big data in the field of internet of things and so algorithms as well as prospects to handle those huge volume of data and extract required knowledge. Here for this chapter some significant dynamic topics associated with big data and

Figure 3. Overview of actor-based IoT Architecture
Source: Ngu et al, 2017.

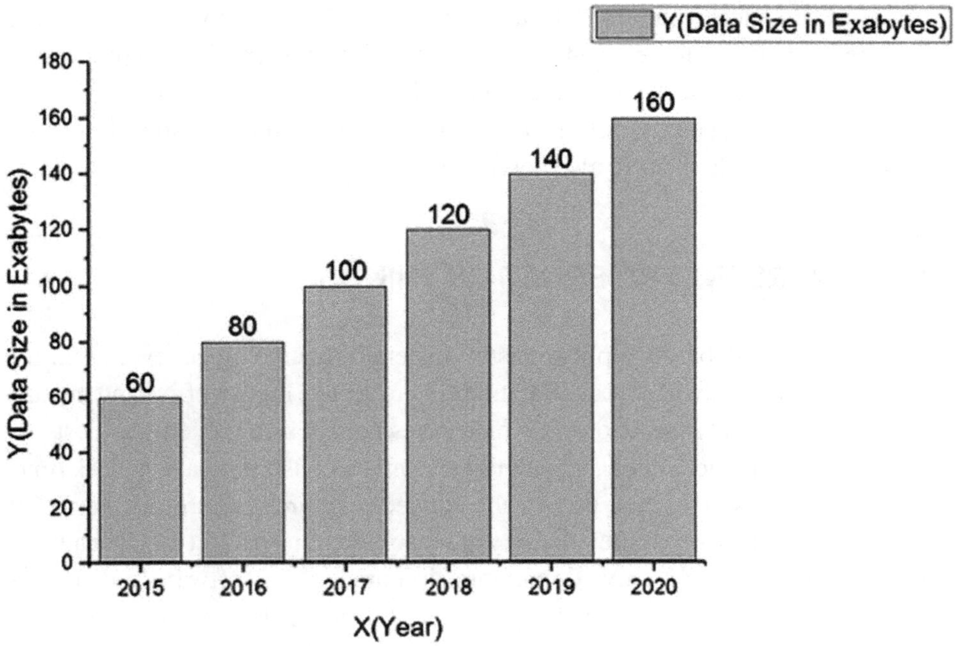

internet of things (IoT) have been integrated within specific context. Since present world is on the edge of machine to machine communication technology, lots of data are transferred within shortest possible time. Thus, knowledge of IOT along with big data processing is mandatory for further advancement in this area. For making this particular things happen this chapter comprehensively illustrates some basic knowledge along with available algorithms in the literature to make familiar with recent improvement in this field. Middleware for internet of things, platforms for socially connected devices are some of the basic level IOT concept. Further portion of this work create a clear vision about high performance computing, location spoofing with the help of IOT infrastructure, clustering tool for IoT data streams. Moreover, methods and prospects for large volume of data have been also demonstrated here along with their performance evolution. However, various big data processing manipulation mechanism including MapReduce, Hadoop, sklar theorem for big data manipulation have been also taken into account here for better clarification regarding the concept of big data and internet of things. Therefore, initially this work clears the perspective of readers regarding the basic concept of middleware for IOT along with the concept of various platforms through which IOT devices and sensors are connected with each other. Those portion illustrates clearly about the protocol, hardware acceleration, manipulator, database system, IOT embedded middleware etc. information in a nutshell. Lots of industrial tasks are handling through the use of various IOT devices. Thus, industrial messaging mechanism between IOT devices need to be emphasized to be invented and improved. However, big data processing and data clustering briefly illustrates our main motive of this chapter which is handling big data embedded with IOT. Various developed mechanisms along with their working procedures have also counted here.

MIDDLEWARE FOR INTERNET OF THINGS

There are lots of IoT based applications in this era of technological evolution. It's become familiar worldwide because of its ability to reduce human effort and manual processing of various aspects of life which saves human's valuable time, cost etc. doing the task more accurately, efficiently as well as consuming a very less time. Here middleware is actually a part of the connectivity architecture of Internet of Things which ensures the accurate activity of sensors creating perfect network among them and also ensures effective communication among the software and sensors to provide exact information transferred by the sensors or devices. Moreover, IoT middleware provide API management system for transferring information, messages, routing and message transformation facilities and so on. Currently, various world class companies like Mulesoft, Oracle, RedHat and WSO2 are providing IoT middleware for this purpose. Therefore, middleware basically perform three types of communication process which are connecting and accessing physical devices, flexible combination of various IoT devices and the support of information discovery along with maintaining security and privacy by using tool kit training. Various types of IoT middleware is shown below graphically.

Currently a lot of IoT middleware is in action. Hydra (Eisenhauer et al, 2009; HYDRA, 2010). is one of them. Basically hydra is a short term project developed and maintained by European Union to improve the middleware embedded network. The main focus of this project concentrated on incorporating various devices with physically connected devices within a network range and internally controlling the devices using internal frequency mostly Wi-Fi, Bluetooth, Zigbee etc. Moreover, the devices under this project is able to exceptionally maintain more secure and trustworthy privacy among the users. Hydra basically enables some features of a device to be activated and response accordingly. In general, the concept of IoT middleware is applicable for three types of spheres such as human healthcare, smart cities and agriculture. This middleware generally operate with SDK tool kit for dynamically work within the specified range to a given architecture. With the

Figure 4. Various types of IoT Middleware

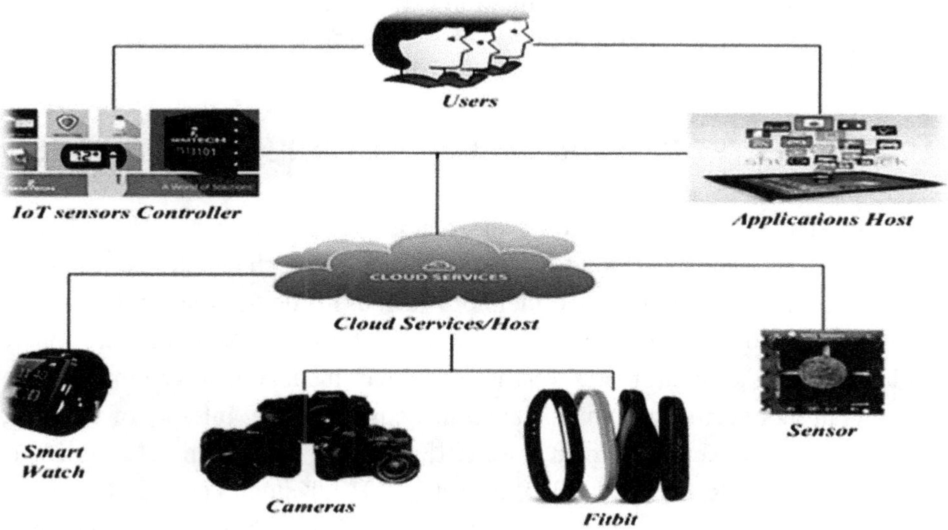

association of SDK tool kit this middleware usually advocates standard web services for addressing all the IoT devices under the supervision. For clarification a watch can be contributed as smart watch by modifying it with a new sensor connecting with it internally can detect the data associated with the sensors. Here for streaming the data transferred by the sensors an integrated network system is need to be constructed and the middleware is up to this significant proceeding.

Global sensor networks ("Global Sensor Networks", 2011). is another IoT middleware system that is aimed at improving IoT based technological devices and sensors working under a certain network or domain by providing flexible integration. The main perspective of this existing system is virtual sensor depredation which ensures the developers to be specified XML based feature used previously in J2EE server (Burke et al, 2006).This system helps to control multiple tasks like notifications, resources, output data structure, SQL processing data etc. all the input as well as output streams react like an envelope which identify the task need to be done to interact or connect with another sensors or devices. It also indicated what should be done for reading data from the sensors and emphasis the need of taking decisions from output data transferring as streams from various sensors. From real world perspective this system is preferable as it has the ability to combine various types of IoT sensors and devices. Moreover for addition of any newly created sensors the users need to know only the procedure to work on XML descriptor. Therefore, it keeps the privacy along with perfect maintenance authorization of every user associated with Global sensors network system. Like stated above IoT based systems Google Fit ("Google Fit", 2015) is similar in patterns though it has some exception like it's an open source system for all.

VARIOUS PLATFORM FOR SOCIALLY CONNECTED DEVICES

For advancement of next generation data streaming along with faster and effective analysis Lysis is one kind of cloud computing based approach which ensures faster, effective, visualization of working principle of various IoT devices currently working for data transaction. Presently, there's a lot of platforms working simultaneously with IoT devices like smart home, smart cities, smart management system, eHealth environment monitoring system, social security and alliance, Internet of vehicles, mine security system etc. (Soliman et al, 2013; Jin et al, 2014; Amaral et al, 2011; Doukas et al, 2012; Riazul et al, 2015; Rao et al, 2012; Kantarci et al, 2014; Sun et al, 2012). Despite all of this available platforms there's a problem occurs in regular basis as these processes only follow vertical implementations and have a lack of horizontal large scale ability to manipulate data.

In recent era the virtualization process is being introduced in various research based IoT projects although most of the IoT platforms usually in lack of this significant feature. Additionally, many devices are connected together and so a strongly effective platform is badly needed to virtualization of this connection and addressing all the IoT devices with unique schemes. This virtual illustration of various devices is considered service objects that help in a great context to handle communication among various connected devices along with sensors. Except this, the reusability of information collected from IoT devices is major concerns for the researchers to be succeeded in this area. Moreover, lack of reusability feature may results in waste of resources as well as bandwidth. The difference of internet enabled devices with IoT devices can be illustrated as shown in figure 5.

From Figure 6, the hardware abstraction layer indicates how the internal connection should be developed in the virtualization level. Making standard communication among the devices and platforms are the major contribution of this Hardware abstraction level (HAL). Thereafter the data manipulator is assigned to check whether the data collected from the devices need to be processed before sending it to device manager.

In general the lysis platform consists of four specific level which are real world level, virtualization level, aggregation level and application level. Basically in real world level all the devices which are connected is indicated by this level. Therefore, this level includes smart cities, IoT devices, IoT sensors, Smart society, smart vehicles etc. additionally all this elements Transfers data from one device to another and in

Figure 5. (a) Physical Devices (PDs) able to communicate with the platform and (b) objects that need a gateway to interact with the platform

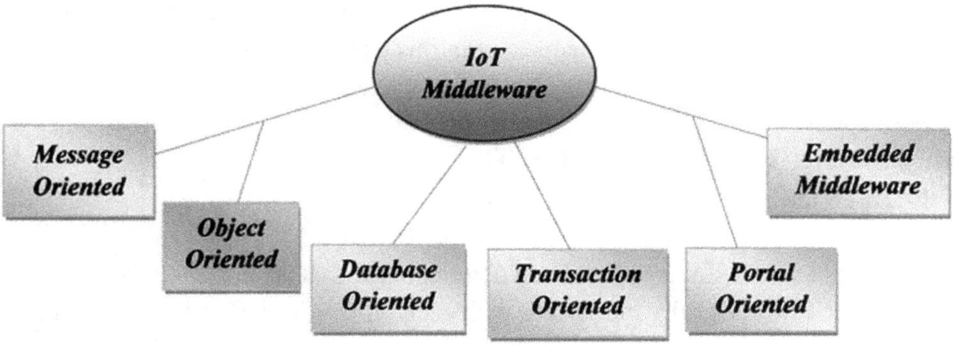

fine towards virtualization level. The virtual level interacts with real world objects creating social virtual objects (SVO). After that the level of aggregation mainly focus on the streaming SVO objects to compose them within Micro engines (ME). The last level is application level where all the user defined macro based services are executed. The basic structure of Lysis is illustrated in the figure 6.

Figure 6. Overview of lysis infrastructure

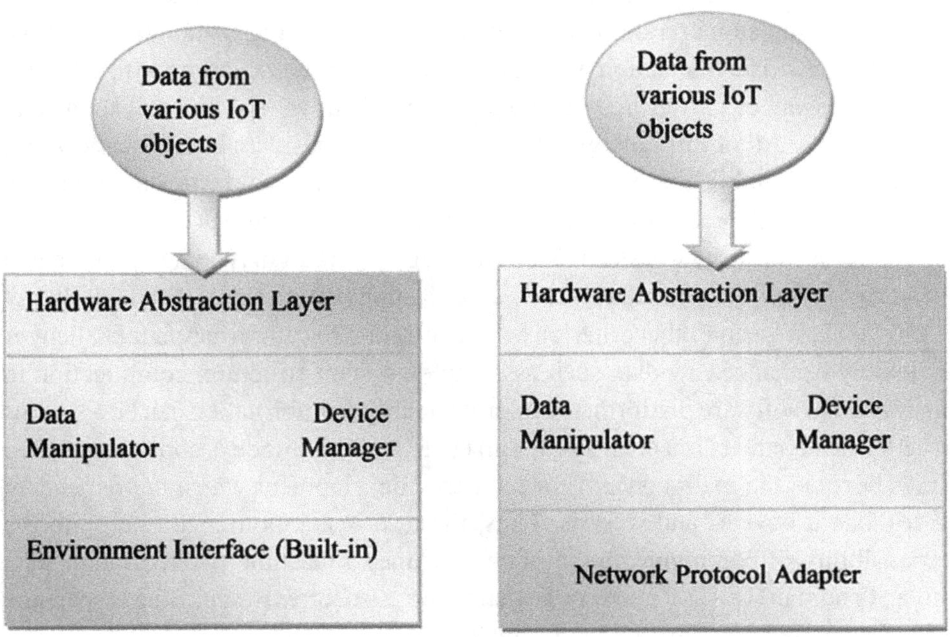

The main obstacle for IoT devices systems nowadays is to create a virtualization environment internally while transferring data from one phage to another phage. Here this phenomenon has been controlled using the social virtual object stage. That is why Lysis platform performs well in this current situation of IoT systems.

MESSAGING MECHANISM FOR INDUSTRIAL IOT APPLICATIONS

According to recent studies it has been demonstrated that within 2020 there will be 50 billion IoT devices fully interactive with each other within a network (Feki et al, 2013; Lumpkins, 2013). In this new IoT era of communication all the devices as well as sensors keep connecting with each other automatically and intelligently for taking any kind of decisions needed and in this context the M2M technology paves the way of communication in a top level to provide disturbance less communication. This messaging system also provides an automatic way to exchange large volume data by storing them in a storage system for human analysis, observation and further knowledge discovery (Schniderman, 2013; Jiang et al, 2014). In this storage system there's a data of data associated with human health, various industrial manufacturing, hone automation, supply chain, retailing etc. from last decade use of machine learning based process to discover data from large volume data becomes threatened for the researchers, students but with the invention of new cloud based IoT data computing paradigm which supports the features of ubiquitous sensing, data normalization, data collection, data interaction with various devices creates a new hope to establish an effective way of clustering big data streaming from various IoT as well as real world devices. All of this emerging technologies are considered to make the world even better by establishing smart manufacturing, digital manufacturing, industrial 4.0 privileges and factory of things (Savazzi et al, 2014; Perera et al, 2014).

In various industrial works this M2M messaging paradigm is being implemented because of its ability to ensure secure data transaction (Perera et al, 2014; Fadlullah et al, 2011). There's some other criterion for which the M2M system may face challenges to flexibly exchange raw data such as complex system structure, complication in hardware and software platform, the internal quality of information can be a serious issue for this context (Liu et al, 2014; Sun et al, 2012). Nowadays above mentioned issues become the main concern for industrial development which are dependent on IoT based devices and sensors. Thus, the next generation is coming with the responsibility of communication among machine to machine (M2M) rather than human to human (H2H). The overall existing infrastructures for machine to machine messaging system is illustrated in figure 7.

Figure 7. M2M network architecture

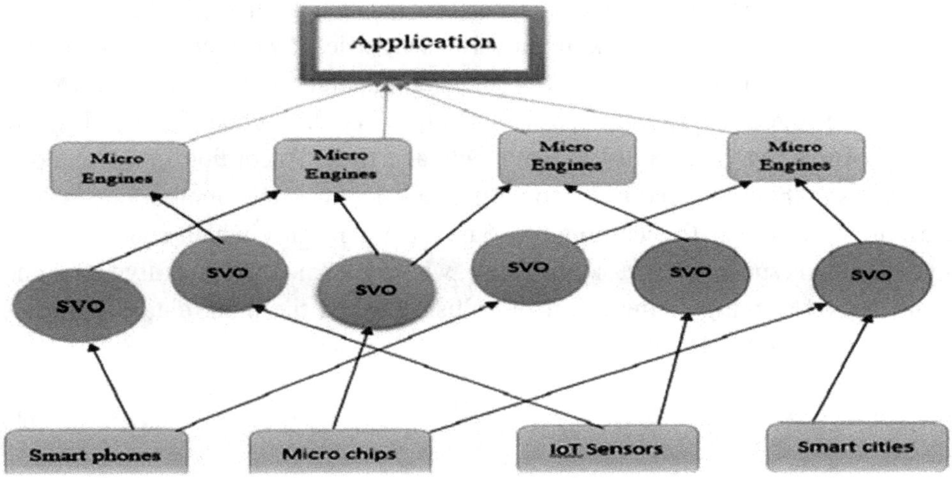

Despite all the research held on M2M technologies till now, no exceptionally better result achieved. All the researches obtain same estimation. For enabling this machine to machine communication process in reality combination of various IoT based devices basis on their work is must for the overall infrastructure. Initially there is a need of sensors, devices, actuators which generally work physically. Thereafter, sensor module, controller, barcode reader is mandatory to collect the information or data transferred by the physical devices placed in initial stage. To serve the streaming data any platform or operating system needs to be present like Ubuntu, windows, OS etc. these embedded operating system provide an environment for all the physical devices to communicate within their range. All of these operating systems are basically micro kernel architecture developed using C/C++ along with standard API system. Additionally, all the existing operating system is built-in with the ability to support standard physical devices or IoT based protocols like HTTP, IPv6, IPv4, UDP, TCP etc. To establish M2M communication system some optional protocols, API, Domains, middleware etc. The basic protocols, middleware, API which appears most for developing M2M system are message queue telemetry transport (MQTT), constrained application protocol (CoAP), Data Distribution Service (DDS), Extensible Messaging and Presence Protocol (XMPP), Advanced Message Queuing Protocol (AMQP) etc. (Sun et al, 2012; Tan et al, 2011; Kraginnis et al, 2015) are some of them. Thus, combination of all this objects along with various IoT devices results in a meaningful machine to machine communication. This promising process are increasing in such a trend that the next generation cannot be thought without its existence.

Recent years IoT indicates automation in various aspects treating all the objects in virtual behavior. Various recent research find that around 8 billion cellular network based devices are working on various significant work now. However, for a possible solution of big data in M2M messaging paradigm the authors of (Meng et al, 2017) suggested a solution which enhances the throughout for M2M traffic in case of massive requests from the clients at a time. According to the authors, the users should be divided into two parts one is learning and another one is non-learning. Now, if α is the learning rate, and r is the reward, which equals 1 if the transmission is successful or –k otherwise, where k is known as penalty factor and is introduced to mitigate the effect of collisions, then the input data or requests should be updated as follows:

$$Q \leftarrow (1-\alpha)Q + \alpha r \tag{1}$$

Moreover, the coexistence problem for both H2H and M2M has been formulated in (Bello et al, 2014). Here game theoretic approach has been implemented successfully. If mixed strategy of M2M and H2H is defined as α_i, probability of M2M as well as H2H are $P_{i,M}, P_{i,B}$ and cost of every transmission is defined as $\lambda \in [0,1]$ then gains (g_i) can be illustrated as,

$$g_i = \begin{cases} 1-\lambda & Successful \\ -\lambda & not\ successful \\ 0 & not\ attempted \end{cases} \tag{2}$$

Therefore, according to game theory scheme the success probability of messaging mechanism for action a_i taken by each player I is determined as,

$$\mathrm{E}[g_i] = \sum\nolimits_{a_i \in \{M,B\}} \sigma_i.(a_i).(P_s(a_i) - \lambda) \tag{3}$$

Simulations represents that the performance of both M2M and H2H is improved with respect to the baseline scheme of delivered message from device to device.

In healthcare technology, lots of used IoT devices mobility progress is quite similar as the Google glass as well as apple watch. In terms of data transformation, a M2M devices transfer small size of data but while vast amount of devices and sensors are connected there must be a verdict or discontinuation may happen in the network radio as well as in the core network. Recently, some significant work is going on for a possible effective solution for this types of problems. Consequently, the recent project titled 3rd Generation Partnership Project (3GPP) is concentrated on to address this problems towards the world for obtaining a solution of cellular protocols. With the process of upgrading cellular network from 2G to 3G some of the M2M concept has been implemented to the core networking system using for long term evolution. Generally, all the M2M devices are compulsorily need high long life cycle guarantee to efficiently handle the communication system to be established.

SMART PHONE OR HIGH PERFORMANCE COMPUTING WITH INTERNET OF THINGS (IOT)

After the evolution of technological advancement with the invention of Internet of things (IoT), nowadays technologies like sensors, short length wireless networks (Yang et al, 2018), local area connection networks, radio frequency identification (RFID) have permitted internet of things to implant in cloud based embedded computing which is the most adequate topic for discussing about in this technological era. In addition, the internet of things presently in its way to cover all spheres of human life keep in touch with physical sensors, devices, smart phones along with connecting network protocols. Most probably, one day it will be the reality of future generation not even thinking the world in their imagination without internet of things. This evolutionary technology boosts after the invention of network protocol version IPv6. Previous version was IPv4 which has some constraints to keep track vast volume of devices which IPv6 can easily and currently machine to machine technology indicates vast amount of connected devices at a time within a local or huge range of networks. Moreover, for this era communication, keeping concentration with internet of things technologies to be shifted later web of things (WoP) is currently providing the connected devices all possible necessary virtual protocols as well as domains to connect through internet or web. From all the actual application of IoT and WoT, IoT basically deals with various heterogeneous devices while the WoT is a platform where devices connection paradigm is flexible compared to IoT. Recently the promising rate of use of internet enabled devices reduce the internal barrier to relate mobile computing with IoT or WoT networking system. Moreover, smart phone of this era support various exceptional sensing capabilities like proximity sensors, gravity sensors, location tracker, audio or videos recorder detector etc. all of these services along with internet connections can enhance the overall capability of mobile phones which is defined as smart phones nowadays because of its involvement with machine to machine technology. Improvement of this technology enlarge security, privacy as well as enhance involvement of user with more flexibility, efficiency, less error rate, more accurately etc. Additionally all of this new technological progress helps the user to get desired information faster since mobile computing is now supporting analysis of vast amount of data following WoT access (Pang et al, 2014; Elmesalawy, 2015; Jahn et al, 2010; Kamilaris et al, 2011). Since, smartphone is everywhere nowadays, thus performance evaluation of it is important to estimate weakness. However, according to (Bello et al, 2014) average power needed by any processor while executing a job L with a frequency f is,

$$P(L,f) = a + b.f + c.f^2 + d.f^3 \tag{4}$$

Where a, b, c, d are the constraints. Following similar approach execution time associated with performance evaluation when α, β, γ be the regression coefficients can be interpreted as follows.

$$P(L,f) = \alpha + \beta.f + \gamma.f^2 \tag{5}$$

However, total energy consumption can be determined through,

$$e(L,f) = n_j.p(L,f).t(L.f) \tag{6}$$

Meanwhile optimization problem for internet of things can be formulated as,

$$\sum_{L\in R}(x_L.e_X(j))$$

With respect to

$$\sum_{L\in R}(x_L.P(L,f_{\min})) \leq \hat{P} \tag{7}$$

Moreover, simply by estimating the energy overall energy cost along with job's completing percentage α_j the projected power consumption by a job can be estimated using the equation mentioned below,

$$e_x(L) \approx a_{L.}n_L.P(L,f_L).t(L,f_L) \tag{8}$$

In addition, we can apply 0-1 knapsack on overall findings we collect from the equations mentioned above.

Basically the relation between mobile computing and IoT or WoT indicates the relations between various domains, protocols, demonstrations, case studies, application and business cases of IoT or WoT. This comparison is generally executed using mobile phones and IoT devices as well as sensors. Lots of recent work find ten different application categories associated with mobile computing as well as internet of things. Those categories are illustrated graphically below in figure 8.

Therefore, these ten different categories perform ten kinds of different task. First one is initial stage in which generally involves while recording some audio or video clip on mobile with the help of developed sensors connected with the smart phone. Thereafter, eco-feedback usually indicates the use of those mobile applications through which the users are able to track the current condition of the phone they are using like battery percentage, energy, and processor or ram condition etc. automation and control refers to maintaining various external devices associated with the phone by means of various sensors for example the smart home management can be controlled using smart phones to control each and every single object present in that home. For healthcare analysis the mobile phone acts as a communication bridge between the devices as well as sensors installed in user's body and web access. Those sensors collect current information of various parts of user's body and thereafter for next analysis as well as feedback those information are automatically sent to the web and after better manipulation immediately transfer the status to the user's smart phones to make aware about his health basis on the result processed. Meanwhile, the sports sensors basically accommodate with the user in times of sport activities. This is generally the combination of various sensors, devices which remain active while user engage with any kind of sporting competition. Usually the athletes use this feature.

Figure 8. Application categories of IoT and mobile computing

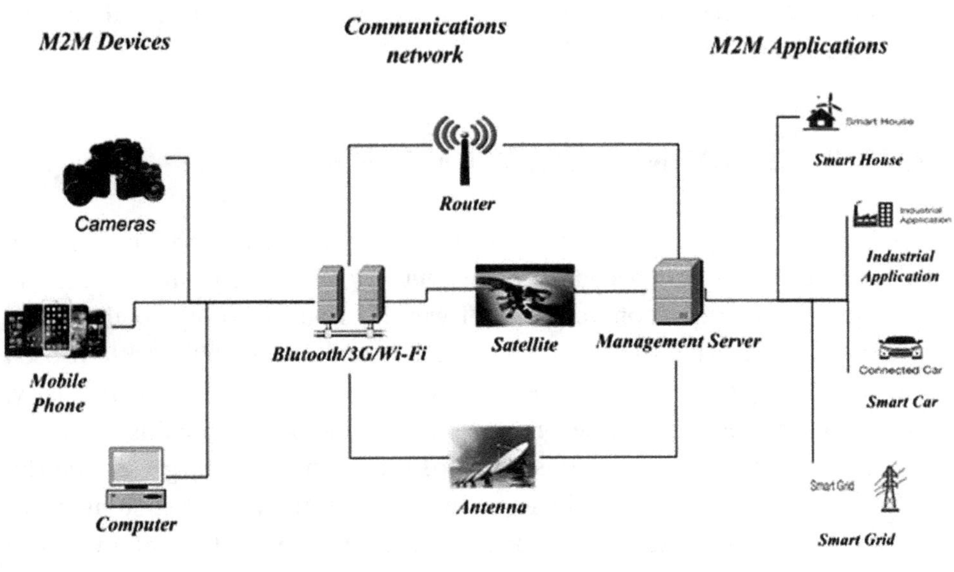

This process is finally executed by means of mobile application connected with appropriate application in the user's phone and the sensors gather logical necessary information from the sports activities after which those collected information are delivered to the web for possible manipulation and analysis of the user's status. In fine, the final estimation also provided to the phone for user's acknowledgement. Furthermore, the mobile computing and web of things are associated with agriculture for improving the condition of productivity, livestock etc. by making the agricultural system smart as well as improved than before. Gaming category illustrate the virtual gaming status to enhance the performance of the mobile device by the user. Since the mobile is connected through web of things therefore every single time the status of internal accessories of the phone delivered to web for better solution as well as satisfaction of the user. Immediately the user got feedback of the instant status and react like that situation. Another category titles "transportation" basically related to smart cities management. Under this phenomenon all the vehicles remain connected internally automatically and intelligently with one another as well as permits the driver to know current location, where to go, how far the destination is, what the shortcut way etc. is. Thus, making the transportation harmless, secure as well as faster is the significant theme of this category taking presence of both mobile communication system along with web of things technology. Large social interaction among the people living in the society and interaction among the things depicts the same consciences as first one always deals with the people, at the same time the second one deals with machine to machine communication within a society, city or whole country through mobile computing as well as web of things technology. This is the broaden application of WoT with mobile computing technology which ensure upcoming evolution of internet of things (IoT) and web of things (WoT) simultaneously. Therefore, for next generation these phenomenon will be the indispensable part for human civilization.

LOCATION SPOOFING DETECTION FOR IOT

With the evolution of technology and internet of things, wireless exact localization of any kinds of things have become a significant way of research to the researchers because of its importance on various real world serious spoofing condition the general people faces. Thus from the last decade thousands of research work have been executed and analyzed for making the process even better as it is now. Particularly, location spoofing detection has become significantly momentous for this new era to get rid some serious condition. In this era of IoT the common people get satisfied by securing privacy, expect more reliable as well as trustworthy for continuing better life as before. With increasing rate of improvement in the field of technology as well as internet of things (IoT) the common users are expecting high reliable information, location based services to their system. Absence of reliable and trustworthy estimates of internet of things as well as machine to machine communication based system the continuous location based system can be disturbed. For getting rid of this critical situation there needs some anchors to be present to enable the overall procedure. Those anchors are usually used for determining the distance of the desired object from the device we are operating through internet. The overall architecture for detecting trustworthy location services is illustrated in figure 9.

Here for location detection purpose the anchors play vital role for determining exact position of that particle which is being searched by user. Initially the anchors send request packet to confirm the distance of the object from that anchor. Generally the signal is sent by the anchors are bi-directional for which it becomes possible for the system to measure the actual distance by measuring the delay period after which the requested signal returns back to the anchors. After gathering all the information returned from the object. All the anchors transfer the distance measured values towards Fusion center where after enormous manipulation the data is send towards internet protocols or domains to transfer clear estimation of findings transferred by the overall system to represent the estimation using various devices or sensors to make user informed regarding the result found after multiple analysis. Additionally,

Figure 9. Overview of trustworthy location tracker

the more the number of anchors the detection is faster and accurate than using less than three anchors. For meaningful estimation of a particular object in a specific location of two dimensional plane more than three or four anchors are mandatory for overall errorless estimation of current location. Furthermore, all the five anchors shown in figure.9 can detect signals transferred to each other and make the signal strength within its limit for make it audible to them. From these audible signals the exact location of an object though it's critical task to transmit the exchanged signal by each anchors to audible sound. Therefore, from the perspective of this audibility concept the overall procedure of object detection is implemented and for this significant task to be executed strong protocol based communication process along with various cloud computing (Zhang et al, 2017; Sarkar et al, 2017; Matallah et al, 2017) along with mobility features are necessary enough to be completed. One of the contribution (Kamilaris et al, 2011; Prinsloo et al, 2016; Kamal et al, 2017) associated with location detection using internet of things represents a universal way of location detection through which an object is currently passing. The study interprets that two types of antennas are in action for data collection and analysis. Initially both the voltages and current flown through the antenna is calculated,

$$voltage = \frac{2}{f_0 C_{tune}(R_t + 6)} \tag{9}$$

$$nt = \frac{6.37}{(R_t + 6)} \tag{10}$$

Here voltage is the peak to peak voltage and f_0 is the operating frequency along with tuning capacitor C_{tune} .Meanwhile, R_t is internal resister. However, circular coil as well as square coil parameters need to be determined for tracking location. The basic operation which is in need of determination of those parameters are as follows,

For circular coil,

$$L = \frac{0.31 * (aN)^2}{6a + 9ha + b} \tag{11}$$

For this circumstance, L is considered as the conductance of the coil, a is average radius, N is the number of windings, h is the height whereas b is the thickness. Moreover, the inductance of an N-dimensional multilayer square coil is depicted in equation 12.

$$L = 0.001aN^2\left(2.303\log_{10}\left(\frac{a}{b+c}\right) + 0.2235\left(\frac{b+c}{a}\right) + 0.726\right) * \mu H \tag{12}$$

Thus, using these equations some estimation results are extracted from the reader and transponder. Analyzing those findings the exact location is usually determined.

CLUESTERING TOOL FOR DYNAMIC IOT DATA STREAMS

Since the next generation is appearing with enormous use of internet of things, the faster manipulation strategies are mandatory for next generation information extraction from large volume of data streaming from the IoT devices as well as sensors. Various recent researches find lots of clustering components for dynamically streaming data. In spite of being available clustering tools next generation faster analysis of higher dimensional data is mandatory because of upcoming data from the devices that will be used for analysis of various aspects of data from human life as well as developed civilization properties like smart cities, smart homes, smart agricultural system etc. For being able to cluster large amount of data various data clustering components are nowadays enabling current technology to be attached with vast amount of data from eHealth, smart phones and smart civilization consequences are measurable conditions here. Therefore, for grouping huge volume of data streaming from the devices which are connected internally through the help of various new technologies like cloud computing technology, mobile computing technology, cellular networking technology, etc. moreover, all the classification approaches need special types of data manipulation ability for efficiency, less time consumption, less memory consumption etc. Since there will be a technological evolution specially IoT or Web of things evolution, therefore, the analysis of the streaming data from various sources of devices will be hot topic for next generation to discover gist knowledge from there. Therefore, research on data clustering technologies should be emphasized more than ever to deal with particular size of data. Various unsupervised as well as automated process are adorable for this context. However, for the purpose of data distributions from any dynamic data streams it is mandatory to estimate a probable function which will interpret overall data distribution process. Kernel density estimation (KDE) (JooSeuk et al, 2012) function is suitable for this circumstance. If $data(w)$ is considered as the stream of data for window size w, $W(.)$ is weighting function, $func_m(x)$ can be interpreted as data samples of size m, bandwidth is $h(m)$ then the overall density estimation function can be illustrated as,

$$func_m(x) = \int_{-\infty}^{\infty} \frac{1}{hW(\frac{x-y}{h}) * data * func_m(y)} = \frac{1}{mh}\sum_{j=1}^{m} W(\frac{x-X_i}{h}) \tag{13}$$

and

$$\int_{-\infty}^{\infty} W(y)dy = 1 \tag{14}$$

Meanwhile, estimating local maxima and merging centroids is mandatory for clustering purpose. If x is taken into account as a specific portion of data stream and $data_i = \left(x_i 1, x_i 2, x_3 3.........x_i m\right)$ be the ith cluster where $data_i$ a point which minimizes mean squared Euclidean distance. Thus, the equation to minimizes sum of Euclidean point can be depicted as follows,

$$data_i = \frac{(x_i 1 + x_i 2 + ... + x_i m)}{m} \tag{15}$$

Since, this equation determines local maxima as well as adjust centroids, merging this centroids seems to be important to the purpose of clustering data. Meanwhile, if σ is the standard variance, μ is the expected outcome, d is a small outlier distance and z is the normal distribution then merging centroids can be done through,

$$out(\beta, \mu, \sigma^2) = \left\langle d_i :| d_i - \mu |> z_1 - \frac{\alpha}{2} * \sigma \right\rangle$$

However, sometimes the outliers distance seems to be appeared far away from the centroids. Consequently, just to check whether the outliers are within our consideration we check outliers validity using the property mentioned in equation 17.

$$d_i < -(z_1 - \frac{\alpha}{2} * \sigma + \mu)$$

The data is usually clustered following various criterions which must be while implementing approaches to cluster or analysis of data. Basically, the main criterion depending on which the data are partitioned as well as interpreted is similarity and homogeneity. Sometimes, the dimension along with data size becomes obstacle to manipulate data easily. For adapting appropriate clustering methods this obstacle need to be taken on concentration for avoiding further analytical obesity problem. Various data mining models have been already discovered by lots of researchers to make advancement in the field of clustering or analysis of data. In addition, some of the developed models are multi-layer data mining model for internet of things, distributed data mining model for internet of things, grid based data mining model for internet of things, Data mining model for IoT from multi-technology integration perspective etc. various models follow unique process to analysis large volume of data. For example, according to multi-layer model for data mining technology executes whole process within four phages which are data collection layer, data management layer, event processing layer and data mining service layer. Data collection layer basically collects data from the devices which are connected instantly along with various sensors and obviously within a network range. Usually the satellite data, GPS data, positional data, sensors data are collected through this phage. Moreover, various types of data need different types of data collection strategies to gather data within specified range. There sometimes seems a lot of problems while collecting data among them energy efficiency, misreading, repeated reading, fault tolerance, data filtering etc.

After execution of the task operated by data collection layer, the data management layer deals with the collected data to make it noise or irreverent information free data for next phage. Thereafter, the next stages handle data storage process to the storage system. After which finally the resultant data is able to be manipulated and illustrate in a desirable way.

In general compared with the general data, the data streamed from IoT devices are quite massive, complex, complicated, distributed, time related, position related. Meanwhile, at the same time the data collected from IoT devices is heterogeneous

Figure 10. Multi-layer data mining model for IoT

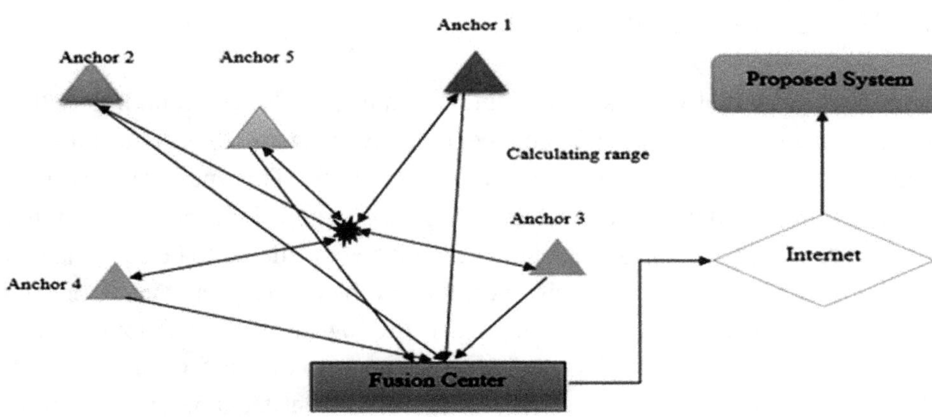

and the resources by which it needs to be analyzed is limited which creates a contradiction towards finding of resultant information mandatory to work with. Moreover, IoT data are massive in size, therefore preprocessing is mandatory for further future process. While depending on internet based preprocessing tools it's unfortunately consumes a lot of time uncover the gist knowledge. For a possible solution of this problem the data distributed system can be applied here for better performance of classification. If the use of this process might not be effective for IoT data then another model titled grid based model for analysis large volume of data. Recent study finds that grid based computing is an exceptionally outperforms the other common algorithms for computation of vast amount of data. Moreover, this grid based computational approach is able to handle heterogeneous, large-scale, complicated data that's why grid based computational approach draws extra attention from the industrial perspectives as well as research community. The main theme of grid based computing is the overall system can use the grid like a resource. These resources indicates various types of computational data collection tools like internet of things devices, sensors, computational resources like data resources etc. Thus, the key concept of grid based computing is connecting various devices through faster internet connection protocol. Because of this consequence the recent available IoT devices internally connected become more intelligent, automatically decision undertaken ability, context awareness, long range operable. Therefore, data mining grid for distribution of devices will cause great improvement to the next generation data mining technology. However, this smart grid technology nowadays creates an evolutionary advancement in the field of internet of things and predictably it will be a great clustering tool for upcoming next generation large streaming data for ever growing data produced from various always connected devices through high level internet connection.

METHODS AND PROSPECTS FOR LARGE VOLUME OF IOT DATA

The massive growth of various technological data is nowadays surrounding us that has been considered nowadays a key element for advancement of recent technology towards next generation analytics (Solanki et al, 2016). This massive growth of data can be illustrated using high velocity, complex, large volume as well as higher dimensional data. Thus, this types of data cannot be considered to be easy manual analysis by human. Because of this enormous problem the data science or data mining is on the way of downward slopping. Thus, nowadays the traditional way of manipulation of data analytics algorithms are fully incapable of performing analytical task on large volume of data in this evolution of current technology (Dey et al,

2018; Elhayatmy et al, 2018). Desired knowledge as well as information cannot be easily extracted neither can be analyzed easily. In this context the rising of various approaches like cloud computing based approaches to reduce data size, volume, dimension, and velocity is now mandatorily important for analytical consciences. However, in this era of technology a distributed as well as faster and efficient way has become mandatory to handle this types of situation. From the field of large data processing technology various platforms have been tried to implement for this problem as well but every times they failed to handle using those traditional analysis paradigm (Bhatt et al, 2017; Dey et al, 2017). For this problem, the most effective first steps was map reduce process Nimmy et al, 2017) which was proposed on 2003. Moreover, this evolutionary invention was designed to work with large volume of data to acquire knowledge from there. For most of the cases data prepossessing is must to step ahead towards big data analysis. Overview of data preprocessing paradigm is illustrated in figure 11.

Additionally, a lots of work has already taken part towards data manipulation this era. Among which kNN-IS: An Iterative Spark-based design of the k-Nearest Neighbors classifier for big data, Fuzzy Rule Based Classification Systems for Big Data with MapReduce, A View on Fuzzy Systems for Big DFata, Data discretization: taxonomy and big data challenge, An Effective Big Data Supervised Imbalanced Classification Approach for Ortholog Detection in Related Yeast Species, Evolutionary Feature Selection for Big Data Classification, The winner algorithm for the ECBDL'14 Big Data Competition (Maillo et al, 2012; Fernandez et al, 2017; Fernandez et al, 2016; Ramirez-Gallego et al, 2016; Galpert et al, 2015; Peralta et al, 2015; Triguero et al, 2015; Fernandez et al, 2014). Big Data with Cloud Computing: An Insight on the Computing Environment, MapReduce and Programming Frameworks. Big Data and Internet of Things – Fusion for different services and its impacts (Fernandez et al, 2014).

Figure 11. Apreprocessing overview of huge datasets

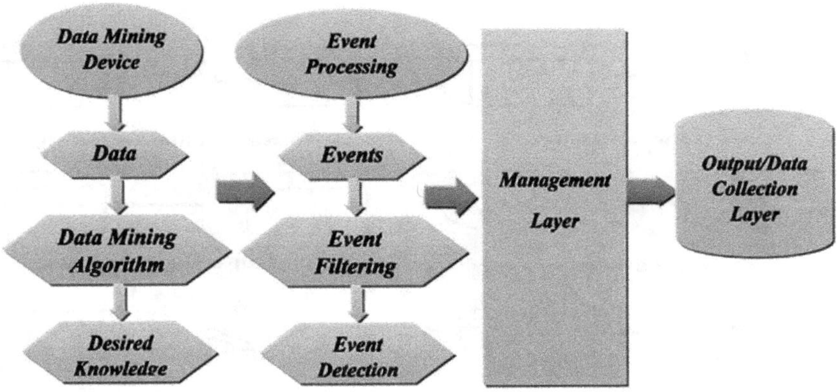

Although thousands of methods and prospects are available in current literature MapReduce (Kamal et al, 2016) is the evolutionary approach to analyze large volume of data. This approach is seemed to be appeared quite worthy for large volume data manipulation for its ability to reduce data complexity applying internal mechanism on input data. Moreover, robustness, dimension problems can also be diminished through MapReduce Approach. Basic working mechanism of MapReduce approach is illustrated below:

Here, some sample medical big data (Jain et al, 2017) have been used to explain basic structure of MapReduce approach (Kamal et al, 2017). After invention of this evolutionary approach in 2003; this approach is most used and popular approach for big data analysis. Lots of research have been conducted based on this method. However, the k-Nearest Neighbors classifier is a simple yet effective widely renowned method in data mining. The actual application of this model in the big data domain is not feasible due to time and memory restrictions. Several distributed alternatives based on MapReduce have been proposed to enable this method to handle large-scale data. Therefore, from the perspective of time and space complexity mapreduce is the best fitted for large volume of data manipulation. All of this works depicts the way how large volume of data can be manipulated for analysis purpose. Therefore, for data extraction from streaming or produced data from billions of IoT devices will be hot topic for next generation and which can be reasonable using various data mining mar reduce process along with cloud computing technology.

BIG DATA PROCESSING FOR IOT

Most of the cases it seems to be appeared that the massive and complex data of internet of things gathered from heterogeneous devices like cameras, vehicles,

Figure 12. Graphical Infrastructure of MapReduce proceedings

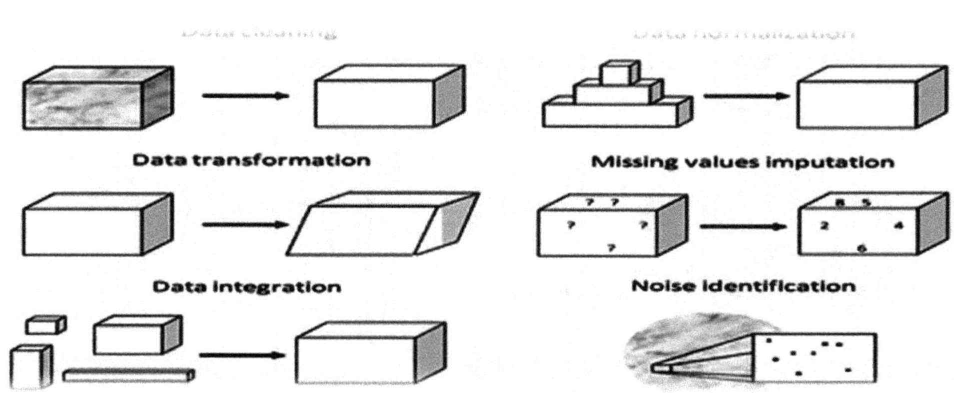

drivers and passengers. In order to obtain specific solution from this sort of data multivariate Gaussian model can be effective which ensures suboptimal solution (Rain et al, 2017). According to Guoru Ding et al. (2001) Sklar' Theorem (Ding et al, 2013) can be implemented as the perfect alternative of Gaussian model to process heterogeneous data. To look forward of processing task if x_n be the data from n-th sensor, $X := \{x_n\}_{n=1}^{N}$ be a heterogeneously distributed dataset and F, C be defined as *N*-dimensional cumulative distribution function (cdf), copulas function (Nelsen et al, 1999) respectively then sklar theorem states that,

$$F(x_1, x_2 ....x_n) = C(F_1(x_1), F_2(x_2), ...F_N(x_N)) \tag{18}$$

Now from the above mentioned equation the probability density function can be interpreted as the N-order derivative of equation.1. Therefore, if f_p be the product of marginal pdf (probability density function) and c () be the copulas density weights then the resultant function should be,

$$\begin{aligned} f(x_1, x_2, x_3, ..., x_N) &= \frac{\partial^N}{\partial x_1 \, \partial x_2 .... \partial x_N} C(F_1(x_1), F_2(x_2), ..., F_N(x_N) \\ &= f_p(x_1, x_2, ..., x_N) c(F_1(x_1), F_2(x_2), ..., F_N(x_N)) \end{aligned} \tag{19}$$

Using the above formulas the data gathered from heterogeneous sensors [89] are processed which creates easier approach for further manipulation of complicated data in an efficient manner which ensures efficient processing mechanism.

However, Massive amount of IoT devices generates higher dimensional as well as large volume of data and further steps may require specific information from those higher dimensional data collected from IoT sensors or devices. Therefore, big data is must for this context to analyze those data and find the desired information. For implementation of big data concept we can interpret the higher dimensional data as an impact matrix which demonstrate higher dimensional data consuming lower system property. The traffic matrix in large scale networks (Iyengar et al, 2011) and image frame matrix in video surveillance (Mardani et al, 2013) are the most effective use of impact matrix for higher dimensional data. Now if we consider M as a large sensing data matrix and Y as a low rank impact matrix then if we decompose matrix and formulate principal component analysis (PCA) then the equation should be,

$$\min_{\{Y\}} \; || Y ||_* \; Subject\ To || M - Y ||_F \leq \varepsilon \tag{20}$$

ε is a noise related parameter and $|| \,.\, ||_*$, $|| \,.\, ||_F$ be various norm of matrix. If more noise or irreverent data defined as A is inserted furthermore then the matrix can be optimized using following equation,

$$\min_{\{Y\}} \; || Y ||_* + \lambda \, || A ||_1 \; Subject\ To || M - Y - A ||_F \leq \varepsilon \tag{21}$$

Here λ is rank-sparsity controlling parameter and $|| \,.\, ||_1$ is 1st norm of the matrix. According to matrix recovery mechanism with higher dimensional data processing approach may outperforms other approaches.

Moreover, future contribution of IoT may create enormous challenges and explore new directions. From the perspective of the contribution (Cand'es et al, 2011) the authors mentioned some future challenges including architecture challenge for wide spreading network of IoT, environmental, technical and hardware innovation challenge, privacy and security challenge may occur because of every connected devices and sensors internally, heterogeneous and big data processing challenges, noisy and massive-intensive data etc. moreover, lots of contributions associated with big data processing conducted among can be taken into account for internet of things big data processing also (Aly et al, 2015; Shanzhi et al, 2014; Shanzchi et al, 2014; "Engineeringforecechange.org, 2015; Kamal et al, 2017). Proposed methods mentioned those contributions seems to outperform lots of big data processing algorithms. Thus, those approaches can be evolutionary resource for internet of things big data processing.

CONCLUSION

Meanwhile, this work basically promotes the main concept of data growth for next generation IoT devices as well as discussion regarding the way that can be emphasized to be proposed to handle those big data. In this purpose big data analytical tools can be effective as well as faster process of extraction of different significant information. Practically, in IoT devices applications, a lot more highly sensible data with mixed characteristics are detected which is considered a great challenging task for computer science as well as improved information technology along with big data analytics for easy, faster, efficient with less memory consumption. From this perspective this work can be effective for brief discussion of mining tools for data from billions of internet of things (IoT) devices, sensors etc by spreading knowledge regarding automatic extraction process, preprocessing mechanism, automatic network process, smart resource allocation, intelligent service providing etc. therefore, cloud computing, mobile computing along with various recently discovered effective technology based big data analytics for IoT can come up with a new evolutionary technological evolution for next generation.

REFERENCES

Ahmed, K., Liu, J., & Wu, X. (2017). An Energy Efficient Demand-Response Model for High Performance Computing Systems, *Proceedings of the 25th IEEE International Symposium on the Modeling, Analysis, and Simulation of Computer and Telecommunication Systems (MASCOTS 2017)*. 10.1109/MASCOTS.2017.25

Ali, A. H., Atia, A., & Mostafa, M. S. M. (2017). Recognizing Driving Behavior and Road Anomaly using Smartphone Sensors. *International Journal of Ambient Computing and Intelligence*, *8*(3), 22–37. doi:10.4018/IJACI.2017070102

Altawy, R., & Youssef, A. M. (2016). Security Tradeoffs in Cyber Physical Systems: A Case Study Survey on Implantable Medical Devices. *Access IEEE, 4*, 959-979.

Aly, H., Elmogy, M., & Barakat, S. (2015). Big Data on Internet of Things: Applications, Architecture, Technologies, Techniques, and Future Directions. *International Journal of Computer Science Engineering, 4*(6).

Amaral, L. A., Hessel, F. P., Bezerra, E. A., Corr^ea, J. C., Longhi, O. B., & Dias, T. F. (2011). ecloudrfid–a mobile software framework architecturefor pervasive rfid-based applications. *Journal of Network and Computer Applications*, *34*(3), 972–979. doi:10.1016/j.jnca.2010.04.005

Apache Hadoop Project. (2015). Retrieved from http://hadoop.apache.org/

Ayagi, S., & Agarwal, A. (2016). A conceptual framework for IoT based healthcare system using cloud computing. *Cloud System and Big Data Engineering (Confluence), 2016 6th International Conferenc*.e

Bello, L. M., Mitchell, P., & Grace, D. (2014). Application of Q-learning for RACH access to support M2M traffic over a cellular network. *European Wireless Conference*, 1–6.

Bhatt, C., Dey, N., & Ashour, A. S. (Eds.). (2017). Internet of things and big data technologies for next generation healthcare. Academic Press. doi:10.1007/978-3-319-49736-5

Bill Burke, R. M.-H. (2006). *Enterprise Java Beans 3.0*. O'Reilly.

Cand'es, E., Li, X., Ma, Y., & Wright, J. (2011). Robust principal component analysis? *Journal of the Association for Computing Machinery*, *58*(3), 1–37. doi:10.1145/1970392.1970395

Chaturvedi, K. K., & Sing, V. B. (2013). Tools in Mining Software Repositories. *Computational Science and Its Applications (ICCSA) 2013 13th International Conference on*, 89-98.

Chen, S. (2014). A Vision Of Iot: Applications, Challenges, An Opportunities With China Perspective. *IEEE Internet of Tings Journal, 1*(4), 349–359. doi:10.1109/JIOT.2014.2337336

Dean, J., & Ghemawat, S. (2008). MapReduce: Simplified dataprocessing on large clusters. *Communications of the ACM*, *51*(1), 107–113. doi:10.1145/1327452.1327492

Dean, J., & Ghemawat, S. (2004). Mapreduce: Simplified data processing on large clusters. OSDI 2004, 137–50.

Dey, N., Ashour, A. S., & Bhatt, C. (2017). Internet of things driven connected healthcare. In *Internet of things and big data technologies for next generation healthcare* (pp. 3–12). Cham: Springer. doi:10.1007/978-3-319-49736-5_1

Dey, N., Hassanien, A. E., Bhatt, C., Ashour, A., & Satapathy, S. C. (Eds.). (2018). *Internet of Things and Big Data Analytics Toward Next-Generation Intelligence*. Springer. doi:10.1007/978-3-319-60435-0

Ding, G., Wang, L., & Wu, Q. (2013). *Big Data Analytics in Future Internet of Things*. arXiv:1311.4112v1 [cs.DC]

Dinh, T., Kim, Y., & Lee, H. (2016). A location-based interactive model for Internet of Things and cloud (IoT-cloud). *Ubiquitous and Future Networks (ICUFN) 2016 Eighth International Conference on*, 444-447.

Doukas, C., & Maglogiannis, I. (2012). Bringing iot and cloud computing towardspervasive healthcare," in Innovative Mobile and Internet Servicesin Ubiquitous Computing (IMIS). *Sixth International Conferenceon. IEEE, 2012*, 922–926.

Eisenhauer, M., Rosengren, P., & Antolin, P. (2009). A development platform for integrating wireless devices and sensors into ambient intelligence systems. In *Sensor, Mesh and Ah Hoc Communications and Networks Workshops, SECON 2009, IEEE Communication Society Conference*. Rome: IEEE. 10.1109/SAHCNW.2009.5172913

Elhayatmy, G., Dey, N., & Ashour, A. S. (2018). Internet of Things Based Wireless Body Area Network in Healthcare. In *Internet of Things and Big Data Analytics toward Next-Generation Intelligence* (pp. 3–20). Cham: Springer. doi:10.1007/978-3-319-60435-0_1

Elmesalawy & Ali. (2015). A Grouped System Architecture for Smart Grids Based AMI Communications Over LTE. *International Journal of Wireless & Mobile Networks, 7*(6).

Engineeringforchange.org. (2015). *Engineering for Change*. Available at: http://www.engineeringforchange.org/home

Fadlullah, Z. M., Fouda, M. M., Kato, N., Takeuchi, A., Iwasaki, N., & Nozaki, Y. (2011). Toward intelligent machine-to-machine communications in smart grid. *IEEE Communications Magazine, 49*(4), 60–65. doi:10.1109/MCOM.2011.5741147

Feki, M. A., Kawsar, F., Boussard, M., & Trappeniers, L. (2013). TheInternet of Things: The Next Technological Revolution. *Computers, 46*(2), 24–25. doi:10.1109/MC.2013.63

Fernandez, Río, & Herrera. (2017). Fuzzy Rule Based Classification Systems for Big Data with MapReduce: Granularity Analysis. *Advances in Data Analysis and Classification, 11*(4). doi: 10.100711634-016-0260-z

Fernandez, A., Carmona, C. J., del Jesus, M. J., & Herrera, F. (2016). A View on Fuzzy Systems for Big Data: Progress and Opportunities. *International Journal of Computational Intelligence Systems, 9*(1), 69–80. doi:10.1080/18756891.2016.1180820

Fernandez, A., Río, S., López, V., Bawakid, A., del Jesus, M. J., Benítez, J. M., & Herrera, F. (2014). Big Data with Cloud Computing: An Insight on the Computing Environment, MapReduce and Programming Frameworks. *WIREs Data Mining and Knowledge Discovery, 4*(5), 380–409. doi:10.1002/widm.1134

Galpert, Río, Herrera, Ancede-Gallardo, Antunes, & Agüero-Chapin. (2015). An Effective *Big Data Supervised Imbalanced Classification Approach for Ortholog Detection in Related Yeast Species*. BioMed Research International. doi: 10.1155/2015/748681

Global Sensor Networks. (2011). *International Conference on Computational Intelligence and Communication Networks*. Author. doi: 10.1109/CICN.2011.172

Google Fit. (2015). Retrieved from https://developers.google.com/fit/

Gupta, R., Gupta, H., & Mohania, M. (2012). Cloud computing andbig data analytics: what is new from databases perspective? *1st International Conference on Big DataAnalytics (BDA)*, 42–61.

Höller, J., Tsiatsis, V., Mulligan, C., Karnouskos, S., Avesand, S., & Boyle, D. (2014). *From Machine-to-Machine to the Internet of Things: Introduction to a New Age of Intelligence*. Elsevier.

HYDRA. (2010). Retrieved from http://hydramiddleware.eu

Iyengar, S. G., Varshney, P. K., & Damarla, T. (2011, May). A parametric copula-based framework for hypothesis testing using heterogeneous data. *IEEE Transactions on Signal Processing*, *59*(5), 2308–2319. doi:10.1109/TSP.2011.2105483

Jahn, Jentsch, Prause, Pramudianto, Al-Akkad, & Reiners. (2010). The Energy Aware Smart Home. *Proceedings of the Fifth International Conference on Future Information Technology (FutureTech),* 1–8.

Jain, A., & Bhatnagar, V. (2017). Concoction of Ambient Intelligence and Big Data for Better Patient Ministration Services. *International Journal of Ambient Computing and Intelligence*, *8*(4), 19–30. doi:10.4018/IJACI.2017100102

Jiang, L., Da Xu, L., Cai, H., Jiang, Z., Bu, F., & Xu, B. (2014). "An IoTOriented Data Storage Framework in Cloud ComputingPlatform," Ind. Informatics. *IEEE Trans.*, *10*(2), 1443–1451.

Jin, J., Gubbi, J., Marusic, S., & Palaniswami, M. (2014). An information framework for creating a smart city through internet of things. *InternetofThings Journal, IEEE*, *1*(2), 112–121. doi:10.1109/JIOT.2013.2296516

Kamal, M. S., Dey, N., & Ashour, A. S. (2017). Large Scale Medical Data Mining for Accurate Diagnosis: A Blueprint. In *Handbook of Large-Scale Distributed Computing in Smart Healthcare* (pp. 157–176). Cham: Springer. doi:10.1007/978-3-319-58280-1_7

Kamal, S., Ripon, S. H., Dey, N., Ashour, A. S., & Santhi, V. (2016). A MapReduce approach to diminish imbalance parameters for big deoxyribonucleic acid dataset. *Computer Methods and Programs in Biomedicine*, *131*, 191–206. doi:10.1016/j.cmpb.2016.04.005 PMID:27265059

Kamilaris, A., Iannarilli, N., Trifa, V., & Pitsillides, A. (2011). Bridging the Mobile Web and the Web of Things in Urban Environments. *First Workshop on the Urban Internet of Things, in Proc. of IoT 2010.*

Kantarci, B., & Mouftah, H. T. (2014). Trustworthy sensing for public safety in cloud-centric internet of things. *Internet of Things Journal, IEEE, 1*(4), 360–368. doi:10.1109/JIOT.2014.2337886

Karagiannis, V., Chatzimisios, P., Vazquez-Gallego, F., & Alonso-Zarate, J. (2015). *A Survey on Application Layer Protocols for the Internet of Things. Transaction on IoT and Cloud Computing.*

Kim & Scott. (2012). Robust kernel density estimation. *Journal of Machine Learning Research, 13*(1), 2529–2565.

Kuprianov & Petukhov. (2016). Distributed data mining based on actors for Internet of Things. *2016 5th Mediterranean Conference on Embedded Computing (MECO),* 480-484. doi: 10.1109/MECO.2016.7525698

Lam, C. (2011). *Hadoop in Action* (1st ed.). Greenwich, CT: Manning.

Li, Xu, & Zhao. (2014). *The Internet Of Things: A Survey.* Springer Science+Business Media.

Liu, C. H., Fan, J., Branch, J. W., & Leung, K. K. (2014). Toward QoI and Energy-Efficiency in Internet-of-Things Sensory Environments. *IEEE Transactions on Emerging Topics in Computing*, *2*(4), 473–487. doi:10.1109/TETC.2014.2364915

Lumpkins, W. (2013). The Internet of Things Meets Cloud Computing. *IEEE Consum. Electron. Mag.*, *2*(2), 47–51. doi:10.1109/MCE.2013.2240615

Maillo, J., Ramírez-Gallego, S., Triguero, I., & Herrera, F. (in press). kNN-IS: An Iterative Spark-based design of the k-Nearest Neighbors classifier for big data. *Knowledge-Based Systems.* doi:10.1016/j.knosys.2016.06.012

Mardani, M., Mateos, G., & Giannakis, G. B. (2013, February). Dynamic Anomalography: Tracking Network Anomalies Via Sparsity and Low Rank. *IEEE Journal of Selected Topics in Signal Processing*, *7*(1), 50–66. doi:10.1109/JSTSP.2012.2233193

Mari, D., & Kotz, S. (2001). *Correlation and Dependence.* London, UK: Imperial College Press. doi:10.1142/p226

Marotta, Both, Rochol, Granville, & Tarouco. (2014). *Evaluating management architectures for Internet of Things devices*. 2014 IFIP Wireless Days (WD), Rio de Janeiro, Brazil. doi: 10.1109/WD.2014.7020811

Matallah, H., Belalem, G., & Bouamrane, K. (2017). Towards a New Model of Storage and Access to Data in Big Data and Cloud Computing. *International Journal of Ambient Computing and Intelligence*, *8*(4), 31–44. doi:10.4018/IJACI.2017100103

Meng, Wu, Muvianto, & Gray. (2017). A Data-Oriented M2M Messaging Mechanism for Industrial IoT Applications. *IEEE Internet of Things Journal, 4*(1), 236-246. doi: .10.1109/JIOT.2016.2646375

Nelsen, R. B. (1999). *An Introduction to Copulas*. New York: Springer-Verlag. doi:10.1007/978-1-4757-3076-0

Ngu, Gutierrez, Metsis, Nepal, & Sheng. (2017). IoT Middleware: A Survey on Issues and Enabling Technologies. *IEEE Internet of Things Journal, 4*(1), 1-20. doi: 10.1109/JIOT.2016.2615180

Nimmy, Kamal, Hossain, Dey, Ashour, & Shi. (2017). Neural Skyline Filtering for Imbalance Features Classification. *International Journal of Computational Intelligence and Applications, 16*(3).

Niyato, D., Xiao, L., & Wang, P. (2011). Machine-to-Machine Communications for Home Energy Management System in Smart Grid. *IEEE Communications Magazine*, *49*(April), 53–59. doi:10.1109/MCOM.2011.5741146

Pang, Y.-C., Chao, S.-L., Lin, G.-Y., & Wei, H.-Y. (2014). Network access for M2M/H2H hybrid systems: A game theoretic approach. *IEEE Communications Letters*, *18*(5), 845–848. doi:10.1109/COML.2014.6845397

Peralta, Río, Ramírez-Gallego, Triguero, Benítez, & Herrera. (2015). Evolutionary Feature Selection for Big Data Classification: A MapReduce Approach. *Mathematical Problems in Engineering*. doi: 10.1155/2015/246139

Perera, C., Liu, C. H., & Jayawardena, S. (2015). The Emerging Internet of Things Marketplace from an Industrial Perspective: A Survey. *IEEE Transactions on Emerging Topics in Computing*, *3*(4), 585–598. doi:10.1109/TETC.2015.2390034

Perera, C., Liu, C. H., Jayawardena, S., & Chen, M. (2014). A Survey on Internet of Things from Industrial Market Perspective. *IEEE Access: Practical Innovations, Open Solutions*, *2*, 1660–1679. doi:10.1109/ACCESS.2015.2389854

Prinsloo, J., & Malekian, R. (2016). Accurate Vehicle Location System Using RFID, an Internet of Things Approach. *Sensors (Basel)*, *16*(6), 825. doi:10.339016060825 PMID:27271638

Raggett, D. (2015, May). The Web of Things: Challenges and Opportunities. *IEEEComputer*, *48*(5), 26–32.

Ramírez-Gallego, S., García, S., Mouriño Talín, H., Martínez-Rego, D., Bolón-Canedo, V., Alonso-Betanzos, A., ... Herrera, F. (2016). Data discretization: Taxonomy and big data challenge. *Wiley Interdisciplinary Reviews. Data Mining and Knowledge Discovery*, *6*(1), 5–21. doi:10.1002/widm.1173

Rani, Ahmed, Talwar, Malhotra, & Song. (2017). IoMT: A Reliable Cross Layer Protocol for Internet of Multimedia Things. *IEEE Internet of Things Journal*. doi: .10.1109/JIOT.2017.2671460

Rao, B., Saluia, P., Sharma, N., Mittal, A., & Sharma, S. (2012). Cloud computing for internet of things & sensing based applications. *Sensing Technology (ICST), 2012 Sixth International Conference on. IEEE*, 374–380. 10.1109/ICSensT.2012.6461705

Riazul Islam, S., Kwak, D., Humaun Kabir, M., Hossain, M., & Kwak, K.-S. (2015). The internet of things for health care: A comprehensive survey. *Access, IEEE*, *3*, 678–708. doi:10.1109/ACCESS.2015.2437951

Salahuddin, M. A., Al-Fuqaha, A., & Guizani, M. (2015). Software-defined networking for rsu clouds in support of the internet of vehicles. *Internet of Things Journal, IEEE*, *2*(2), 133–144. doi:10.1109/JIOT.2014.2368356

Sarkar, M., Banerjee, S., Badr, Y., & Sangaiah, A. K. (2017). Configuring a trusted cloud service model for smart city exploration using hybrid intelligence. *International Journal of Ambient Computing and Intelligence*, *8*(3), 1–21. doi:10.4018/IJACI.2017070101

Savazzi, S., Rampa, V., & Spagnolini, U. (2014). Wireless CloudNetworks for the Factory of Things: Connectivity Modeling andLayout Design. *IEEE Internet Things J.*, *1*(2), 180–195. doi:10.1109/JIOT.2014.2313459

Schniederman, R. (2013). Machine-to-Machine Connectivity isbooming. *IEEE Signal Processing Magazine*, *30*(4), 10–13. doi:10.1109/MSP.2013.2255940

Solanki, V. K., Katiyar, S., BhashkarSemwal, V., Dewan, P., Venkatasen, M., & Dey, N. (2016). Advanced automated module for smart and secure city. *Procedia Computer Science*, *78*, 367–374. doi:10.1016/j.procs.2016.02.076

Soliman, M., Abiodun, T., Hamouda, T., Zhou, J., & Lung, C.-H. (2013). Smart home: Integrating internet of things with web services and cloudcomputing. *Cloud Computing Technology and Science (CloudCom), 2013 IEEE 5th International Conference on,* 2, 317–320.

Sun, E., Zhang, X., & Li, Z. (2012). The internet of things (iot) and cloud computing (cc) based tailings dam monitoring and pre-alarm system in mines. *Safety Science, 50*(4), 811–815. doi:10.1016/j.ssci.2011.08.028

Sun, Z., Liu, C. H., Bisdikian, C., Branch, J. W., & Yang, B. (2012). QoIaware energy management in internet-of-things sensory environments. *2012 9th Annual IEEE Communications Society Conference on Sensor, Mesh and Ad Hoc Communications and Networks*, 19–27.

Tan, Sooriyabandara, & Fan. (2011). M2M Communications in the Smart Grid: Applications, Standards, Enabling Technologies, and Research Challenges. *International Journal of Digital Multimedia Broadcasting*. doi:10.1155/2011/289015

Triguero, I., Río, S., López, V., Bacardit, J., Benítez, J. M., & Herrera, F. (2015). ROSEFW-RF: The winner algorithm for the ECBDL'14 Big Data Competition: An extremely imbalanced big data bioinformatics problem. *Knowledge-Based Systems, 87*, 69–79. doi:10.1016/j.knosys.2015.05.027

Want, R., Schilit, B. N., & Jenson, S. (2015). Enabling the Internet of Things. *IEEE Computer, 48*(1), 28–35. doi:10.1109/MC.2015.12

White, T. (2012). *Hadoop, The Definitive Guide*. Sebastopol, CA: O'Reilly Media, Inc.

Yang, W., Wang, X., Song, X., Yang, Y., & Patnaik, S. (2018). Design of Intelligent Transportation System Supported by New Generation Wireless Communication Technology. *International Journal of Ambient Computing and Intelligence, 9*(1), 78–94. doi:10.4018/IJACI.2018010105

Yuvaraj, S., & Sangeetha, M. (2016). Smart supply chain management using internet of things(IoT) and low power wireless communication systems. *Wireless Communications Signal Processing and Networking (WiSPNET) International Conference on*, 555-558. 10.1109/WiSPNET.2016.7566196

Zhang, W., Qi, Q., & Deng, J. (2017). Building Intelligent Transportation Cloud Data Center Based on SOA. *International Journal of Ambient Computing and Intelligence, 8*(2), 1–11. doi:10.4018/IJACI.2017040101

Section 2
The IoT and Industry Expectations

Chapter 6
Industry 4.0:
The New Industrial Revolution

Krishnan Umachandran
Nelcast Ltd., India

Igor Jurčić
Mostar University, Bosnia and Herzegovina

Valentina Della Corte
University of Naples Federico II, Italy

Debra Sharon Ferdinand-James
The University of the West Indies, Trinidad and Tobago

ABSTRACT

Industry 4.0 can be considered the 21st century's industrial revolution and will soon be the new form of manufacturing delight. The definitive customer would experience manufacturing requests determined by artificial intelligence, machine learning, and automated technologies linked with data science support for gauging customer necessities. Phenomenally, Industry 4.0 is rapidly changing the firm's management and organizational systems, and competencies, as well as making its environment much more explored, even if more complexed than in the past. This new industrial revolution would possess systems with transformative technologies for managing interconnected systems between its physical assets and computational capabilities. Such enterprises would require skilled workforce to improve and operate advanced manufacturing tools and systems, and investigate the machine data, clients, and global capitals, resulting in an escalating need for trained employees proficient in cross-functional capacities and with competencies to cope new processes and IT systems.

DOI: 10.4018/978-1-5225-6207-8.ch006

INTRODUCTION

Unlike the previous three industrial revolutions (18th, 19th, and 20th Centuries), the 4th will be more decentralized, automated, and controlled interdependently (Qin, Liu, & Grosvenor, 2016). In the first industrial revolution, the factory achieved production primarily through machines powered by water and steam and heavy manpower. In the second, operations became slightly more complexed through machines powered by electricity supported by mass production and division of labour. The third industrial revolution ushered in the use of electronics and information technology, adding more complexity to the production process in making it more automated (Brettel, Friederichsen, Keller, & Rosenberg, 2014; Wolfgang, 2016). Undoubtedly, these three industrial revolutions would have impacted their countries' economies. For example, through their relative contribution from the manufacturing sector, especially in European countries where the GDP accounts for 17% and creates about 32million jobs (Qin, Liu, & Grosvenor, 2016). Other such evidence can be seen in industrial landscape of factories and activities in the physical environments as well as improvement in living standards (Calvert & Simandan, 2010; United Nations Industrial Development Organisation, 2015).

Introduced by Germany at the Hannover Fair in 2011, the world's leading trade show for industrial technology (Qin, Liu, & Grosvenor, 2016), Industry 4.0 can be seen as an amalgamation of tangible outputs from integrated information technology systemic processes with mostly interdisciplinary engineering facilitations coordinating into a seamless connect (Sniderman, Mahto, & Cotteleer, 2016). Consequently, the future of the industrial revolution would demand more mobile technology. Vision 2020 in ICT will be the base for Industry 4.0, especially producers of 5G mobile networks seen as the locomotive for Industry 4.0 (Jurčić & Gotovac, 2016) and other modern mobile technologies such as Li-Fi technology. According to IOTUK (2016), Li-Fi is "a new technology with speeds that surpass Wi-Fi hundreds of times over by utilizing light instead of radio waves (Transferring data using fluorescent lights" para 1). Of course, existing modern mobile telecom operators, or more so their key managers, will have to have a "shift in mindset" before they start playing one of the most important roles in this new business order. Subsequently, the integrating processes will include sensor networks (IoT/IIoT), computing and sharing data and finally, a collaborating community. Regarding integrating processes mentioned above, it is good to note that in many literature the terms Internet of Things and Industrial Internet of Things replaced Internet of Everything (IoE). Briefly, Internet of Everything (IoE) was introduced by Cisco for bringing together people, processes, data and things to make networked connections more relevant and valuable than ever before – turning information into actions that create new capabilities, richer experiences, and unprecedented economic opportunity for businesses, individuals and countries (Evans, 2012).

Evidently, the growth and impact of this new industrial revolution depends on highly automated manufacturing processes that require less traditional manufacturing jobs and more modernized ones for growing the manufacturing sector and, by extension, the economy. High expectations abound for Industry 4.0 that will impact key stakeholders favourably: the ultimate customer would experience manufacturing applications driven by artificial intelligence, machine learning, and automated technologies linked with data science support for gauging customer requirements. Thus, this modern version of manufacturing would transform the existing infrastructure-people-systems to experience a more extrapolative and natural way for revolutionizing the manufacturing industry of the future. Telecommunication technology developers continue to revolutionize the manufacturing industry with predictive modeling of data streams and forecast flow of people-process-systems through the following: Internet of Things (IoT), Big Data and additive manufacturing for supporting the evolution of new solutions, which helps to increase the uptime for utilizing these opportunities. In this new era, the term "telecommunication technologies" will very often be changed by the term "Information and Communications technology (ICT)", which shows that telecom operators will have to have, as previously mentioned, a "shift in mindset" in order to sustain growth in this market in following years (Jurčić & Gotovac, 2016).

The new industrial revolution is demanding more innovative production for firms to gain competitive advantage but also needing to build capacity for dealing with industrial hazards that come with such innovation. The central industrial sectors of Europe housing manufacturing operations are prompted to innovate products with new materials, functionalities, and manufacturing approaches for improving functionality and versatility of design (European Commission, 2013). This new or 21C (21st Century) industrial-technological revolution is fuelled by the Internet and cyber-physical systems (cps) through large-scale networks. Scholars have defined the third industrial revolution as that of digitalization and the fourth as the Smart one (Lasi, Fettke, Kemper, & Hoffman, 2014). The latter differences suggests that the new 21C industrial phenomenon is still evolving but developing rapidly, changing the firms' management systems, organizations and competencies and making the environment much more explored, even if more complex, than in the past. Learning from the experience of traditional market economies that may at times bring environmental damages, more focus is now placed on increasing industry share and output to inclusive and sustainable industrial development. For instance, the agro-industry and pharmaceuticals have revolutionized these sectors. In agro-industry, the latest advances are oriented towards safety in the use of fertilizers, growing demand for"ethical food" (e.g., hal al or kosher productions) and the new ICT applications in the process have determined a profound transformation in the value chains with an increasing attention to health and the environment. This approach is fundamental to finding the most appropriate relationship between the tradition connected with local identities and innovation (Della Corte, Zamparelli, & Micera, 2013). In the pharmaceutical sector, recent innovations are driving both to an increasing attention to human health and wellbeing and to increase the capacity to deal with industrial hazards. This capacity includes the ability to efficiently process healthcare information and automated analysis of patients ubiquitously using different Internet of Things (IoT) technologies (Bhatt, Dey, & Ashour, 2017; Jain & Bhatnagar, 2017). As in other industries, a lot of important controls to protect workers, workplaces and entire locations have been put in place. Started in US, this process is signaled by increased attention on the prevention and control of hazards through different actions such as in the following: worker involvement in policy making; developing a hierarchy of controls and connecting plans for implementation during emergencies and non-routine activities; a continuous review of the process and of its technologies in order to make it more and more protective, reliable, and efficient (United States Department of Labor, n.d.).

Therefore, Industry 4.0 is an ongoing process that still in need of a unique and shared definition. In this paper, we aim to identify the main drivers of industry 4.0., explain the implications for firms, and analyse some interesting data that show what is happening in two very different countries with very different histories in terms of innovation: India and Germany.

THE MAIN DRIVERS OF INDUSTRY 4.0

Today, the manufacturing environment for Industry 4.0 is mostly characterized by fast changing processes, short development periods, abrupt technological changes and a growing necessity for individual demand and customized products. Consequently, significant changes are occurring in firms not only for the physical plants but also for future technological manufacturing skills and competencies required in these firms for driving Industry 4.0 These significant changes will happen in companies from all business fields and expressed from one business field to another one. But definitely, any company form any business field will feel these changes in the following years and decades. The emerging technological applications and studies that have been developed and characterizing the new Industry 4.0. manufacturing system of the future are as follows:

Cyber-Physical Systems (CPS)

By definition, CPS's are "transformative technologies for managing interconnected systems between its physical assets and computational capabilities" (Viles, Ormazabal, & Lleo, 2018 p. 346). These systems are conceived to manage big data through algorithms and technologies that allow each system layer to cooperate with a unified structure and get an appropriate overall functionality in terms of efficiency, reliability and product quality. Lee, Bagheri, and Kao (2014) proposed a 5-level architecture as shown in Figure 1 and iterated as follows: 1. A smart collection level finalized to acquire accurate and reliable data; 2. A data-to-information conversion level, based on tools ad methods to convert available data into useful information (in this case, human contribution is very important); 3. A cyber level that gathers all information and central hub to spread the relevant information to the functions that are interested in them; 4. A cognition level important for generating knowledge from the monitored system, giving decision makers useful information for decision-making; and 5. A configuration level that is a feedback from cyber space to physical space and is a self-control system (Wang, Wan, Zhang, Di., & Zhang, 2016). In addition, one of the main features of Industry 4.0 is the integration of machines and software and within sectors (from commerce and logistics to finance), new production processes/distribution (Sniderman, Mahto, Cotteleer, 2016) and new platform architecture.

Figure 1. Architecture for implementation of cyber-physical system
Source: Wang et al. (2016)

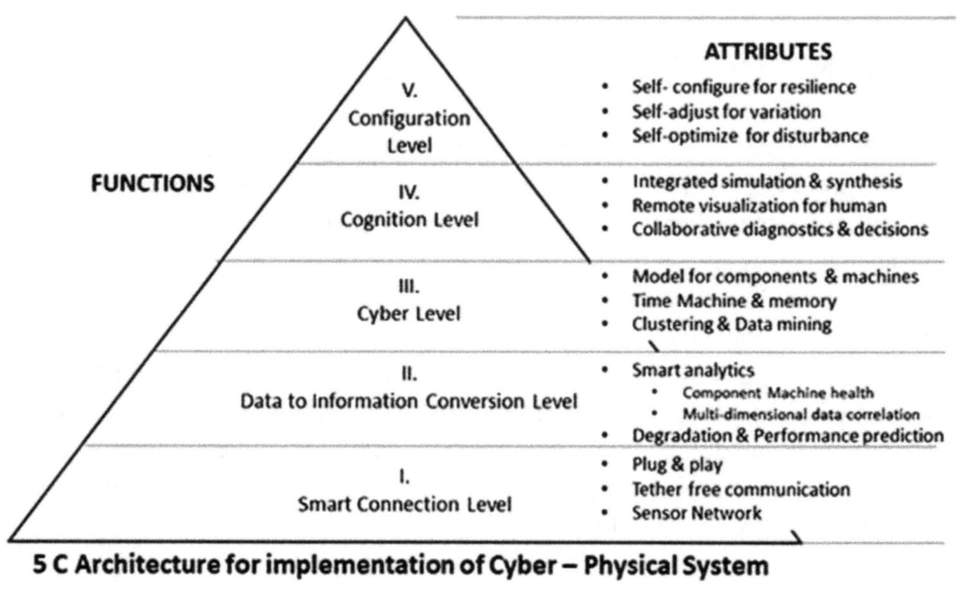

Virtualization and Decentralization

The huge amount of data acquired through smart systems provides knoweledge about customerization and manufactturing processes with impacts on the overall value chain. This new direction of individualized production (Brettel, Friederichsen, Keller, & Rosenberg, 2014) affords the advantage of ecocnomies of scale. Thanks to Mass Customization (Fogliatto, Da Silveira, and Borenstein, 2012), such production is based on a strategy that focuses on personalized mass products through flexible processes, modularized product design, and shared producers within the value chain. As such, modularization becomes a product of architecture that divides the production process into modules which can be combined differently. This process increases flexibility for launching new products as well as shortening time to market, significantly. In this scenario of a smart factory, Reconfigurable Manufacturing Systems (RMS) are arranged, allowing to manage change while still respecting cost effciency.

Open Networks' Building

Previously, human decisions were mainly based on the personal experience of decision makers, but in the near future, Industry 4.0 will assist decision makers with dynamic "self optimizing and knowledgeable manufacturing systems" (Brettel et al., 2014, p. 38) that help managers to follow market trends. From this point of view, the architectures described above are a great advantage for small and medium enterprises (SMEs) that can have access to systems, tools and techniques absolutely not affordable individually. One of the main strengths of Industry 4.0 is the possible horizontal integration of collaborative networks and environments (Brettel et al., 2014) with increasing complexity of products and processes for collaborative manufacturing (Bohu et. al., 2010). The availability of product data for the whole network favours global optimization of the procecces going much further than the boundaries of the single firm. This favourable condition provides allows for SMEs to increase their marketing sales opportunities, especially for modern mobile operators, who with "shifts in mindsets", can become ICT providers, providing secured "services in clouds" and many new other products and services (Jurčić & Gotovac, 2016; Sarkar, Banerjee, Badr, & Sangaiah, 2017). Regrettably, without such a shift in mindset, they will slightly dissapear with other companies siezing these opportunities.

Therefore, SMEs can concetrate on their core competencies and outsource the other activities to the members of the network that are more specialised in those specific activities. This concentration represents business district logic for core competency and market feasibility, which are highly favoured for technologies that include the following components: cloud computing, big data through information and knowledge flows, providing deep market information with a common infrastructure that the single firm could never afford. The combination of cloud computing and big data storage can be effected with the Hadoop Distributed File System (Matallah, Belalem, & Bouamrane, 2017). Overall, these open networks interact with other sectors with a more elevated level of competitiveness.

Servitization

The advent of Industry 4.0 brings a focus on the customer that blurs the boundaries between manufacturing and services. Baines, Lightfoot, Benedettini, and Kay (2009) have characterized servitization as the organization's ability to transform its skills, capabilities, processes, and products into a bundle of services as proven by the Product Service System (PSS). All these drivers push the firm towards big challenges with inevitable serious management changes, strategies, organization, marketing choices, and new approaches to customers, which include pre-sales, sales, post-sales processes and customer-care approaches. If all the above mentioned factors have the advantage of opening companies to huge and sometimes unexpected opportunities, guiding decision makers with deeper and more precise information, then new profiles are required within organizations with different and specific skills becoming necessary. Besides, Industry 4.0 is not a source of competitive advantage *per se*: it depends on the single firm's resources and capabilities (Della Corte & Sciarelli, 2012) as well as on its relational capabilities in taking the initiative to a network with counterparts in related sectors.

A FIRST EMPIRICAL ANALYSIS: A COMPARISON BETWEEN GERMANY AND INDIA

Industry 4.0 organizations are in fact world-class with a global presence that every organization needs to become a preferred leader to global customers coupled with achieving par excellence in manufacturing standards and market reach. Bringing customer delight is possible only when employees are engaged and empowered in a healthy loyal business. Factories of the future will have the ability to do the following: enhance flexibility and increase production lines, take advantage of installed capabilities, and improve workplace ergonomics and safety by means of advanced materials and power micro-electronics, etc. Upgraded re-cycling or re-manufacturing solutions would ease the shortage of valuable materials through cost-effective conversions and less wastage. India and Germany as manufacturing-based economies share the view that the manufacturing sector is very important for future economic growth. Hence, for these global multinationals to succeed and remain competitive, they need to be innovative in concentrating on manufacturing, performing differently in order to increase their market share and employ Industry 4.0 technologies to achieve continued growth. (Hiskey, 2017). As the world becomes a global village, companies must understand that their markets are not in a county, a country, or in a region – their real market is becoming the entire world.

Industry 4.0: Manufacturing Base Tilted

In the current global value chains, many organizations are positioning various activities in different geographic locations to take advantage of the specific strengths of particular local labour markets (World Economic Forum, 2016). Proceeding with a comparison between India and Germany, some interesting hints can be derived. The Indian Manufacturing Industry needed to have policies in place which encourage and attract other said sectors such as in Japan, South Korea, China, Germany and the U.S. many years ago. They should now cover the infrastructural deficit in the following: in transport, power, urban development, oversight of direct investment funds, governance, simplification of procedures for land acquisition, people skill development through partnership, innovation incubation, and capacity expansion (Umachandran, 2012).

Although India has issues with inadequate infrastructure, thick bureaucracy and inconsistent supply of resources, it is obsessed with abundant supplies of skilled human resources, the vital link for achieving the position of a globally preferred manufacturing destination of the future (Sagar & Jadhav, 2017). The dominant technologies of Industry 4.0 are inter-disciplinary, including mechanics, ICT's, electronics and robotics, and other knowledge areas such as biotech and nanotech. Therefore, the business expectations of Industry 4.0 needs both enhance social and technical skills. In the Indian industry, robots have already been positioned as part of the latest revolution. In the future, robots would be become intelligent, adapting, communicating, and impacting cost structures, skills supply/demand, and production sites. Thus, emerging robotics in manufacturing can yield significant increases in productivity as organizations replace human-structured and closed-looped workflows, giving way to more complex and intelligent tasks in manufacturing (Aulbur & Singh, 2014).

Industry is a major pillar of the European economy as shown in Figure 2. Europe has no choice but to partner with industry now more than ever in order to remain competitive in the global economy (European Commission, 2013). Arguably, in the European context, Industry 4.0 will be the new industrial revolution, which should include reindustrialization for the changing environment and realignment of processes, production and products to the new situation (Berger, 2014).

Figure 2. Capital intensity vs profitability
Source: Berger (2014)

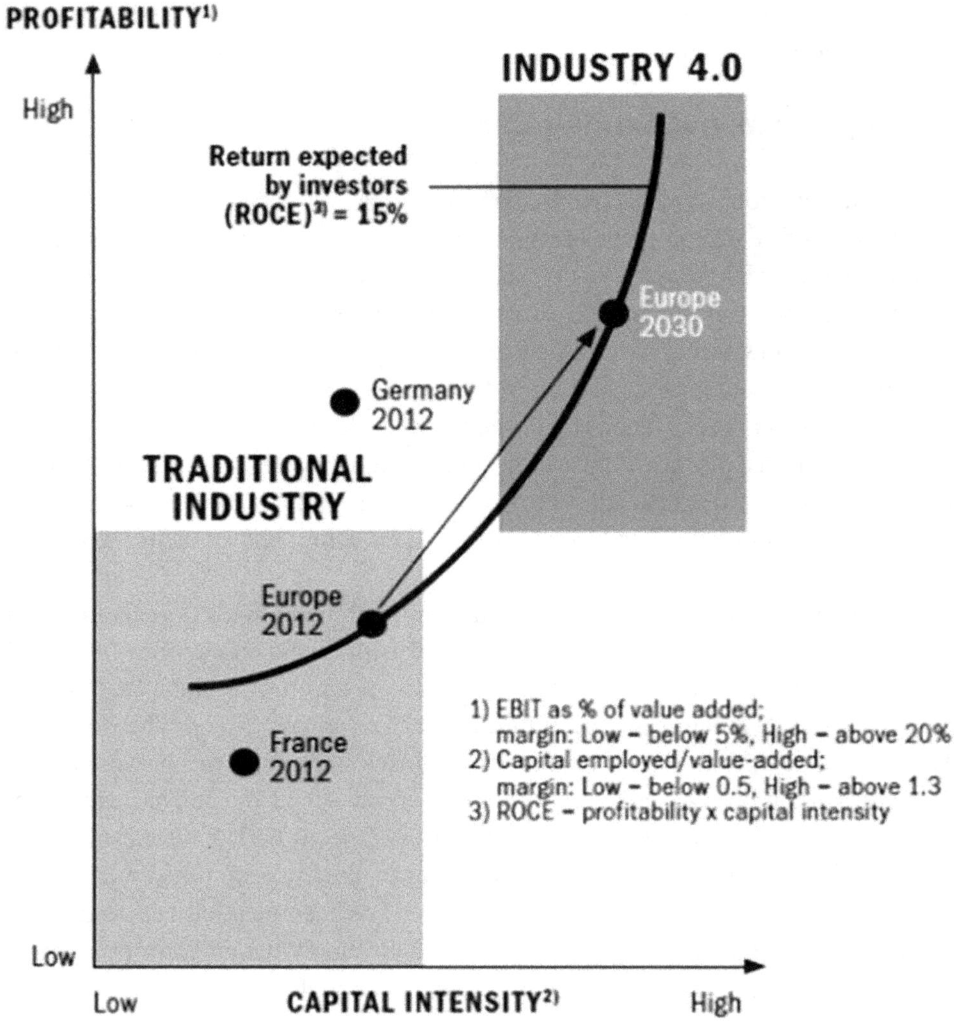

Global economies advance and become wealthier through their manufacturing sector by growing gross domestic product significantly more towards innovation, productivity, efficiency, and sustainability (World Economic Forum, 2013).

Industry 4.0 will bring the artificial intelligence with speed at affordable cost to the shop floor, automating manufacturing solutions, while fully exploiting the information and communication technology (ICT) C21 revolution. The data-capturing technologies and industrial automation controls will be more reliable and feed real-time information to the manufacturing systems with highly improved process utilizations, and effectiveness and efficiency of all resources in the organization. Enterprises such as those in Germany and India would require skilled workforce to improve and operate advanced manufacturing tools and systems. Of paramount importance is investigating machine data, clients, and global capitals that would result in an escalating need for trained employees who are proficient in cross-functional capacities and with competencies to cope new processes and ICT systems. The demand for technical talent would determine the alteration of job creation in the manufacturing sector, necessitating additional competent employees for the shop floor activities. Machine vision technologies has been on a continuous upgrade where coordination between work centres are completely numerically controlled with sensing inputs of pervious and successive operation being closed looped.

Changing Demands on Manufacturing

The advanced material processing and technological solutions will help businesses to collaborate and make the transition towards high value manufacturing (HVM), which must be supported by sustainable industrial development strategies for long term benefits. In a global context, Agenda 2030 recognizes the role of technology and innovation in industrialization and has as its 9th Goal – "Build resilient infrastructure, promote inclusive and sustainable industrialization and foster innovation" (United Nations Industrial Development Organization [UNIDO], 2015, p. xiii). Strategies for achieving this goal include setting related policy supported by resolute implementation by policy makers; planned educational drive for raising awareness of this policy to sustainable industrial development, and sponsorship by key stakeholders for supporting these latter efforts that should include poverty eradication (UNIDO, 2015). Predominantly, biosciences–medicines–healthcare will have a significant demand on the following sectors: healthcare (Bioactive and compatible drug delivery; dressings and scaffolds; anti-microbial materials and surfaces); energy generation and distribution (secure, clean affordable energy usage); defense (self-sensing and self-repairing composite structures in aerospace); and transport (lightweight materials and structures and towards zero carbon emissions (Technology Strategy Board, 2010).

Moreover, modern technologies have helped to create a safer and healthier living and working environment. As industrialization and urbanization give rise to overcrowding and other ill effects, ICTs have been used to secure city resources such as power supplies; ease road traffic congestion through advance traffic channeling; and safeguard cities with the use of security cameras (Solanki, Katiyar, BhashkarSemwal, Dewan, Venkatasen, & Dey, 2016). Considerations for a centralized transportation cloud data center and use of a MapReduce-oriented framework have also enhanced operational systems efficiency (Kamal, Ripon, Dey, Ashour, & Santhi, 2016; Zhang,

Qi, & Deng, 2017). Smartphone sensors have been used effectively in intelligent transport systems to alert aggressive driver behaviour and hazardous road conditions for reducing road accidents (Ali, Atia, & Mostafa, 2017). Data mining for diagnosing and decision-making on widespread diseases through machine learning have enhanced disease information processing greatly (Kamal, Dey, & Ashour, 2017).

A research analysis on future of manufacturing as an opportunity or challenge stated that "the quality and skills of the workforce will be a critical factor in capturing competitive advantage" (Foresight, 2013, p. 6). Essentially, policy makers should focus on the supply of skilled workers, including apprenticeship schemes, support for researchers, and the supply of skilled managers. As shown in Figure 3, firms will need to pay much more attention to building multidisciplinary teams for developing increasingly complex products, and also innovative business models (Foresight, 2013). Pervasive technology which includes ICT, integration of sensors, advanced materials, bio-tech, intelligent transportation system, and sustainable clean technological applications would be facilitated by Big data, IoT, robotics, additive manufacturing, and cloud- based mobile computing (Dey, Ashour, & Bhatt, 2017; Yang, Wang, Yang, & Patnaik, 2018).

European Union, Trade With India

According to the European Commission (2017), the Directorate-General for Trade, as shown in Figure 4, projected that embarking upon a process of reformation for an integration with the global economy would mean infusing a new path of rapid growth.

Figure 3. Phase shifts to sustainable manufacturing
Source – Foresight (2013)

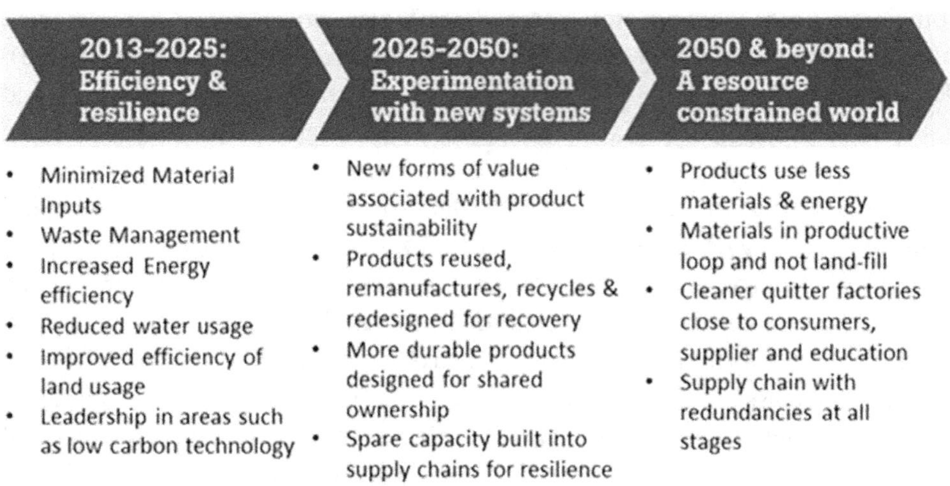

Figure 4. EU trade flow and balance
Source – Trading Economics (2017)

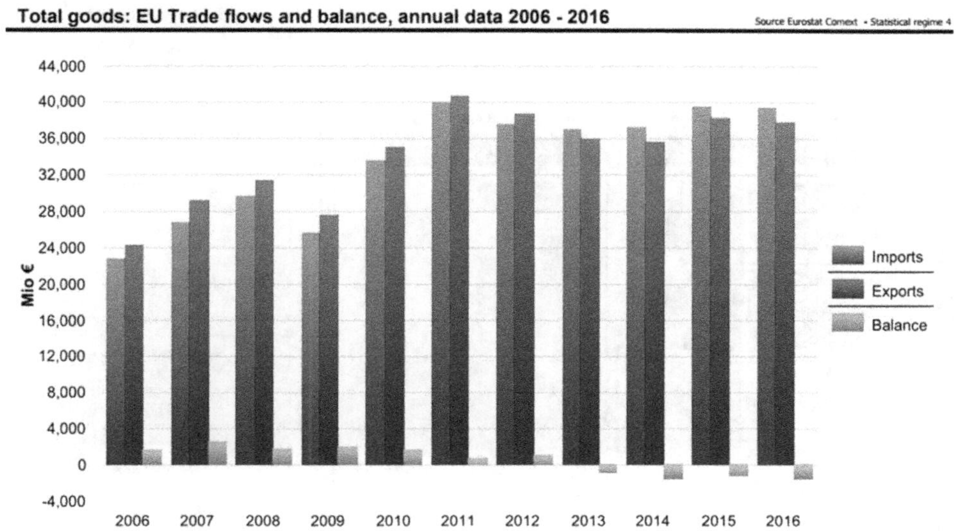

These data explain the changing demand of competencies in the manufacturing world with implications for the education systems. The Indian, German and US main educational systems reflect the following profiles: the average profile in Germany is more concentrated on undergraduate programs; both in India and the U.S., there is a greater focus on post graduate programs, which could mean that there is a higher demand for Post-Graduate and PhD students. The American and German educational systems are more closely aligned, but there is need for a closer alignment between these and the Indian educational system as shown in Figure 5. Improved quality, access, and ascendency in technical education will also promote strong Indo-EU partnerships and collaborations.

Simultaneously, the more skillful market in Germany remains more ambiguous. Employment statistics show that 577,000 persons were employed as manufacturing engineers paying social security (Brenke, 2012). Further, (Brenke (2012) also reported on a 2008 microcensus survey conducted in Germany and found that 738,000 persons admitted to working as engineers. However, "this is a third more than the total number of engineers in employment and paying social security contributions (June 2008: 551,000). This can be partially accounted for by the fact that the self-employed are also recorded in the Microcensus" (Brenke, 2012, p.4). A closer examination of this one-third of engineers reveal that nine percent were self-employed while another four percent were those who are not required to pay social security contributions. However, the latter differences in the employment and

Figure 5. Comparison of existing educational systems: US-Germany-India
Source: Umachandran (2016) and Institute of Education Science (2015)

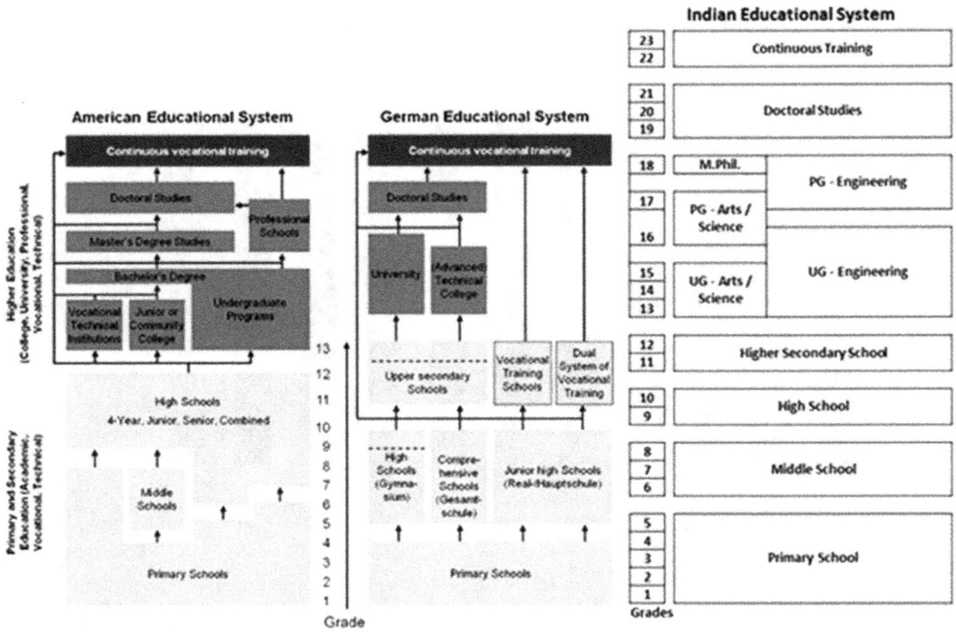

microcensus survey groups have not been reconciled or fully explained. Perhaps, a more transparent educational system that would be able to evaluate the impact of educational and training programs on the labor market is necessary. In addition, internationalization of educational programmes bears significance here as the total number of foreign national students enrolled in Indian institutions for the academic year 2015-2016 was 45,424. This figure represents foreign students from 165 different countries who show the following course enrollment breakdown: 78.5% or the highest number are enrolled in Undergraduate courses, while 14.1% are enrolled in Post-Graduate courses (MHRD, 2016). Emerging from this analysis is that the post graduate educational system is richer in the U.S. and India than in Germany, signaling that there is a higher demand for more skilled human resources outside of Europe.

CONCLUSION

Industry 4.0 would be used to demarcate countries and continents by their manufacturing and trade quotients with efficient supply chain management supported by strategic industrial development efforts. Failing health of people will increase when the human resources are not properly managed, and the quality of work life is overlooked, which will eventually have a long term impact on the family and society, especially employees. Proper skill development and training need not be emphasized as requirements for Industry 4.0 but need to be taken as very much essential with technical schools being an extension of industrial establishments. Educational institutions need to be sponsored and adopted by neighboring industries so that requisite skills needs are made within the local community. Harnessing of large-scale data for responding to changing skill needs can be achieved through modern technologies for informing effective decision-making. The former requires knowledge of Big Data management and analytics (Umachandran & Ferdinand, 2017) and would help industries to be more proactive in contributing to societal initiatives, thus minimize governmental intervention towards people welfare.

Ultimately, Business is about the bottom lines and deliverables. India is poised to take a leap over rest of the world with more skilled people available in its trained pool. Trading Economics (2017) indicate that India reflects monthly wages as low at less than $1000 per month (High Skilled is paid $ 782, INR 50300), while the German monthly wages are 4.3 times the outflow compared to India ($ 4354, EURO 3745). These two factors are cutting edge for India becoming a base for manufacturing in the new Industry 4.0 era. Of course, it would be interesting to compare more than just two countries for an overall view to verify which system appears to be most appropriate for the new skills required on the job market. Such a review would make it easier for millennials to move across countries with an average international approach, prepared to valorize their specific competencies and skills, granted that countries change their modus operandi

Another interesting consideration is that the profile of human resources connected with Industry 4.0 is different from the past as several processes within firms have changed. Nowadays, more specific skills are required in terms of data analysts and web experts as the firm-market relationship has changed, significantly. Moreover, thanks to the new tools, firms can get prompt and deep information on consumers' profiles that a marketing manager can use to take decisions and instruct designers on what product or service to create. But, the traditional process of creativity is significantly changing, especially the capacity for creating opportunities in developing new ideas that generate new markets niches. Therefore, the risk is that of going back to a context of dependence of the firm from the market. This return should be the object of reflections both for entrepreneurs and decision makers in safeguarding the *human touch* and *insight* in strategic decisions. Besides, it also explains why some typical activities like craftsmanship, although at a very high level, are disappearing.

The expertise revealed in such reflections that open up new insight to the entrepreneur seems no longer important. Further reflections on the prospective development of human resources in firms' management is a topic that both scholars and decision makers should consider in order to safeguard the basis of firms' competitiveness (Della Corte, Zamparelli, & Micera, 2013).

REFERENCES

Ali, A. H., Atia, A., & Mostafa, M. S. M. (2017). Recognizing Driving Behavior and Road Anomaly using Smartphone Sensors. *International Journal of Ambient Computing and Intelligence*, *8*(3), 22–37. doi:10.4018/IJACI.2017070102

Aulbur, W., & Singh, H. V. (2014). *Next gen manufacturing: Industry 4.0: A look at changing landscapes in manufacturing*. Confederation of Indian Industry. Retrieved from https://www.gita.org.in/Attachments/Reports/CII%20Roland%20Berger%20 Industry%20 40.pdf

Baines, T., Lightfoot, H., Benedettini, O., & Kay, J. M. (2009). The servitization of manufacturing: A review of literature and reflection on future challenges. *Journal of Manufacturing Technology Management*, *20*(5), 547–567. doi:10.1108/17410380910960984

Berger, R. (2014). THINK ACT - Industry 4.0. The new industrial revolution How Europe will succeed.

Bhatt, C., Dey, N., & Ashour, A. S. (Eds.). (2017). *Internet of things and big data technologies for next generation healthcare*. Gewerbestrasse, Switzerland: Springer. doi:10.1007/978-3-319-49736-5

Bohu, L., Lin, Z., & Shilong, W. (2010). Cloud Manufacturing: A New Service-oriented Networked Manufacturing Model. *Jisuanji Jicheng Zhizao Xitong*, *16*(1), 1–7, 16.

Brenke, K. (2012). Engineers in Germany: No Shortage in Sight. *DIW Economic Bulletin, 5*. Retrieved from https://econpapers.repec.org/article/diwdiwdeb/2012-5-1.htm

Brettel, M., Friederichsen, N., Keller, M., & Rosenberg, M. (2014). How virtualization, decentralization and network building change the manufacturing landscape: An industry 4.0 perspective. *World Academy of Science, Engineering and Technology*, *8*(1), 37–44.

Calvert, K., & Simandan. (2010). Energy, space, and society: A reassessment of the changing landscape of energy production, distribution and use. *Journal of Economic and Business Research, 1*, 13–37.

Della Corte, V., & Sciarelli, M. (2012). Can coopetition be source of competitive advantage for strategic networks? *Cooperate Ownership & Control, 10*(21), 363–379.

Della Corte, V., Zamparelli, G., & Micera, R. (2013). Innovation in tradition-based firms: Dynamic knowledge for international competitiveness. *European Journal of Innovation Management, 16*(4), 405–439. doi:10.1108/EJIM-06-2012-0065

Dey, N., Ashour, A. S., & Bhatt, C. (2017). Internet of things driven connected healthcare. In *Internet of things and big data technologies for next generation healthcare* (pp. 3–12). Cham: Springer. doi:10.1007/978-3-319-49736-5_1

European Commission. (2013). Factories of the future - Multi-annual roadmap for the contractual PPP under. *Horizon, 2020*. Retrieved from https://www.scribd.com/document/271903700/Factories-of-the-Future-2020-Roadmap#

European Commission. (2017). *European Union, Trade in goods with India*. Author.

Evans, E. (2012). *Internet of everything: How more relevant and valuable connections will change the world*. Retrieved from https://www.slideshare.net/CiscoIBSG/internet-of-everything

Fogliatto, F. S., Da Silveira, G. J. C., & Borenstein, D. (2012). The mass customization decade: An updated review of the literature. *International Journal of Production Economics, 138*(1), 14–25. doi:10.1016/j.ijpe.2012.03.002

Foresight. (2013). *The future of manufacturing: A new era of opportunity and challenge for the UK: Summary Report*. The Government Office for Science.

Hiskey, T. (2017). Preparing for Manufacturing's Future with Industry 4.0. *Industry Week*.

IOTUK. (2016). *Li-Fi Internet of Things*. Retrieved from https://iotuk.org.uk/li-fi-internet-things/

Jain, A., & Bhatnagar, V. (2017). Concoction of Ambient Intelligence and Big Data for Better Patient Ministration Services. *International Journal of Ambient Computing and Intelligence, 8*(4), 19–30. doi:10.4018/IJACI.2017100102

Jurčić, I., & Gotovac, S. (2016). New approach in mobile telecom analysis – The eight key field analysis. *WICT III, Workshop on Information and Communication Technologies III, SoftCOM 2016*. Retrieved from http://marjan.fesb.hr/SoftCOM/2016/files/apk/final_program_2016.pdf

Kamal, M. S., Dey, N., & Ashour, A. S. (2017). Large Scale Medical Data Mining for Accurate Diagnosis: A Blueprint. In *Handbook of Large-Scale Distributed Computing in Smart Healthcare* (pp. 157–176). Cham: Springer. doi:10.1007/978-3-319-58280-1_7

Kamal, S., Ripon, S. H., Dey, N., Ashour, A. S., & Santhi, V. (2016). A MapReduce approach to diminish imbalance parameters for big deoxyribonucleic acid dataset. *Computer Methods and Programs in Biomedicine*, *131*, 191–206. doi:10.1016/j.cmpb.2016.04.005 PMID:27265059

Lasi, H., Fettke, P., Kemper, H., & Hoffman, M. (2014). Industry 4.0. Business and Information Systems Engineering. *The International Journal of WIRTSCHAFTSINFORMATIK*, *6*(4), 239–242.

Lee, J., Bagheri B., & Kao H. (2014). A cyber-physical systems architecture for Industry 4.0- based manufacturing systems. *Manufacturing Letters, 3*(2015), 18-23.

Lee, J., Kao, H., & Yang, S. (2014). Service innovation and smart analytics for Industry 4.0 and big data environment. *Procedia CIRP*, *16*, 3–8. doi:10.1016/j.procir.2014.02.001

Matallah, H., Belalem, G., & Bouamrane, K. (2017). Towards a New Model of Storage and Access to Data in Big Data and Cloud Computing. *International Journal of Ambient Computing and Intelligence*, *8*(4), 31–44. doi:10.4018/IJACI.2017100103

Ministry of Human Resource Development (MHRD). (2016). *All India survey on higher education (2015-16)*. Government of India.

Qin, J., Liu, Y., & Grosvenor, R. (2017). A categorical framework of manufacturing for Industry 4. 0 and beyond. *Procedia CIRP*, *52*, 173–178.

Sagar, B. S., & Praveen, D. J. (2017). A study on impact of Industry 4.0 in India. *International Advanced Research Journal in Science, Engineering and Technology, 4*(7), 24-28.

Sarkar, M., Banerjee, S., Badr, Y., & Sangaiah, A. K. (2017). Configuring a trusted cloud service model for smart city exploration using hybrid intelligence. *International Journal of Ambient Computing and Intelligence, 8*(3), 1–21. doi:10.4018/IJACI.2017070101

Sniderman, B., Mahto, M., & Cotteleer, M. J. (2016). *Industry 4.0 and manufacturing ecosystems: Exploring the world of connected enterprises*. Deloitte University Press. Retrieved from http://dupress.com/articles/industry-4-0-manufacturing-ecosystems-exploring-world-connected-enterprises/

Solanki, V. K., Katiyar, S., BhashkarSemwal, V., Dewan, P., Venkatasen, M., & Dey, N. (2016). Advanced automated module for smart and secure city. *Procedia Computer Science, 78*, 367–374. doi:10.1016/j.procs.2016.02.076

Technology Strategy Board. (2010). *Accelerating business innovation across the UK: The Technology Strategy Board, regional development agencies and devolved administrations working in partnership*. Author.

Umachandran, K. (2012). *Exploitation of human rights in Indian industrial sector (Automobile Segment)*. Paper presented at the 3rd International Seminar of the Regional Network on Poverty Eradication.

Umachandran, K., & Ferdinand-James, D. (2017). Affordances of Data Science in Agriculture, Manufacturing, and Education. In S. Tamane (Ed.), *Privacy and Security Policies in Big Data*. IGI Global. doi:10.4018/978-1-5225-2486-1.ch002

United Nations Industrial Development Organization. (2015). *Industrial Development Report 2016. The Role of Technology and Innovation in Inclusive and Sustainable.* Industrial Development.

United States Department of Labor. (n.d.). *Recommended practices for safety and health programs – Hazard prevention and control*. Retrieved from https://www.osha.gov/shpguidelines/hazard-prevention.html

Viles, E., Ormazabal, M., & Lleo, A. (Eds.). (2018). *Closing the gap between practice and research in industrial engineering*. Gewerestrasse, Switzerland: Springer. doi:10.1007/978-3-319-58409-6

Wang, S., Wan, J., Zhang, D., Li, D., & Zhang, C. (2016). Towards smart factory for industry 4.0. *The International Journal of Computer and Telecommunications Networking, 101*(C), 158–168.

Wolfgang, W. (2016). *Industrie 4.0: Cyber-Physical Production System for Mass Customization.* Retrieved from 16/Industrie_4_0_Cyber-Physical_Production_Systems_for_Mass_Customizations.pdf

World Economic Forum. (2013). *Manufacturing for growth strategies for driving growth and employment - Volume 1: Globally competitive policy - A World Economic Forum Report in collaboration with Deloitte Touche Tohmatsu Ltd.* Author.

World Economic Forum. (2016). *Global Challenge Insight Report - The Future of Jobs, Employment: Employment, Skills and Workforce Strategy for the Fourth Industrial Revolution.* Author.

Yang, W., Wang, X., Song, X., Yang, Y., & Patnaik, S. (2018). Design of Intelligent Transportation System Supported by New Generation Wireless Communication Technology. *International Journal of Ambient Computing and Intelligence, 9*(1), 78–94. doi:10.4018/IJACI.2018010105

Zhang, W., Qi, Q., & Deng, J. (2017). Building Intelligent Transportation Cloud Data Center Based on SOA. *International Journal of Ambient Computing and Intelligence, 8*(2), 1–11. doi:10.4018/IJACI.2017040101

Section 3

Safety and Security Systems

Chapter 7
An Integrated Digital Authentication Mechanism for Intrusion Detection System

Ch Rupa
V.R. Siddhartha Engineering College, India

ABSTRACT

The internet of things is the internetworking of physical devices, vehicles, buildings, and other items embedded with electronics, software, sensors, and network connectivity that enable these objects to collect and exchange data. Security has become an important issue everywhere. In current days, security is becoming necessary as the possibilities of attacks and threats are increasing day by day. In this situation, specific sensitive premises should monitor by a secure alert system with IoT-based advanced technology in order to prevent the threats and attacks on persons or system assets by intruders. The purpose of this system is to notify the use of the intruder's presence at premises and send alert messages to the authority people who help to take prevention actions as well as detection if in misfire situations. This notification will be helpful to know about intruder's presence even if right persons are away from the location.

INTRODUCTION

Security is an important concern in our daily life. It wanted by everyone as much as possible. People premises are protected and provide peace of mind when they are away from it. Different security agencies were working against the threat agencies like masquerades, misfeasors and clandestine in the society. Whereas intruders

DOI: 10.4018/978-1-5225-6207-8.ch007

have tried to attack the information as well as user's assets frequently. A high - end updated security systems were proposing and designing by the security agencies to protect the assets and privacy of the end users. It extends that an alert message send by advanced security systems to the users which can able to reduce the crime rate. People rely on intelligence based machines to get security from the organised crimes and criminals in this modern world.

There are different types of security systems available for the protection of the home/organisation. Some of the systems use buzzers and alarms for alerting the user about the theft at the home/organization. And some of the systems use the arrangement of password checking systems at the home for preventing the unauthorized persons entering into the home. Internet of things (IoT) with advanced techniques is used by all sensitive zones. Especially it has been showing more impact on healthcare system by its quick and efficient results [Elhayatmy 2018& Bhatt, 2017].

This chapter holds a sample real time application process of IDS that holds an access control for specified area forming an important secure connection. The Digital authentication based security system that can adopt at Home/ organizations or any specified areas by the authorities. It has maintained by various types of sensors like PIR and others like a microcontroller and a web camera. The Passive infrared sensor (PIR), Vibration sensor, web camera and the microcontroller were at the transmitter end which are will continuously monitor the premises. If any security problem has found then immediately receiver end microcontroller would switch on the web camera and then sent an alert event to an owner entity through image and voice. This equipment can work with low power and it can able to function in real time scenarios.

In global market, simple alert message security systems to fully monitored security systems by using advanced techniques were available. These can provide full security to the coverage premises as well as keep away from the threats on human assets. The problems come when the users need to spent a lot of money for setup, equipment, monthly monitoring fees, and even warranties. This secure premises system does not need to be monitored by premises that offer such services because the alert will be send directly to higher officials. In a normal commercial alarm system creates a loud blast sound which is noticeable to the intruder or create disturbance to the neighborhood. This will create a panic situation and no one knows that the intruders could be armed with guns or any type of light weapons, this could be dangerous to people around. To avoid this situation, this secure system will not create any sound. A text message and the image (face recognition) of the intruder will be sent to the users if the system triggered. This kind of digital authentication improves the efficiency of the system to detect the intruders. In this system each and every system maintains with unique ID using Hash values that no one has to be utilized or tampered or spoofed the value. Only authorized persons can able to access the ID to allow seeing or sending the alert events.

RELATED WORK

In order to detect and prevent the threats from various threat agents many security techniques were developed. Most of the techniques focused on how to notify the intruder's presence to the authority. Andi, et. al [Andi, 2015] developed a home security system with low cost by using various sensors and Arduino microcontroller. It generates one time password for user's entry and sends alert messages via GSM and alarm signal. The main drawbacks of the system it was restricted only to GSM which will works based on SIM card as well as alarm triggers sound which one creates disturbance.

Suryawansi, et. Al [Suryawansi 2016] system objective is that to automate the devices such as Arduino and Raspberry Pi based home appliances, and by using sensors like PIR, LPG, temperature and uses Mail communication system for alerting the end users. The system sends the alerts to E-mail ID. Here, LEDs are used to indicate the switching action of the devices and sensors. The major drawback of this system is (1) Intruder alerts have to be sent to E-mail (2) Internet needs to be presence at all the situation.

Y. Yang, et. al [Yang 2013] from University of China proposed a system using smart Grid protocols. It was helpful to reduce the attacks on networking but no alert message to the clients and server regards the intruders.

U. Patil et. al [Patil 2010] from C-DAC, worked on advanced technologies which can monitors and detect the sensory data. In this clearly explained about main causes of an Intruder attack and how can it identify. But the main drawback of this system is IDS analysis not reported. Their system unable to give alert even to the main system regards IDS.

Luigi Coppolino et. al [Luigi 2010] from University of Italy, describes about protection of critical infrastructure from intruders. This Intrusion Detection System has given an accurate periodic report on malicious activities by the threat agents. It helps to counter the attacks as well as to tolerate them.

Dr. Anurag Kumar, et. al from DRDO proposed a system for human intruder detection using wireless networks. In that, they have applied this project only for hot spots. The main drawback of the system is unable to send the information about the intruder. It sends only alert message to the owner. L. M. Patnaik, et. al from

IISC, Bangalore designed a system for detection of an Intruder. It can applicable in Homogeneous wireless networks. This approach is not sending any alert events to the owner. As well as, no ID maintenance system is used for intruder's detection. It is difficult to analyze the allowing rate of intruders.

Solanki, et. al [Solanki, 2016] designed a framework for securing the main resources such as electricity, traffic control system and public security. This system describes that how to uses advanced Information and communication technologies to stop the drained recourses. It has maintained the records of them. The main drawback of this system is alert system to concern resource authority by sending the immediate notifications.

Yang et.al [Yang, 2018] discussed an intelligent transportation system by using wireless communication technology. It designed by applying process oriented structure analysis on intelligent transportation system (ITS). This paper consists of the process of structural system analysis based ITS with logical structures and Physical structures. Jain et.al [Jain 2017] extends his work by associating intelligence with the Patient Services. It helps to detect heartbeat blood pressure, etc. variations automatically and update it in the report by its intelligence feature.

SAMPLE APPLICATION WITH ANALYSIS

N. dey et. Al [N. Dey, 2017& N. Dey 2018] has shown about the controlling of Home appliances using BASCOM microcontroller. In this, N. day team clearly has shown that bluetooth range problem can overcome by Wi-Fi. In addition, in our proposed system uses Wi-Fi to send alert messages on threats to the authorised persons at any time and from anywhere. This application can able to stalk the intruder with the help of Infra Red transceiver and camera that can able to surveillance specific space continuously. The camera captures the stills and records the images as soon as the intrusion occurs. Images are stored with Unique ID by Hashing Techniques into the cloud it has connected through the dual-radio board based central control authorities to take actions efficiently and immediately [Zhang et.al 2017 & Sarkar 2017]. This system uses threshold based intrusion detection for updating the incidents log in day and night. Statistical based intrusion detection helps to modify and analyze the log in the cloud database for reporting periodic report on incidents [Matallah 2017]. Notify effectual detection intrusion through E-mail and SMS to the concerned authorities. This system proposes a security solution to detect intruder as well as to reduce the attacks and threats by illegal stalkers. PIR sensor plays a vital role in the proposed system.

The PIR sensor set up is that consists of four sensing elements that have arranged in 2×2 array and it was enclosed by multilane plano convex lenses. Each pixel can able to capture temperature variations. In this process, all the pixels wired in differential mode to away sensor trigger problems due to changes in the temperature. Otherwise, it reflects on noise ration of the images. If any two pixels generate common variation in the temperature then immediately set off the sensors. Plano convex lens based IR radiation keeps alert on sense the elements of PIR sensor. This can creates a virtual beam array in the field of sensor view. The functioning of the proposed system is shown as a Figure 1 and shows in form of steps.

DETECTION ALGORITHM

Step 1: Before owner wants to go out from premises, he needs to initialize the secure system by switching it on. Now system is in active state and starts monitoring of the surroundings. It can detect any intruder at Specified Premises.

Step 2: If system observing that any of the intruder trying to enter into the specified location activates the camera and captures the image through web camera and sends them to authority with the help of GSM.

Step 3: The captured images are stored into the memory with the unique ID generated by Hashing algorithm.

Step 4: Then this information will be send to the owner's mobile as in form notification. So this notification contains the alert message and image of the intruder with Unique hash value as ID.

Figure 1. Process of intruder detection

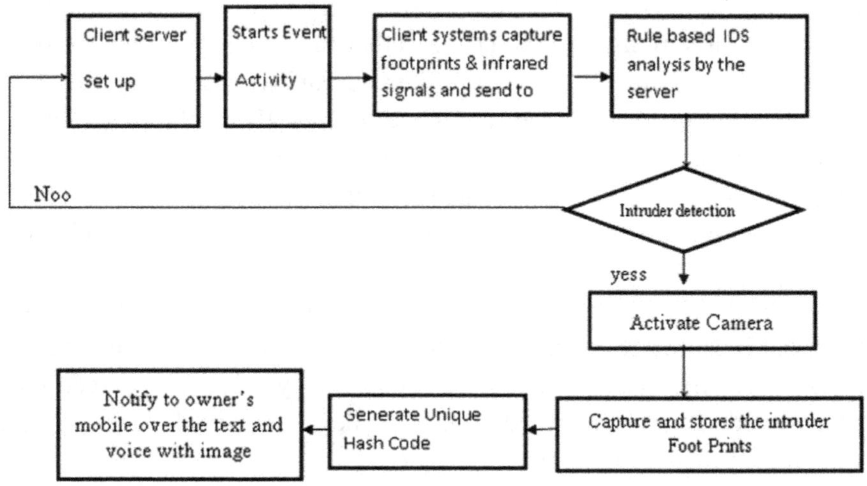

The Unique ID of the data has generated by hashing algorithm. While storing in the database, the verified data will be associated with hash value. It can help to reduce the redundancy of the data as well as improves quality of the system. In this proposed system for validating the captured image data before stores at databases, evaluate the PSNR, MSE and WPSNR and WMSE values. If the results will satisfies the throughput then will move to next step.

RESULTS AND ANALYSIS

This IDS system tested on various environments like classrooms, home doors and Hospital Wards as shown in Figure 2 and Figure 3. In this approach, metrics like Unique Code [Rupa, 2016; Tiwari, 2010] and Quality of captured Images are considered by the metrics such as Peak signal to noise ratio (PSNR) [Nyeem, Boles, et.al, 2013] and weighted peak signal to noise ratio (WPSNR) [Ponomarkenko, et.al, 2014]. Mean Square Error is used in the calculation of Peak Signal Ratio. It has measured in decibels (dB) by the Formula (1) and (2) [Rupa, 2016]:

$$PSNR = 10\log_{10}\left(\frac{255^2}{MSE}\right) \tag{1}$$

$$MSE = \sum_{k=1}^{m}\sum_{l=1}^{n}\frac{\left(p_{kl} - q_{kl}\right)^2}{mn} \tag{2}$$

whereas ‘m’ and ‘n’ are the width and height of the image. Weighted Peak Signal Ration is an additional property of PSNR. Textured and Flat regions were unable to considered by Peak Signal Ration (PSNR):

$$WPSNR = 10\log_{10}\left(\frac{255^2}{WMSE}\right) \tag{3}$$

$$WMSE = \frac{MSE}{1 + var\left(i, j\right)^2} \tag{4}$$

where i, j are local variance of image.

Figure 2. System connections

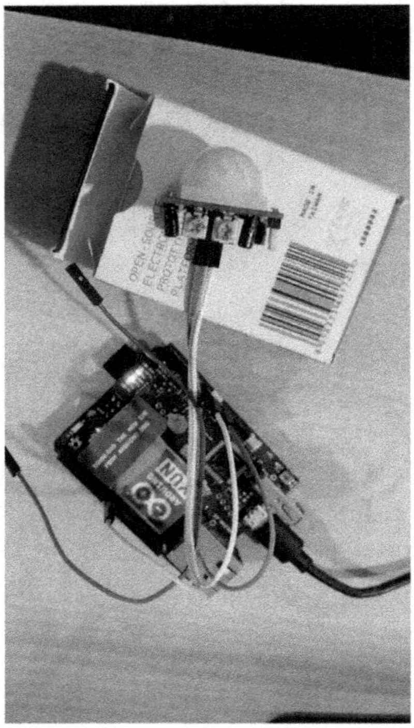

Figure 3. SMS alert

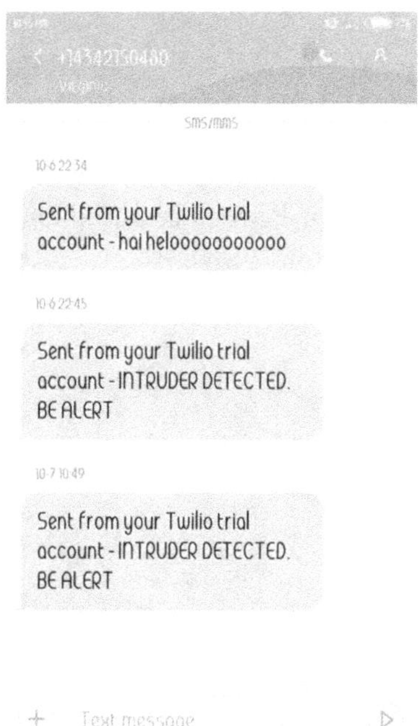

Table 1. Analysis report on reviewed schemes and proposed scheme

Scheme Developers	Redundancy	SMS Alert	E-Mail Alert	Unique ID
Andi et.al	High	Yes	No	No
Yauz et.al	High	Yes	No	No
Musab et.al	High	Yes	No	No
Yang	High	Yes	No	No
Zhihong	High	Yes	No	No
Proposed Method	Less	Yes	yes	Yes

If any image quality metrics are greater than 45 dB then only can allow to stores the footprints at database as a log and otherwise, discarded them. It can acts as threshold value as per statistical IDS system. This value is updated based on statistical reports on intruder threats activities. This approach can helps to improves efficiency of the log maintenance and saves space at logs by controlling the redundancy data.

An analysis report has shown in Table 1 by considering various features like Redundancy, SMS alert, E-mail alert, Unique ID with reviewed schemes and proposed scheme. The proposed method developed on the assumption that the intruders generally try to modify only the significant part.

CONCLUSION

We have reported on the objectives, scope and future work of the proposed chapter. Here we have demonstrated an application on IDS with analysis reports. This system helps for prediction purpose from the illegal stalkers. The main scope of this work is that the information technology world, organisations move their business to outsource through cloud. This can be used at specified locations of the *defence*, army and societal need locations such as hospitals, Houses, organizations etc. It is applicable to army activities by doing environmental surveillance and sending alert messages on threat agents' actions to the concerned officers. Advanced Secure Intrusion Detection System (IDSS) is required to the applications of this digital world. This work can enhance into make critical for determining the bearing of incoming data as well as outgoing data by intruders. It extends to that IDSS prototype needs to check the possibilities of implementation by using other scientific conventional sensors and Highend infrared beams. This work can enhance in the perspective where the significant commercial importance is there in the industry. It needs to design with low cost and low power wireless mesh networks with reliability feature. This is one of the challenges that needs to pursue currently.

REFERENCES

Adrian, A., & Dani, A. W. (2014). *Design of small Home system based on Arduino*. IEEE Electrical Power, Electronics, Communications, Control and Informatics.

Alaa. (2017). A review of smart home applications based on Internet of Things. *Journal of Network and Computer Applications, 97*, 48-65.

Bhatt, C., Dey, N., & Ashour, A. S. (Eds.). (2017). Internet of things in Health care. Internet of things and big data technologies for next generation healthcare.

Chen, S., Xu, H., Liu, D., Hu, B., & Wang, H. (2014). A Vision of IoT: Applications Challenges and Opportunities With China Perspective. *IEEE Internet Of Things Journal, 1*(4), 349–359. doi:10.1109/JIOT.2014.2337336

Coppolino. (2010). An Intrusion Detection System for Critical Information Infrastructures Using Wireless Sensor Network Technologies. *IEEE Critical Infrastructure*.

Dey, N., Ashour, A. S., & Bhatt, C. (2017). Internet of things driven connected healthcare. In *Internet of things and big data technologies for next generation healthcare* (pp. 3–12). Cham: Springer. doi:10.1007/978-3-319-49736-5_1

Dey, N., Hassanien, A. E., Bhatt, C., Ashour, A., & Satapathy, S. C. (Eds.). (2018). *Home Automation using IoT. In Internet of Things and Big Data Analytics Toward Next-Generation Intelligence*. Springer. doi:10.1007/978-3-319-60435-0

Elhayatmy, G., Dey, N., & Ashour, A. S. (2018). Internet of Things Based Wireless Body Area Network in Healthcare. In *Internet of Things and Big Data Analytics Toward Next-Generation Intelligence* (pp. 3–20). Cham: Springer.

Furestenau, N., Schmidt, M., Horack, H., Goetze, W., & Schmidt, W. (1997). Extrinsic fabryperot interferometer vibration acoustic senso systems for airport ground traffic monitoring. *Proc. IEE on Optoelectronics*, 144, 134–144. 10.1049/ip-opt:19971268

Gamba, M., Gonella, A., & Palazzi, C. E. (2015). Design issues and solutions in a modern home automation system. *Proceedings of International Conference on Computing Networking and Communications (ICNC 2015)*, 1111-1115. 10.1109/ICCNC.2015.7069505

Hougen, D. F. (2000). A miniature robotic system for reconnaissance and surveillance. *Proc. 2000 IEEE International Conference on Robotics and Automation*, 501–507. 10.1109/ROBOT.2000.844104

Jain, A., & Bhatnagar, V. (2017). Concoction of Ambient Intelligence and Big Data for Better Patient Ministration Services. *International Journal of Ambient Computing and Intelligence*, *8*(4), 19–30. doi:10.4018/IJACI.2017100102

Jassas, Qasem, & Mahmoud. (2015). A Smart System Connecting e-Health Sensors and the Cloud A Smart System Connecting e-Health Sensors and the Cloud. *Proceeding of the IEEE 28th Canadian Conference on Electrical and Computer Engineering Halifax*, 712-716.

Kholidy, H. A., & Baiardi, F. (2012). CIDS: A framework for intrusion detection in cloud systems. *Proceedings of 9th IEEE International Conference on Information Technology-New Generations*, 379–85. doi: 10.1109/ITNG.2012.94

Liao, Y. (2014). SnowFort: An Open Source Wireless Sensor Network for Data Analytics in Infrastructure and Environmental Monitoring. *IEEE Sensors Journal*, *14*(12), 4253–4263. doi:10.1109/JSEN.2014.2358253

Matallah, H., Belalem, G., & Bouamrane, K. (2017). Towards a New Model of Storage and Access to Data in Big Data and Cloud Computing. *International Journal of Ambient Computing and Intelligence*, *8*(4), 31–44. doi:10.4018/IJACI.2017100103

Miles, R. N., Robert, R., & Hoy, R. R. (1995). Mechanically coupled ears for directional hearing in parasitoid fly ornia ochracea. *The Journal of the Acoustical Society of America*, *98*(6), 3059–3070. doi:10.1121/1.413830 PMID:8550933

Nyeem, H., Boles, W., & Boyd, C. (2013). Utilizing least siginificant bit-planes of RONI pixels for medical image watermarking. In *Proceedings of the International Conference on Digital Image Computing:Techniques and Applications* (pp. 1–8). IEEE.

Patchava, V., Kumar, S. D., Hussain, S. R., & Naidu, K. R. (2015). Raspberry PI controlled SMS-Update-Notification (Sun) system. *Proceeding of IEEE International Conference on Electrical Computer and Communication Technologies (ICECCT 2015)*, 1-4.

Patil. (2010). Intelligent Intrusion Detection System (In2DS) using Wireless Sensor Networks. *IEEE Networking, Sensing and Control*.

Ponomarkenko, N., Keivenko, S., Egiazarian, K., Astola, J., & Lukin, V. (2010). Weighted MSE based metrics for characterization of visual quality of image denosing methods. *Proceeding of the 8th Inte.Workshop on Video Processing and Quality Metrics for consumer Electronics*.

Rupa. (2016). Performance Evaluation of Message Security Using Fractal Sieve with MMD. *International Journal of Digital Crimes and Security, 34*(4).

Rupa, C. (2010). Performance Evaluation of Message Security Using Fractal Sieve with MMD. *International Journal of Security, Privacy and Trust Management, 2*(4), 27–32.

Sarkar, M., Banerjee, S., Badr, Y., & Sangaiah, A. K. (2017). Configuring a trusted cloud service model for smart city exploration using hybrid intelligence. *International Journal of Ambient Computing and Intelligence*, *8*(3), 1–21. doi:10.4018/IJACI.2017070101

Shruti, G. (2016, March). Implementation of Smart Home Automation System through E-mail using Raspberry Pi and Sensors. *International Journal of Innovative Research in Electrical, Electronics, Instrumentation and Control Engineering*, *4*(3).

Solanki, V. K., Katiyar, S., BhashkarSemwal, V., Dewan, P., Venkatasen, M., & Dey, N. (2016). Advanced automated module for smart and secure city. *Procedia Computer Science*, *78*, 367–374. doi:10.1016/j.procs.2016.02.076

Speller, K. E., & Yu, D. (2004). A low-noise MEMS accelerometer for unattended ground sensor applications. *Proc. SPIE 5417, Unattended/Unmanned Ground. Ocean, and Air Sensor Technologies and Applications*, *4*. doi:10.1117/12.540337

Tiwari. (2010). A Secure Hash Function MD-192 With Modified Message Expansion. *International Journal of Computer Science and Information Security, 7*(2).

Vujoic, V., & Makshimovic, M. (2015). Raspberry Pi as a sensor Web node for home automation. *Elsevier Transaction on Computer and Engineering*, *42*, 153–171.

Yang, W., Wang, X., Song, X., Yang, Y., & Patnaik, S. (2018). Design of Intelligent Transportation System Supported by New Generation Wireless Communication Technology. *International Journal of Ambient Computing and Intelligence*, *9*(1), 78–94. doi:10.4018/IJACI.2018010105

Yang, Y. (2013). A Self-Learning Sensor Fault Detection Framework for Industry Monitoring IoT. *Mathematical Problems in Engineering*, *15*, 2013.

Yang, Z. (2011). *Study and application on the architecture and key technologies for IOT*. IEEE Multimedia Technology.

Yavuz, Hasan, Serkan, & Duygu. (2007). Safe and Secure PIC Based Remote Control Application for Intelligent Home. *International Journal of Computer Science and Network Security, 7*(5).

Zhang, W., Qi, Q., & Deng, J. (2017). Building Intelligent Transportation Cloud Data Center Based on SOA. *International Journal of Ambient Computing and Intelligence*, *8*(2), 1–11. doi:10.4018/IJACI.2017040101

Section 4

Applications in Electronic and Digital Technologies

Chapter 8
Smart Traffic System Operations

Neelu Khare
VIT University, India

Shruthy Bhavanasi
VIT University, India

ABSTRACT

Traffic congestion is a growing problem in the world. It causes waste of time and fuel, increases pollution and stress. This is especially of road traffic in India, which faces problems like congestion, unpredictable travel time, urban chaos and noise. A smart traffic system automates the traffic control activity in order to obtain preference for the lanes on the basis of certain parameters. Hence, it provides an automatic control on the traffic in an optimized manner. It takes inputs from various sensors (utilising ultrasonic sensors, light sensors, motion sensors, cameras and IoT devices) and sends interrupt signals to the controlling devices. It consists of the IoT (Internet of Things) coupled with smartphone technology, RFID, sensors and a LADAR system. The system deals with the major issues of traffic congestion and collision avoidance and suggests remedies to tackle the same. In this chapter, various methodologies and architectures for smart traffic system operations are described.

INTRODUCTION

Smart Traffic Management is a system in which midway controlled movement signs and sensors direct the activity stream through the city because of interest.

Redesigning and coordinating every one of the signs on the fundamental streets in the city will:

DOI: 10.4018/978-1-5225-6207-8.ch008

- Diminish regular clog particularly, by smoothing movement streams and organizing activity because of interest continuously;
- Diminish contamination all through the city: stop-begin driving is wasteful and dirtying;
- Offer need to transports moving toward intersections, staging lights to give activity streaming with transports a 'green wave' through the city;
- Empower a significantly more successful reaction to activity occurrences. The framework can be pre-modified to deal with a sudden increment in activity on any of the ten radials.

It will likewise give an ideal chance to introduce observing hardware to gather considerable information better than which we have at present. Each arrangement of movement lights will have correspondence gear that can be utilized to transmit (unknown) vehicle information, either from ANPR cameras or Bluetooth locators, and also CCTV sustains. It will also reduce everyday congestion markedly, by smoothing traffic flows and prioritizing traffic in response to demand in real time. Pollution throughout the city is also reduced.

ARCHITECTURE

1. It will likewise give an ideal chance to introduce checking gear to gather considerably more point by point movement and voyage information than we have now. Each arrangement of movement lights will have correspondence gear that can be utilized to transmit (unknown) vehicle information, either from ANPR cameras or Bluetooth indicators, and CCTV bolsters.

ANPR (Automatic Number Plate Recognition) programming utilizes OCR (Optical Character Recognition) abilities for programmed acknowledgment of vehicles number plates. ANPR permits you dependably recognize vehicles with confinements (stolen auto, not authorized, unpaid fines, and so forth, cautioning the experts when they have gone through review, permitting prompt activity, for example, seizure of the vehicle, surrender of travelers and some other activity.

When going by the camera, ANPR gets a photograph of the vehicle's number plate, enrolling it on a pictures database, with date, hour and camera data, permitting a lot of counsels and a more powerful movement administration.

Frameworks normally utilize Infra Red (IR) lighting to enable the camera to take the photo whenever of the day. ANPR innovation has a tendency to be area particular, attributable to plate variety from place to put.

Activity control frameworks utilize ANPR units which give insights about the development and stream of vehicles around the street organize. These points of interest can feature the issue zones and help to settle on episode administration choices. The photo of the auto and the driver can be put away and recovered if there is a requirement for proof in debate or wrongdoing circumstances.

2. If we consider, a Radio Frequency Identification (RFID) system has both RFID controller and RFID tag.

RFID Controller

The RFID controller comprises of RFID examiner. It examines the utilized correspondence apart from the RFID tag. The RFID controller at that point receives the signs/information which the questioner got before. Informing obstruction is utilized to transfer summons and information messages through the controller segments. The controller center will be available on the RFID controller. This tunes into the cross-examiners by relying on the arrangement; now the controller center can do read/compose tasks on the RFID tag or it can perform both tuning along with the operation performance. It will also have an order interface from which outer GSM/GPRS gadgets can be interfaced in order to perform a double radio gadget.

RFID Tag

RFID labels are remote gadgets that perform utilization of radio recurrence electromagnetic fields to exchange information, which is utilized for distinguishing and following of the articles. RFID labels consists of two types: Active and Passive. Dynamic RFID has a battery introduced, in which the latent RFID doesn't have. Latent RFID needs to rely upon outside hotspot for working. Labels data can be put away in a non-unstable memory. The label comprises of a Radio Frequency transmitter and recipient. Each tag can be doled out an interesting serial number.

SYSTEM OVERVIEW

Every vehicle will be introduced with a RFID tag. The tag will save all the data once it watches the vehicle, for example, the vehicle number, and so forth. RFID labels can be utilized as a part of recognizing every vehicle exceptionally and furthermore helps the driver to get some movement alerts in the form of mobile notifications. The current flagging framework can be combined with the RFID controller. As portrayed in figure 1, each flag can have the data with respect to each vehicle that goes by it.

Figure 1. Architecture using RFID

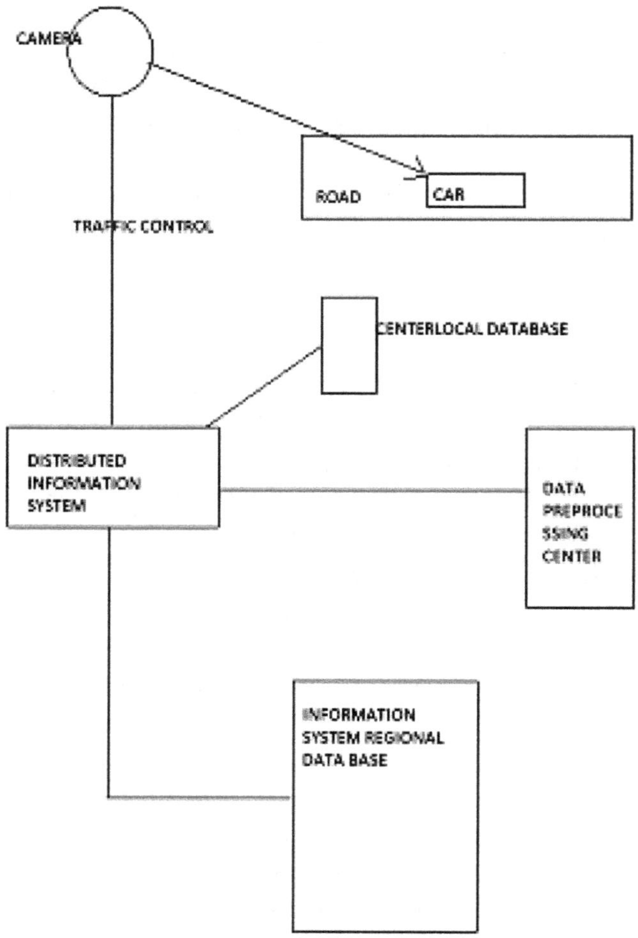

In this way whenever a vehicle goes by a flag, the flag will consequently note the tally of the vehicles going by, and helps in the location of movement blockage. Each flag ought to be put away with a limit an incentive for which it ought to be red and green. Presently relying on the recurrence of the vehicles going by the flag every second, the clock can be progressively minimized. Every controller of the flag ought to be kept away with an estimation of least recurrence of the vehicles moving by the flag. When this base recurrence is achieved, the controller ought to transfer a charge to the flag to turn red. In this way, the flag is minimized progressively. For instance, assume a flag, most extreme time of which it can be read is set for about 30 seconds and greatest time of which the flag can be green is set as 20 seconds.

The controller is put away with an estimation of least recurrence of vehicles going through every second as 5. Presently assume the flag turns green, the clock begins with the greatest estimation of 20. At first the recurrence of the vehicles moving by the flag every second is 10, following 10 seconds this recurrence diminishes to 5, and after that consequently, the RFID controller gives a summon to the flag to turn red. Along these, the flag hands red and its contiguous flag over that intersection turns green. This procedure proceeds in a cycle. Therefore, powerful controlling of the flag is useful in decreasing the wastage of time. This additionally helps in staying away from activity clog as a need is given to a high vehicular movement street. The framework is useful in recognition of movement blockage. On the off chance that the recurrence of the vehicles moving by the flag every second stays higher than the esteem set despite the fact that the maximum value of the time is reached, then the congestion has occurred at that point.

Once the congestion is recognized, the RFID controller can make an impression on its first flag's controller informing it to incidentally stop movement along that extent. In the wake of accepting the message from its next flag, the RFID controller will turn ON the red flag for that extend towards that congested intersection point for a predefined day and age. At the point when the clog is discharged at the intersection, the individual flag's controller will send another message to its prior controller showing to continue the activity stream again towards the path. Tolerating this message the controller of the previous flag put the red light OFF and the green flag ON and then it restart the flag cycle as some time recently.

Figure 2. System Architecture

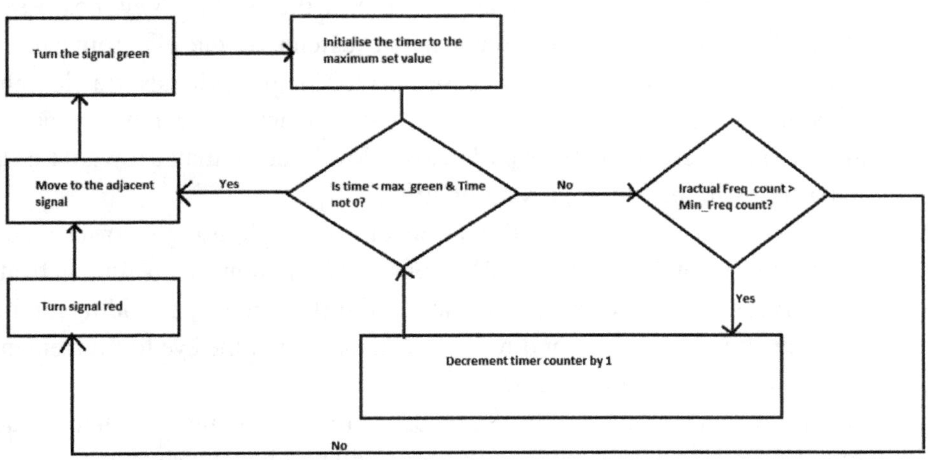

DESIGN OF SMART TRAFFIC SYSTEM OPERATION

1. The utilization from claiming the traffic lights to dispense with confusion, chaos, time-wasting furthermore mishaps over street junctions curvets have seen a considerable measure of transforms in quite some time despite those fundamental color indications held;
2. The vast majority of these progressions have been kept tabs on the kind of control utility. Others bring been concentrated on the red, yellow also green sessions timing, Furthermore in the similarity or state of an organization;
3. Most L-junctions, T-junctions and 4- manner intersection bring different traffic lights in most urban areas on the planet these days. These traffic control lights which formerly utilized radiant lights for watts going starting with 20 should 100 w need bit by bit been displaced with light emitting diodes (LEDs). These LEDs which diminished those force, consumption of the traffic lights should a couple of watts additionally made another problem, which will be drivers' inability of seeing or translating these lights as they approach of the traffic lights on ostensible pace;
4. This might be brought about concerning illustration an after effect of a few variables including at not restricted to sun-oriented glare, separate reflections also refractions happening previously, ongoing in front of the driver. These variables tend with help the drivers' decisions, activities furthermore inactions. Virtual pass out of driver's dream when faced with the floodlights of approaching vehicles (light glare) is an additional motivation behind of drivers not having the ability should decipher traffic indications;
5. This will be in light of it takes a few occasions when to a driver who encounters light glare to conform should typical lighting since the mankind's eye possesses differed light of light in the unmistakable extend of the electromagnetic range. However, in the greater part cases, drivers' slip previously, traffic light interpretations might prompt accidents, a significant part camwood makes finished to decrease these mishaps due to the low lumen radiated toward LEDs comparing with the encompassing lighting;
6. The LEDs intensities contrasted with encompassing lighting's power made Toward that sun and the various reflections and refractions happening in front of the driver need aid precise low and not solid sufficient with providing for the driver the obliged summon or direction, as the retina of the eye for movement will just react will brighter light;
7. Also, those pupils of the eyes have a tendency to widen in place with translate dimmer lights. Furthermore, this expansion will require the driver on the back off alternately much totally prevent to completely the way they decipher those signet cetera make fitting activities;

8. This is practically difficult as most drivers would in scurry or during secondary velocity particularly for exactly nations, the place speed breaking points are not directed alternately authorized. Those mixture lighting innovation organization (HLT) suggested in this fill in trusts to dispose of these issues and likewise give a choice for drivers fighting for different sun-oriented glares throughout that daytime also reflections starting with floodlights in the night.

METHODOLOGY

- The Hybrid Light System is the mixture framework permits double operation. For every arrangement of red, yellow furthermore green (RYG) sign periods, a third of the lighting span may be permitted will a chance to be powered toward the halogen radiant lights same time those remaining time of every move may be powered by those LEDs;
- The three 100 w halogen Lights comparing of the RYG circuits utilized 220 v mains voltage and the LEDs utilization 12 v full wave rectification of the mains supply utilizing a venture down transformer mounted on the mainboard. The control supply unit additionally gives the 24 v full wave rectification, which will be important should force the unit's cooling fan. Those controller circlet utilization 12 v immediate current (DC);
- TLCS could use fixed timed sessions same time canny TLCS plans utilization element timing of the RYG circuits dependent upon reaction control frameworks that are dependent upon the traffic thickness during any. Specific chance previously, in spite of the last will be utilized to this research, those outcomes would pertinent with dynamic- reaction frameworks;
- The required exchanging of the Different circuits which furnished Eventually Tom's perusing bipolar transistors and electromechanical all transfers on the TLCS mainboard previously. An approach that those whole unit fits under t-intersection unit. It may be guaranteed that the plan doesn't make "false directions" and no two lights work in those same duration of the time as stated by Q2 those safety rules. In the rules "a traffic light control framework model must need 50certain characteristics to legitimate furthermore protected the operation;
- For example, that controller ought to not bolt up (deadlock) because of a portion unforeseen consolidation from claiming actions, it ought to not permit conflicting developments will need right of way simultaneously, it ought to have the ability to serve constantly on sign periods Furthermore come back on the percentage starting state". The timed sessions in the HLS are furnished by

the counter out with stable created pulses. Furthermore, reflectors would use to focus those beams to a purpose as stated by seeing a point about drivers. Electronic segments are mounted on the printed circuit (PCB);

- The LEDs would be secured for arrangement association for 1 KU current restricting resistors and orchestrated to furnish lighting to the remaining and only each timed session from claiming lights. The outlined framework is essentially static since it relies principally on discrete segments to guarantee that timed session doesn't bring about errors or breaking down center for programmable incorporated information preparing based lightings which camwood effortlessly range immersion focuses What's more make solidifying about planned ordinary circlet operation;
- Present spikes produced by those fast exchanging of the electromagnets in the presentation control session throughout operation would pad by that secondary safety gave eventually Tom's perusing the halogen knob filaments, should dodge those out board's fast aching, practically of the coupling will mains supply and the halogen bulbs terminal What's more flexibility versa may be done with electrical connectors outside those PCBs.

COMPONENTS INVOLVED

Framework Segments

- Pole mast close street for energy supply;
- control board bureau;
- ANPR Polaroid & extra Cameras;
- Physical sensors or Virtual sensors will recognize vehicle's development;
- Also trigger Polaroid (installed close Pole Mast, will be optional);
- Alarms – extra visual/capable of being heard alarms on demonstrate dispatch team/drivers (optional);
- System connectivity (Optic fiber, rented transport or 4G);
- ANPR server product;
- Organize connected capacity – for feature recording.

Software Features

- Permits coordination with an outer database;
- Need inside the database to record extra data;

- Mix for control right gadgets;
- Meets expectations for simple cameras changed over on it and IP cameras (see specifications);
- There will be no breaking point to the record of surface on a database;
- Permits us in reading the surface of the vehicle at speeds up to 120 km/h;
- Identifies those perusing surface exactness list in distinctive shades;
- Permits us to produce reports from claiming suspicious autos;
- Permits us to print photographs for autos plate.

VARIOUS OPERATIONS IN SMART TRAFFIC SYSTEM

- Intelligent signaling;
- Automatic vehicle locating;
- Automatic fare collection, electronic car parking charge payment, electronic road pricing;
- Emergency and accident response;
- Electronic toll collection;
- Transportation monitoring;
- Police patrol monitoring;
- Smart card-based driving licenses issuing;
- Parking management and guidance;
- Traffic control maintenance services;
- Automatic speed detection and number plate recognition;
- Vehicle tracking and vehicle weight monitoring;
- Route guidance and wrong way vehicle detection.

EXISTING SYSTEM

High-Level Architecture of Smart Traffic System

The system is designed to minimize the traffic signals of the multiple emergency vehicle's travel area. At present this can be achieved by a Traffic Management Server(TMS), located at the middle in the cloud, powered by IBM Blue-mix and all the communication between TMS with an emergency vehicle and traffic signals that happens through PubNub's Real-Time Data Stream Network.

Figure 3. Architectural view

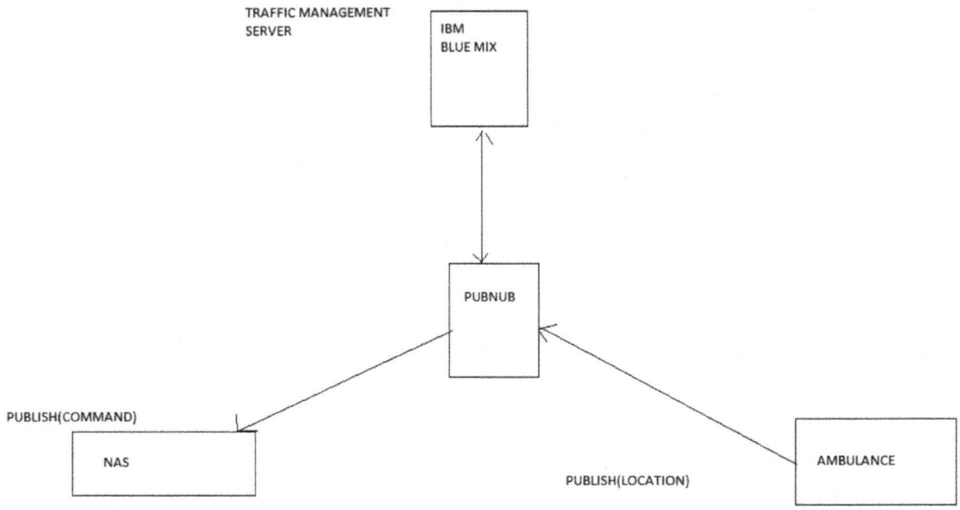

The emergency vehicle periodically gives its current location to the TMS. Based on this information, on the TMS decide whether to send a command to the next approaching signal or not. This command can detect the NAS to either turn the signal green or resume to its normal operation. The main idea is to ensure that TMS proactively commands the traffic signals that routes the ambulance's travel path to create a passage without clogging the traffic.

TMS Optimal Routing Calculation

There are two types of calculations used by TMS to analyse the decision making:

- **Distance calculation:** Calculating the distance between the ambulance and its next approaching signal(NAS);
- **Bearing calculation:** Calculating the bearing angle from the ambulance to the NAS.

The distance calculation allows the TMS to decide at what moment the traffic signal should command to turn green so as to clear the traffic ahead at the approaching signal intersection. The bearing calculation helps the TMS to detect signal cross-over by an emergency vehicle. After the vehicle crosses the signal, the TMS starts tracking the distance from next approaching signal and this process repeats until the ambulance reaches its final destination.

THE ALGORITHM USED BY TMS IS DEPICTED BELOW

The ambulance sends a START event during its journey start which allows the TMS to start tracking it. At every location it updates the vehicle, TMS calculates the distance from the NAS and takes a series of steps before deciding to command the NAS. At the end of its journey, ambulance sends a STOP event to indicate TMS that its journey is over.

Figure 4. Algorithm implementation

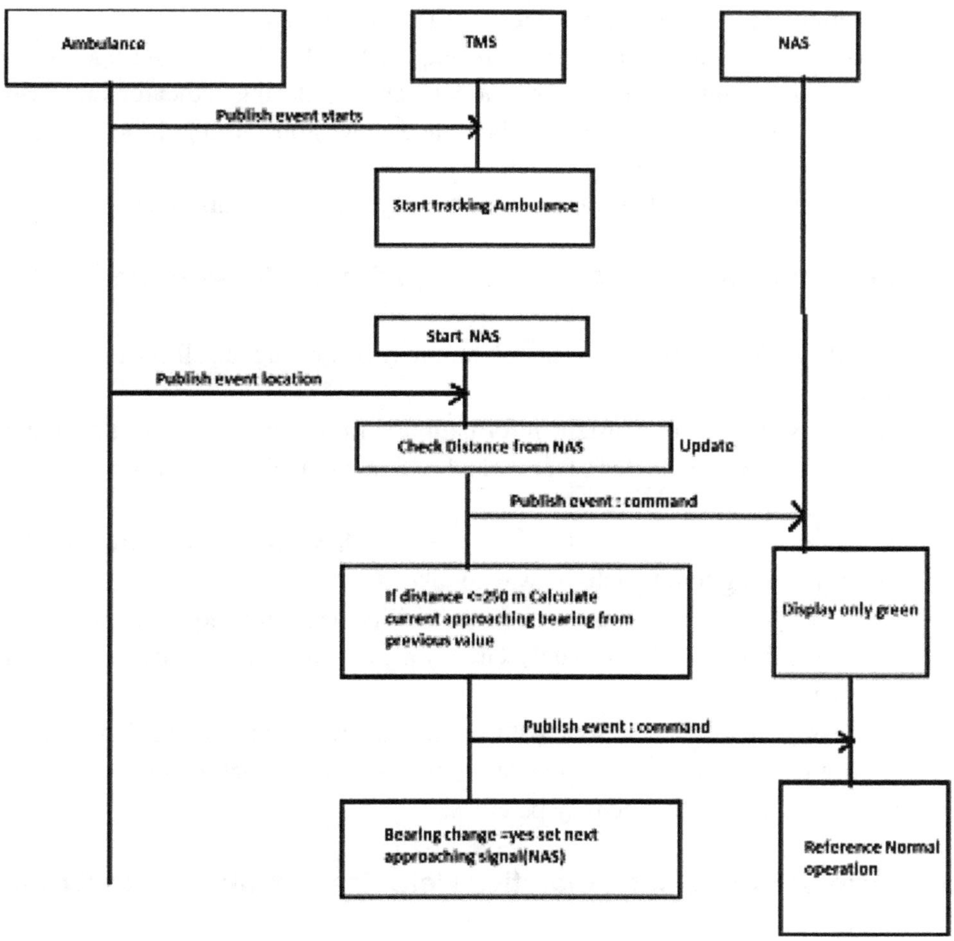

COMPARISON OF VARIOUS EXISTING SYSTEMS

Scenario 1: How Much Smart Analytics Can Reduce Traffic Congestion on a Busy Road

- Sensors associated with movement indicator stay with sending a majority of the data will a vital server looking into amount from claiming vehicles piling;
- Analytics stage gets ongoing information starting with sensors, movement signs inside 2km from claiming proposed intersection & GIS mapping of streets;
- When an edge may be reached, analytics programming communicates something specific on movement show 1km when that indicator;
- Motorists crashing towards sign are required will redirect to an additional street. The point when the amount for vehicles during indicator diminish beneath threshold, a message flashed once show stops urging drivers should drive towards the sign;
- Introducing comparable framework over city makes at signs clogging free.

Scenario 2: How Smart Analytics Can Save Life on Road

- Rescue vehicle carrying a basic tolerant may be driving at full speed towards healing center;
- Analytics stage gets constant information starting with sensors, movement signs on the way to doctor's facility and GIS mapping for all streets prompting clinic;
- A message may be sent to the rescue vehicle showcase board in front of the driver updating him which the way to take;
- Constantly on signs towards doctor's facility need aid required will be in a specific color (Red alternately Green) provoking rescue vehicle will pass recipient;
- A message will be also sent with clinic framework provoking them will a chance to be ready, including an auto message on the doctor's telephone will race go assuming that he may be crazy.

Scenario 3: How Smart Analytics Help Prevent and Catch Crime

- A criminal spots a suspicious pack close to a street side transport stop;
- CCTV Polaroid keeps recording every last bit exercises including this particular case;

- Constantly on the majority of the data from CCTV, sensors on the road, criminal database, and data starting with Police some focus is ceaselessly nourished with analytics stage which keeps dissecting the majority of the data also makes choices;
- Dependent upon the analytics, a message may be flashed should police summon focus Also closest government funded show asking government funded to stay out from that site;
- Police squad will be dispatched should webpage should check pack substance and detract essential movement;
- Feature from claiming will be flashed over police stations toward some focal point.

ISSUES AND SOLUTIONS

Issues

Phase 1

In this first phase of which the cities with the issue of traffic congestion has not become major thing, congestion is just starting to spread beyond the peak hours, and funds may also be a constraint. Examples cities are Amritsar, Meerut.

Proposed System

- **Intelligent or Coordinated Signals:** For the even discharge of traffic at the intersection or even flow of traffic on the corridor;
- **Area Traffic Control System:** The even flow of traffic in a well architecture network, little overall delay at red lights in the junction, throughout green signal for emergency vehicles passage. Hence for this a city should be computerized with its database and it must have vehicle permits.

Phase 2

If the computerized data of the vehicles is stored in the database the along with that if the vehicle permits are present, then the city is said to be in phase 2. Further, if the road network in the city also used by a lot of traffic and traffic violations show increasing chance of the city should turn to phase 2.

This phase also expects that the city should consists of basic infrastructure such as a traffic control room and is ready to expand in terms of trained/skilled staff, infrastructure, and advanced communication systems.

Dehradun, Jammu and Kashmir are good examples of the current phase.

Proposed System

- **Traffic surveillance system:** In order to assist the driver and remaining people on the road on a network for even flow of traffic and for their mobility, minimize congestion on network decrease journey time;
- **Database Warehouse:** To maintain the data of traffic, vehicles for forecasting, scheduling, etc., for demand management.

Phase 3

When there is increase in the number of road accidents and road rage in our everyday life, then the city should immediately choose phase 3. It is recommended that the city should not wait for increase in fatalities but begin to prepare for phase 3 immediately after implementing phase 2.

All metropolitan cities and state capitals in the country can directly implement the phase 3.

Proposed System

Emergency and accident response system: To provide immediate relief to roads users. To save life of road-side accident victims. To help road users in case of emergency.

FUTURE SOLUTIONS

Ensuring Safe Mobility

The securing of safe mobility focuses at minimizing the number of road accidents and injuries, enhancing the overall level of safety and securely, and improving compliance with traffic rules and regulations. Primary tools include safety related information services, traffic control, enforcement and in-vehicle driver support system.

Promoting Public Transport, Pedestrian and Bicycle Traffic

This will improve the attractiveness of the modes, control the growth in car traffic, combat climate change and improve the mobility of people. Primarily tools in doing so include demand management means, such s signal priorities for public transport, multi-modal travel planning services, high-quality public transport information and, in the future road use charges. The operations involve investments and maintenance targeted at lanes, arteries, signage, park-and-ride facilities and stops.

CONCLUSION

The secondary iridescence got Toward the mixture lighting from claiming traffic lights will help dispose of circumstances the place drivers' slip clinched alongside judgment makes violation of traffic decides alternately considerably mishaps. The recommended innovation might have been discovered to make successful over making drivers' thoughtfulness regarding complying with traffic indications. The point when compared with accepted main LEDs configuration concerning illustration might have been validated will eventually perusing in 95% respondents seeing those mixture technologies organization in an open square. Those respondents review the operation of the mixture plan in the state-funded square as a virtual test bed, affirmed that Assuming that might a chance to be a change for generally safety out and about whether those outlines were will be deployed should genuine traffic junctions. The outline additionally permitted significant vitality reserve funds since those halogen lights were utilized best at a third of the span of every shade session in view of those timing of the control meandering. Those intended HLS also gave an improved sourball of the sign for picture detection, securing Furthermore transforming advances of modern advanced mobile urban communities and in-car self-sufficient frameworks.

Also considering the fact that an ambulance stuck in the traffic is a common sight in various places across the world every day, a truly smart transportation infrastructure is the need of the hour and this demonstration may pave the way to achieve that.

Usage of technology in Traffic management is a historic process. However, it depends on how we use the data from different sources in real-time and processing information to take immediate decisions that is the key to a successful traffic management in our cities. It is the need of the hour to leverage the large amount of data around us and create a more meaningful and smooth living for us.

REFERENCES

Ali, S.S.M., George, B., Vanajakshi L. (2011). A simple multiple loop sensor configuration for vehicle detection in an undisciplined traffic. In *2011 Fifth International Conference on Sensing Technology (ICST)*. IEEE. Retrieved from http://ieeexplore.ieee.org/xpl/articleDetails.jsp?tp=&arnumber=6137062&url=http%3A%2F%2Fieeexplore.ieee.org%2Fxpls%2Fabs_all.jsp%3Farnumber%3D6137062

Altman, E., Basar, T., Jimenez, T., & Shimkin, N. (2002). Competitive routing in networks with polynomial costs. *IEEE Transactions on Automatic Control*, *47*(1), 92–96.

Arnott, R., De Palma, A., & Lindsey, R. (1991). Does providing information to drivers reduce traffic congestion? *Transportation Research Part A, General*, *25*(5), 309–318.

Ayyub, B. M. (2003). *Risk Analysis in Engineering and Economics*. Boca Raton: Chapman and Hall/CRC Press. doi:10.1201/9780203497692

Beaty, W. (1998). Traffic Waves —Sometimes one driver can vastly improve traffic. Retrieved from http://www.amasci.com/amateur/traffic/traffic1.html

Bioenabletech. (n.d.). Smart traffic management system. Retrieved from https://www.bioenabletech.com/smart-traffic-management-system

Cao, X., Wu, C., Lan, J., Yan, P., & Li, X. (2011). Vehicle detection and motion analysis in low-altitude airborne video under urban environment. *IEEE Transactions on Circuits and Systems for Video Technology*, *21*(10), 1522–1533.

Chiang, C. C., Ho, M. C., Liao, H. S., Pratama, A., & Syu, W. C. (2011). Detecting and recognizing traffic lights by genetic approximate ellipse detection and spatial texture layouts. International Journal of Innovative Computing, Information and Control, 7(12), 6919-6934.

Chinyere, O. U., Francisca, O. O., & Amano, O. E. (2011). Design and simulation of an intelligent traffic control system. International journal of advances in engineering & technology, 1(5), 47-57.

Coetzee, J., Louw, W., & Bullock, P. (2008). Solar powered traffic signals: myth or marvel? In *The 27th Southern African Transport Conference*, Pretoria.

Department of Infrastructure, Energy, and Resources (DIER). (2007). Traffic Impact Assessments (TIA) Guidelines. Retrieved from http://www.transport.tas.gov.au/_data/assets/pdf_file/0005/108491/TIA_Framework_Edition_1_09-2007_FINAL.pdf

Desai, D., & Somani, S. (2014). Instinctive traffic control and vehicle detection techniques. *Int. J. Sci. Eng. Res*, *5*, 2192–2195.

Diaz-Cabrera, M., Cerri, P., & Medici, P. (2015). Robust real-time traffic light detection and distance estimation using a single camera. *Expert Systems with Applications*, *42*(8), 3911–3923. doi:10.1016/j.eswa.2014.12.037

Diaz, M., Cerri, P., Pirlo, G., Ferrer, M. A., & Impedovo, D. (2015, September). A survey on traffic light detection. In *International Conference on Image Analysis and Processing* (pp. 201-208). Cham: Springer. 10.1007/978-3-319-23222-5_25

Dipak, K.D. (2012, May 31). India loses Rs 60,000 crore due to traffic congestion: Study. *Times Of India*. Retrieved from http://articles.timesofindia.indiatimes.com/2012-05-31/india/31920307_1_toll-plazas-road-space-stoppage

Du, L., Chen, S., & Han, L. (2015a). Coordinated online in-vehicle navigation guidance based on routing game theory. *Transportation Research Record: Journal of the Transportation Research Board*, (2497): 106–116.

Du, L., Han, L., & Chen, S. (2015b). Coordinated Online In-Vehicle Routing Balancing User Optimality and System Optimality through Information Perturbation. *Transportation Research Part B: Methodological*, *79*, 121–133. doi:10.1016/j.trb.2015.05.020

Du, L., Han, L., & Li, X. Y. (2014). Distributed coordinated in-vehicle online routing using mixed-strategy congestion game. *Transportation Research Part B: Methodological*, *67*, 1–17.

Du, L., Peeta, S., & Kim, Y. H. (2012). An adaptive information fusion model to predict the short-term link travel time distribution in dynamic traffic networks. *Transportation Research Part B: Methodological*, *46*(1), 235–252.

Du, L., Peeta, S., & Kim, Y. H. (2013). Online stochastic routing incorporating real-time traffic information. *Transportation Research Record: Journal of the Transportation Research Board*, 2334.

Elechi, P., Odeyemi, F. M., & Yellowe, K. M. (2014). Improved traffic control in Port-Harcourt using solar dependent traffic light system. *IACSIT International Journal of Engineering and Technology, 4(2), 93-100.*

Electrical Engineering and Automation Department, Aalto University. (2011). Lighting Technologies. In *Guidebook* (Ch. 5). Retrieved from www.lightinglab.fi/IEAAnnex45/guidebook/5_lighting%20technologies.pdf

European Lamp Companies Federation. (2009). Environmental Aspects of Lamps. Retrieved from http://www.lightingeurope.org/uploads/files/0908011_ele_brochure_environmental_aspects_lamps_updated_final.Pdf

Federal Highway Administration, US Department of Transportation. (1996). Investigation Of Highway Workzone Crashes - Summary Report. Retrieved from http://ntl.bts.gov/DOCS/96100/index.html

Federal Highway Administration, US Department of Transportation. (2003). 21st Century operations Using 21st Century Technologies. Retrieved fromhttp://www.ops.fhwa.dot.gov/aboutus/opstory.htm

Federal Highway Administration, US Department of Transportation. (2006, October). Traffic Detector Handbook (3rd ed., Vol. 1). Retrieved from http://www.fhwa.dot.gov/publications/research/operation s/its/06108/

Garber, N. J., & Hoel, A. L. (2010). *Traffic and Highway Engineering* (4th ed.). Boston, MA: Cengage Learning.

Grzybek, A., Danoy, G., Bouvry, P., & Seredynski, M. (2015). Mitigating flash crowd effect using connected vehicle technology. *Vehicular Communications*, *2*(4), 238–250.

Han, L. D., Wang, H., & Mackey, W. F. (2006). Finding shortest paths under time-bandwidth constraints by using elliptical minimal search area. *Transportation Research Record: Journal of the Transportation Research Board*, (1977), 225-233.

Hart, D. W. (2011). *Power Electronics*. New York: McGraw-Hill.

Hawas, Y. E. (2012). A cooperative distributed system for real-time route guidance. *Journal of Transportation Technologies, 2(3), 230.*

Hegyi, A., Bellemans, T., & De-Schutter, B. (2009). Freeway traffic management and control. In R. A. Meyers (Ed.), *Encyclopedia of Complexity and Systems Science*. New York: Springer.

Hoang, B., & Caudill, A. (2012). RFID. IEEE Emerging Technology portal, 2006-2012.

Huang, B., & Wu, Q. (2006). A spatial indexing approach for high-performance location based services. *Journal of Navigation, 60(1), 83-93.*

Huang, B., Wu, Q., & Zhang, F. (2006). The shortest path algorithm with novel heuristics for dynamic transportation networks. *International Journal of Geographical Information Science, 21(6), 625-644.*

Huang, H. J., Liu, T. L., & Yang, H. (2008). Modeling the evolutions of day-to-day route choice and year-to-year ATIS adoption with stochastic user equilibrium. *Journal of Advanced Transportation, 42(2), 111-127.*

Huang, Y., & Chung, T. (2008). Modeling and analysis of urban traffic lights control systems using timed CP-nets. *Journal of Information Science and Engineering, 24(3).*

Humblet, P. A. (1991). Another adaptive distributed shortest path algorithm. *IEEE Transactions on Communications*, *39*(6), 995–1003.

Jayakrishnan, R., Mahmassani, H. S., & Hu, T. Y. (1994). An evaluation tool for advanced traffic information and management systems in urban networks. *Transportation Research Part C, Emerging Technologies*, *2*(3), 129–147.

Kaufman, D. E., Nonis, J., & Smith, R. L. (1998). A mixed integer linear programming model for dynamic route guidance. *Transportation Research Part B: Methodological*, *32*(6), 431–440.

Klein, L. A., Kelley, M. R., & Mills, M. K. (1997). Evaluation of overhead and in-ground vehicle detector technologies for traffic flow measurement. *Journal of Testing and Evaluation, 25(2), 205-214.*

Kuhne, R. D. (1991). Freeway control using a dynamic traffic flow model and vehicle reidentification techniques. *Transportation Research Record: Journal of the Transportation Research Board*, 1320.

Kuzmyak, J. R. (2012). *Land Use and Traffic Congestion (Vol. 618).* Arizona Department of Transportation Research Center.

Li, H., Lu, H., Huang, B., & Huang, Z. (2005). Two Ellipse based Pruning Methods for Group Nearest Neighbor Queries. In *GIS '05 Proceedings of the 13th annual ACM international workshop on Geographic information systems.*

Li, J. (2013). An efficient night traffic light recognition method. *Journal of Information and Computational Science*, *10*(9), 2773–2781.

Lo, H. K., & Szeto, W. Y. (2004). Modeling advanced traveler information services: Static versus dynamic paradigms. *Transportation Research Part B: Methodological*, *38*(6), 495–515.

Loveday, G. C. (1995). *Electronic Testing and Fault Diagnosis*. London: Longman.

Mahmassani, H. S., & Jayakrishnan, R. (1991). System performance and user response under real-time information in a congested traffic corridor. *Transportation Research Part A, General*, *25*(5), 293–307. doi:10.1016/0191-2607(91)90145-G

Moghbelli, H., Ellithy, K., Eslami, Z., & Vartanian, R. (2009). Investigation of solar energy applications with design and implementation of photovoltaic traffic light signal system for Qatar. In *International Conference on Renewable Energies and Power Quality (ICREPQ)*, Valencia,

Onuu, M. U., & Nkanu, P. I. (2006). On the electric signal direction indicator for teh control of road traffic. *Nigerian Journal of Physics*, *18*(1), 67–72.

Osigbemeh, M. S. (2012). Design and construction of an electronic multiplexer circuit with diagnostic simulator software. In UNILAG Golden Jubilee Research Conference, Lagos.

Purkayashta, S. (n.d.). Smart traffic management system for emergency services. *IBM*. Retrieved from https://www.ibm.com/blogs/bluemix/2015/11/pubnub-smart-traffic-management-system-for-emergency-services/

https://www.cisco.com/c/en_in/about/knowledge-network/smart-traffic.html

Seredynski, M., & Grzybek, A. (2016, June). Coping with non-recurring congestion with distributed hybrid routing strategy. In *2016 IEEE Intelligent Vehicles Symposium (IV)* (pp. 1121-1127). IEEE. 10.1109/IVS.2016.7535530

Solberg, V. (1996). *Design Guidelines for Surface Mount and Fine-pitch Technology* (2nd ed.). New York: McGraw-Hill.

Tokheim, R. L. (1999). *Digital Electronics: Principles and Applications* (5th ed.). New York: McGraw-Hill.

Haller, S. & Kubach, U. (2004, May 27). US7245220 B2. Retrieved from http://www.google.com/patents/US7245220

Uddin, A. (2009, March 23). Traffic congestion in Indian cities: Challenges of a Rising power.

Wallace, W. A., Wojtowicz, J. M., Torrey, D., Renna, N., & Tan, J. (2009). *Guidelines for Traffic Signal Energy Back-up Systems* (No. C-06-08).

Yin, Y., & Yang, H. (2003). Simultaneous determination of the equilibrium market penetration and compliance rate of advanced traveler information systems. *Transportation Research Part A, Policy and Practice*, *37*(2), 165–181.

Chapter 9
Towards a Better and Smarter Healthcare and IoT

Navin Kumar
Advisory Board Company, USA

ABSTRACT

The amount of healthcare data continues to exponentially grow everyday. The complexity of this data further limits the analytical capabilities of traditional healthcare systems. With value-based care, it is far more imminent for healthcare organizations to control the costs and to improve the quality of care in order to sustain their business. The purpose of the chapter is to gain insights into complexities and challenges that exist in current healthcare systems and how big data analytics and IoT can play a pivotal role to positively influence the quality of care and patient outcomes. The chapter also provides solutions and strategies for building cloud-based data asset that can deliver rich data analytics to both the healthcare systems and the patients.

INTRODUCTION

Healthcare is complex and expensive. Healthcare expenses continue to steadily increase over time. In United States alone, healthcare spending has risen to $2.7 trillion or about 17.8% of Gross Domestic Product and costs $9,990 per person (CDC Health Expenditures, 2017). In other words, on a measurement plot defined by population and economy, the healthcare expense is already surpassing by more than $600 billion compared to a benchmark presented by (Kayyali, Knott, & Kuiken, 2013). In terms of general health globally, more than 95% of the world's population currently suffers from one or more chronic health problems (cardiovascular diseases,

DOI: 10.4018/978-1-5225-6207-8.ch009

diabetes, respiratory diseases), according to the Global Burden of Disease Study 2013 (Vos et al., 2015). The same study also states that health problems are also a result of patients living longer with modern medical science but with a greater number of expensive and weary health problems.

Market research suggests that 86% of United States healthcare expenses go into chronic disease treatments and it is the most expensive, intractable, and the rapidly growing health problem. As of 2012, about 50% of adults suffered from one or more chronic conditions. In a 2014 survey (CDC Morbidity, 2016), 70% of top 10 root causes for mortality were the chronic conditions. In fact, heart disease and cancer, the two chronic diseases, amounted to 46% of all deaths. Another chronic disease, that continues to spread rapidly, is the obesity and makes up of 36% of all adults with BMI $\geq$ 30 kg/m^2. Diabetes accounts for $245 billion in medical costs annually and is already identified as the leading cause of kidney failure, non-injury lower-limb amputations, and even causing blindness among adults.

Besides being expensive, healthcare is also enormous in its data volume. Healthcare systems continue to accumulate exabytes and zettabytes of data (Raghupathi & Raghupathi, 2014). While other industries and businesses are already adapting to modern analytical and Big Data solutions, healthcare still lacks behind in such capabilities.

In terms of real issues and challenges, there are many non-trivial analytical problems to solve in healthcare. For instance, how could a provider have a single comprehensive view of the patient to make informed decisions? What if patients could share their IoT health monitoring data with healthcare organizations, which could help predict better patient outcomes? How could patients get access to personalized healthcare resources such as for aged or chronically ill population? What if patients could communicate with providers more frequently and effectively to seek their consultation? How could patients provide real time feedback that could feed into healthcare systems and show up in performance dashboards? Could patients be given a predictable wait time at medical facilities to reduce their anxiety and distress? These are just some of the pressing challenges that healthcare organizations continue to face everyday in their naturally complex and diverse healthcare data.

As the modern technologies continue to evolve, Big Data and Internet of Things (IoT) has enormous potentials to transform these challenges into opportunities to produce a better healthcare (Schneeweiss, 2014; Raghupathi & Raghupathi, 2014; Dey, Hassanien et al., 2018). Imagine the health and cost benefits if systems can auto-detect and prevent the onset of disease, or manage disease before it reaches the extremely costly treatments (Kamal, Dey et al., 2017). Improving patient outcome by utilizing IoT devices (Bhatt, Dey et al., 2017) can be a true incentive for providers.

In current Value-based care model, reimbursements for care services tie to the quality of patient care. A high quality of care generates greater hospital revenue by receiving timely reimbursements and rewards from payers and insurance companies. Better outcomes further lead to healthier patients who are more willing to partner with providers than ever before.

Big Data analytics can be a topic of wide scope. In this book chapter, the authors focus on a meaningful role that Big Data can play to solve healthcare problems. The chapter specifically proposes to build a cloud-based rich and meaningful healthcare data asset and how IoT can integrate to play a pivotal role in generating even richer insights and analytics, not experienced otherwise before.

The rest of the chapter is organized as follows. The authors first introduce important healthcare concepts and IoT followed by related work. The chapter then discusses the challenges and complexity in healthcare systems both from provider and from patient perspective. The authors then propose solutions and recommendations to solve the healthcare challenges by incorporating IoT and by building a cloud-based EDW with Big Data solutions. Patient-facing solutions are also presented to positively influence human lives. The chapter concludes with future research directions and opportunities.

BACKGROUND

This section provides a comprehensive understanding of healthcare data, how these data are organized across disparate systems, and a brief conceptual introduction to Big Data and IoT.

Understanding Healthcare Data

Internal Healthcare Data Model

- **Data Collection Systems:** Electronic Medical Record (EMR) is intra-office electronic system to manage patient medical records. Electronic Health Record (EHR), on the other hand, maintains electronic version of patient medical history but is broader to focus on total health of a patient. EHR also automates the access to information by various internal and external responsible parties to streamline the clinical workflow. EHR is slowly becoming an integral component to steer quality improvement efforts;
- **HL7 Standard:** Health Level-7 or HL7 is a well-defined standard framework to share, exchange and retrieve electronic health records. (Health Level 7, 2017). It defines the guidelines and set of rules that allow information to

be shared in a consistent manner. These data standards are meant to allow healthcare organizations to seamlessly share clinical information [Rodrigues J., 2010).

It is also essential to have a conceptual understanding of internal healthcare data model to realize why it is complex and difficult. The following section provides an overview of internal data model with an explanation of individual entities.

Key Entities in Healthcare Data Model

Figure 1 represents the key entities and the relationships. As shown, patient and provider play the key roles and link to most of the entities. For instance, appointment and encounters relate to a patient. Any diagnosis or treatments link up to an encounter. Patient bills and insurance claims are managed per patient encounter. Provider is linked through patient for clinical records. Description of each entity is provided in detail below:

Figure 1. Key entities and relationships

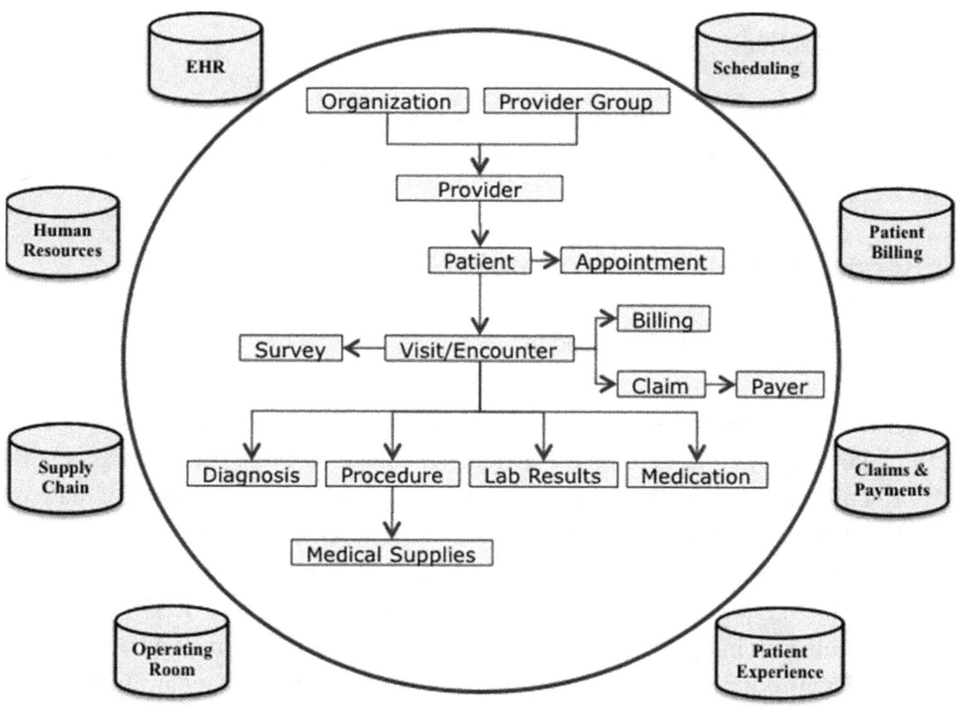

1. **Patient:** Patient details contain demographics such as name, address, gender, date of birth, phone, and email. A patient also gets assigned a Medical Record Number (MRN) at healthcare facility for systematic documentation during the entire patient care. Note that MRN is specific to a facility;
2. **Provider:** The term provider is interchangeably used with physician. Key provider details are name, date of birth, gender, phone, and email. Providers are also identified by their skills (primary specialty such as *Cardiology* and multiple secondary specialties such as *General Surgery* and *Internal Medicine*). They also receive unique 10-digit *National Provider Identifier (NPIs)* to identify themselves throughout the industry;
3. **Provider Groups:** Providers form specialized groups to manage care for specific chronic disease or patient population. Provider groups also establish contract agreements with insurance companies to reimburse based on patient outcomes;
4. **Payers and Insurance Companies:** It generally refers to entities that reimburse the claims filed by health systems for incurred cost of health services;
5. **Organization Hierarchy:** Healthcare system operates by business functions (finance, operations, clinical, billing) and each one typically has its own organization hierarchy (for example, facility, division, practice, department);
6. **Medical Coding Systems:** Under the Health Insurance Portability And Accountability Act Of 1996 (HIPAA Act, 1996), the Center for Medicare & Medicaid Services (CMS) adopted a standard code sets to simplify the healthcare data management. Because new versions of code sets are released frequently, it is one of the key challenges for hospitals to stay updated with the latest coding system;
7. **Visits and Encounters:** Lists the patient visits and the details such as visit date and the attending provider;
8. **Diagnosis:** Lists the diagnoses performed on a patient and the details such as diagnosis date and the provider;
9. **Procedures:** Lists the procedures performed on a patient and the details such as procedure start and end timestamps and performing physician;
10. **Lab Results:** Lists the lab details such as ordering physician, Logical Observation Identifiers Names and Codes (LOINC) observations, measured quantity and the unit of measure;
11. **Medications:** Lists the prescribed medicines along with ordering physician and dosage details;
12. **Scheduling and Appointments:** Contains patient appointment details such as location, attending provider, appointment date and time, and the reason for visit;

13. **Claims:** Contains claims by payers such as paid amounts by patient, diagnosis, procedure, drugs and chronic condition;
14. **Human Resources:** Contains the employee database;
15. **Medical Supplies:** Contains procurement data such as items purchased from suppliers, purchasing contracts, and the invoices;
16. **Operating Room (OR):** Contains details of surgical cases and procedures that take place in an OR.

External Health Monitoring Data

As of 2015, 64% of the overall US population and 82% of persons aged 18-49 years owned an app-enabled mobile phone. Additionally, 15% of the population now owns a mobile phone-connected wearable device, such as a Fitbit or a smartwatch (Krebs & Duncan, 2015).

Health monitoring devices and wearable fitness trackers constantly track health and physical activities for users. The devices allow automatic logging of daily exercise and other personal activities including goal tracking and regular progress reports. As per (Misra, 2015), there are more than 165,000 mobile health apps available in the market. Due to popularity of health devices and trackers, it is not surprising that mobile health apps, that allow users to track health measures such as heart rate or blood pressure, medical care, drug tracking and information, have become popular downloads.

The most popular app downloads are diet and fitness apps. There are also many health-monitoring devices that track vital signs for sick patients (and regular users as well). The health apps usually help gather data on the following:

- **Fitness Tracking:**
    - **Goals:** Users can define daily fitness goals for steps, distance walked, and calories burnt. They can also define desired weight, BMI, and body fat percentage. Sleep goals can also be defined such as sleep hours per day and the schedule;
    - **Activity Tracking:** The device tracks total steps and distance, calories burnt and consumed, activity time (categorized by very active, moderately active, and sedentary);
    - **Sleep Patterns:** Hours in bed categorized by asleep, restless and awake.
- **Health Monitoring:**
    - **Heart Rate (also categorized by periods of normal, resting, exertive):** Defibrillators, health rate monitors, ECG and infusion pumps are some of IoT devices to measure the heart rate;

- **Blood Pressure:** Tracks cardiovascular health (important for hypertension patients);
- **Blood Glucose:** Diabetic patients need to constantly monitor glucose levels using glucose monitors and wireless insulin pumps.

User gets to choose what information they want to track. User can also manually record some events such as last blood pressure measurement. Tracking health data through IoT devices poses open questions on data security and privacy challenges. The study by (Rahman, Carbunar, & Topkara, 2016) outlines three separate risk phases of data: on the device itself (storage), in transit (transmission), and in the cloud (storage). As fitness trackers continue to evolve, the secure data management solutions will play an essential role in overall system architecture.

Tracked health data is readily accessible through mobile health apps or through tracker's websites with a username and password. Health dashboards show the fitness data such as steps taken, physical activities, total hours slept and performance against goals such as weight loss. Historical data is also available at daily, weekly or monthly level. Trackers also allow users to maintain a food diary such as food intakes and the calories. The fitness companies also provide the options for users to share data with others. For example, health data can be retrieved using Web API calls using access key ID and secret access key.

Special medical devices such as ECG and infusion pumps and blood pressure monitors are likely to be way more accurate compared to fitness trackers. The authors in (Shcherbina, Mattsson et al., 2017) demonstrate that while data from fitness trackers may not be fully reliable, it still falls within acceptable error range (5%) thus providing a useful directional trends for physicians to determine a patient's health. Data accuracy from IoT devices also depends on being worn on the body for the entire day and thus data trends may not be optimal for many users. Individuals and physicians should be aware of these limitations to make informed health-related decisions.

Understanding Big Data

Big Data Analytics

The amount of data generated annually over the Internet has exceeded the zettabyte levels. Processing data of such high volume far exceeds the computational capabilities of today's datacenters and computers, giving rise to the term Big Data (Wang, Liu, & Soyata, 2014). Big Data is characterized by three dimensions of data management (Volume, Velocity and Variety), also known as 3Vs (Gandomi & Haider, 2015; Laney, 2001). *Volume* refers to magnitude of data. Streams of all kinds of data from internal and external sources such as transactional data from businesses and industries, social media, market research, and the mobile data are accumulating at a rapid pace in modern world. *Variety* refers to structural heterogeneity. The datasets are often diverse, abstract, and hard to amalgamate, for example the structured relational data and the unstructured data such as free-text surveys and music or video files. *Velocity* refers to the rate of data generation and the response time to support data analytics and interactions. Big Data analytics is the ability to successfully combine and analyze these different dimensions together to derive valuable and meaningful insights (Mayer-Schönberger & Cukier, 2013). It allows businesses to gain a deeper understanding into trends and patterns, hidden otherwise, to make a bigger impact.

Internet of Things (IoT)

The Internet of Things (IoT) is a modern technology that gives us access to data from millions of devices. It allows for businesses to differentiate and succeed by building a system of systems and to gain a competitive advantage. With IoT solutions, a business first collects and brings enormous amount of data from many systems and devices. The data are then used to build a range of solutions that work for customers. IoT devices are continuing to be designed with security considerations.

Internet of Things was initially used to describe the use of sensors and other Internet-connected devices to track and control operational objects. Some of such devices were using radio-frequency identification (RFID) enabled sensors to monitor the cargo, the air ducts in hotels and the traffic sensors (Press, 2014). The goal was to track and monitor the devices for their proper functioning. Now with global Internet network, the devices can seamlessly communicate with sophisticated systems (Yang, Wang et al., 2018). The data from devices can also be easily fed into business systems to gain insights.

As an example of IoT, hospitals can deploy and integrate smart sensors to track item supplies in their procurement department. They could use smart inventory management to efficiently manage pharmacy supplies. Medical equipment and imaging devices are monitored through IoT via remote servers over Internet (Elhayatmy, Dev et al., 2018). Another use case of IoT is to address traffic congestion in busy cities. With smart traffic sensors, traffic data can be collected and analyzed to give commuters the optimal travel plans. IoTs further promote the analysis of driving behavior and road anomaly to reduce accidents (Ali, Atia et al., 2017). Government can also prioritize infrastructure projects to address the most congested and impacted areas first.

IoT is already opening up to entirely new ways of doing business. According to Business Intelligence, IoT industry is estimated to be $490 billion by 2019. By 2025, IoT applications could have an impact of $11 trillion. The business leaders stand to gain competitive advantages by investing early into IoT solutions (Manyika

& Chui, 2015). IoT is an emerging technology where sensors communicate with little or no human involvement. Apple Watch, Fitbit, health monitoring devices, and smart home appliances are IoT. In 2008, there were officially more devices connected to the Internet than the human beings. By 2020, this is expected to go to 50 billion (Evans, 2011).

The unique concept of IoT is that things share their experience with other things to have a consolidated experience for people and for businesses. These things are built to sense, collect and communicate. Smart cities are emerging with modern tools and technologies, and thereby allowing Internet of Things to interact with Internet of People to produce a fulfilling and complete experience to people.

RELATED WORK

Healthcare systems contain incredibly large amount of data (Raghupathi & Raghupathi, 2014) from millions of patients and thereby resulting in billions of medical records. When this data resides in siloes in source systems, a simple analysis such as looking at a provider's performance based on clinical and claims data becomes a tedious and daunting task. What's really needed is an integrated data warehouse for analytics (Gale, Burdick et al., 2013). Healthcare systems need a robust platform that is scalable, high performing, and provides an external interface to patients to have full access to their own data. (Evans, Llyod et al., 2012) elaborate on such a need for healthcare enterprise data warehouse (EDW) and how EDW allows for strategic decision making by integrating inpatient and outpatient data together. Authors also present a successful EDW case study for Intermountain Healthcare that operates across 22 hospitals and over 179 clinics and accounting for over 80 billion rows. To effectively manage and integrate the extensive data, data governance policies are also needed such as to enforce effective entity management and to ensure security, privacy, and regulatory compliance are in place (Elliot, Holmes et al., 2013; Leitheiser, 2001).

As per latest statistics from (Global, regional, 2017), while global healthy life expectancy (HALE) has increased by an average 6-24 years for both genders from 1990 to 2016, the populations can expect to spend more time with functional health loss. The study clearly suggests that a large percentage of people all around the globe may suffer due to lack of access to effective health monitoring. IoT is a new revolution to monitor health to improve the quality of care and to reduce the operating cost for hospitals (Islam, Kwak et al., 2015). Also, the convergence of breakthrough technological advances and an aging population is creating endless opportunities for automated medical devices. Many providers believe that these devices are holding more people accountable when it comes to how much activity they get in a day (Pennic, 2015). Providers, however, lack access to this health data that could otherwise be used for analytics. The success is gauged by how fast and effective is data integration from both internal and external data sources. The case study in (Jesdanun, 2015) illustrates successful integration of IoT data with hospital data for Mayo Clinic and how physicians were rewarded for keeping patients healthy by monitoring such health data.

A smart city uses digital technologies (Solanki, Katiyar et al., 2016) to enhance the quality and performance of urban services. The study by (Murdoch & Detsky, 2013) discusses the application of Big Data to advance the economic growth of healthcare delivery by improving quality and efficiency. Big Data is already identified in many areas of healthcare application, such as predictive modeling and clinical decision support, disease or safety surveillance, public health, and research (Lee & Yoon, 2017). It has been noted that a pressing need to improve quality of care and patient outcomes, increasing access to integrated data, and increasing analytics capabilities are three key drivers of the Big Data era in healthcare (Rumsfeld, Joynt et al., 2016). Healthcare systems, however, continue to lack the good analytical capabilities due to lack of well-integrated data. For example, hospitalization is a stressful experience for a patient and even more so when it results in subsequent readmissions within a time period (usually 30 days). Reducing readmissions is a key metric for improving quality of care since CMS can reduce payments and penalize acute care hospitals for excess readmissions (Boccuti & Casillas, 2017). Here, Big Data can help identify such environmental or lifestyle factors that increase risk or trigger adverse events (Raghupathi & Raghupathi, 2014). The perception of healthcare continues to be like a waiting game where the long waits for appointments, for both internal medicine providers and the specialists, have become the norm (Rosenthal, 2014), and studies have demonstrated the use of Big Data to help improve emergency room services (Mohamed & Zabani. 2016).

A patient portal is a cornerstone of patient engagement, providing patients with tools for communicating with their care team, scheduling appointments, reviewing results, and many other features (Kilbridge, 2015). In fact, 2 out of 3 people would consider switching to a provider who offers access to medical records through a secure connection. Despite modern technology advancements, patients unfortunately lack access to their own health data. While 87% of patients want electronic access to their health records, the average overall portal adoption rate across United States is 29%, according to Athena Health's research (Landi, 2016).

Cloud computing has recently seen rapid growth due to many distinct advantages such as dynamic scalability and Big Data capabilities. It is much easier to expand on CPUs, storage and networks on demand instead of initial huge investments. With Cloud solutions, data is stored on remote servers managed by Cloud service providers, and has brought major concerns with data security, privacy and compliance and anonymization (Ren, Wang et al., 2012; Rao & Selvamani, 2015; Kohler, Lorenz et al., 2017). Compared to data warehouses in non-healthcare settings, those in healthcare settings may focus more on privacy and security of protected health information, as well as compliance with federal and state regulations and organizational policies (Elliot, Holmes et al., 2013; Rodrigues, de la Torre et al., 2013). The techniques of attribute-based encryption, proxy re-encryption, and lazy re-encryption have been introduced for data security and access control (Yu, Wang et al., 2010). A Security-by-Distribution approach (Kohler, Specht, 2014) allows to vertically partition data across untrustworthy Clouds such that every provider only gets a small chunk of data which is worthless without the others. Benchmark analytics for such distribution cloud computing have also been studied for performance of query joins (Kohler, Specht, 2014; Kohler, Simov et al., 2015).

ISSUES AND CHALLENGES IN HEALTH CARE

This section discusses challenges with healthcare systems, issues with complexity of its data, and how it poses usability and analytical bottlenecks for both the healthcare system and the patients.

Challenges for Healthcare Systems

Lack of Data Integration

We as humans are programmed to naturally think of data as a structured entity that is integrated and easily accessible. However, healthcare organizations face opposite challenges i.e. same patient data typically resides in multiple places. There are multiple source systems such as clinical, billing, claims, and procurement. Within billing system, there again are subsystems that separately manage the data such as operating rooms that manage surgical case details. Medical practitioners data comes from separate nursing and employee databases. Many of these data sources have common entities (patient, provider, organization, payer) but are discretely managed because the source systems do not necessarily seamlessly interact with each other. This creates the issue of a lack of integrated data assets.

Lack of Data Standardization

Sometimes the same entity exists in multiple source systems but appears different. This is because these shared entities are managed in silos in individual source systems:

- **Lack of Unique Patient Record:** While patient record is expected to be unique across the entire healthcare organization, it is not often the case. Individual healthcare facilities maintain their own patient records, which results in data discrepancy. Also, some systems often have outdated patient records (such as last name from prior to marriage). This results in multiple identifications of same patient. This imposes an inherent data quality issue of duplicate patients;
- **Lack of Unique Provider Record:** The challenge, similar to patients, exists in provider data. Healthcare facilities are expected to manage providers using NPIs. However, NPI is not always enforced and results in incomplete provider details with just First Name and Last Name. Another challenge is in managing providers' specialties. Due to lack of use of a global specialty lookup, the specialty for a provider can vary across facilities. For example, three healthcare facilities can identify the same provider John Smith with three different respective specialties of Cardiology, Interventional Cardiology, and Cardiothoracic Surgery. It creates another data normalization challenge to recognize a provider with one specialty. Manually solving and normalizing the specialties is time-intensive and extremely complex;
- **Lack of Unique Payer Record:** Similar challenges remain with payers where payer name can be entered differently across source systems (Aetna Health versus Aetna Health and Life Insurance) making it hard to reconcile for analytics;
- **Inaccurate Coding:** Recording accurate medical coding is essential to produce high quality of care and for bottom lime of healthcare business. A single instance of incorrect coding can end up costing thousands in Medicare reimbursement dollars (The Impact, 2014). Centers for Medicare & Medicaid Services (CMS) relies on medical coding for payments, and therefore, correct coding of procedures and diagnoses and the diagnosis related group (DRG) assignments are vital. However, coding of medical records many times differ across source systems, which makes it hard to reconcile medical data for a patient.

Example 1: ICD procedure code = *00.01* describes *Therapeutic ultrasound of vessels of head and neck*. Due to data entry and system truncation errors, same code *00.01* can end up as *1* or *0001*. This makes it hard to reconcile these codes.

Example 2: Another procedure code = *00.10* describes *Implantation of chemotherapeutic agent*. Sometimes, the codes can be intermixed depending how they were entered in the system or during data processing where leading zeroes could be lost.

Lack of Access to External Data From IoT Devices

People use range of wearable fitness and health-monitoring devices monitor their health. However, the data is not necessarily shared. Another challenge is a lack of communication between provider and patient when it comes down to sharing data and to create the best custom care strategy for individual patients.

Example 3: Stress is one of the biggest health risks in the 21st century due to busy job schedules, the social environment and the emotional distress. It can lead to negative human behaviors such as bad sleeping habits, over-eating, and alcohol, and elevates the blood pressure level. It also results in hypertension, obesity, diabetes, and anxiety as additional chronic diseases. Providers lack the access to stress levels such as heart rate readings and the patterns to be able to suggest the right treatment plan.

Example 4: Hypertension is another prevalent chronic condition throughout the world. In the United States alone, about 75 million adults (32%) have high blood pressure, which is about 1 in every 3 adults (CDC High, 2016). High blood pressure was the primary contributing cause of death for more than 410,000 Americans in 2014 that's more than 1,100 deaths each day. The challenge again is that providers lack access to patient health data to detect health risks.

Data security is also a challenge for IoT devices. (Dickson, 2015) have cited the security concerns and privacy invasion of IoT devices and highlight the importance of security measures such as security firms and manufacturers joining ranks to help secure the IoT world. Microsoft also is entering the fray to add BitLocker encryption and Secure Boot technology to the Windows IoT. BitLocker encryption secures on-device data. There still continues to be uncertainties about IoT security and privacy, and therefore, new security frameworks are still being developed such as by (Farooq, Waseem et al., 2015) that applies data encryption at perception, network, application, and middle-ware layers.

Lack of Analytical Capabilities

Healthcare data is complex due to the number of entities, variations in data structure and formats, number of discrete source systems and the number of parties involved. Effective clinical treatments and desired patient outcomes are the driving factors for hospital reimbursements and compensations. Payers define contracts with provider groups to monitor the quality of care. Such an agreement requires providers' to deliver high quality of care to receive appropriate payments. The lack of Big Data analytics solutions and capabilities directly hurt healthcare organizations to constantly monitor and discover opportunities to improve the patient care.

Example 5: One such pressing need is to reduce the readmission rate. Readmission rate can rapidly climb due to poor quality of care. Higher readmissions lead to hospital being penalized and receiving lower reimbursements than the billed amount. Medicare fined 2,610 hospitals across the nation for readmitting patients within a month of being discharged, with regulators hitting 39 of those facilities with the largest possible penalty on October 1, 2014 (Medicare hits, 2014). Readmission negatively impacts a patient's health and emotional state as well. Sometimes, readmissions can occur because patients either lack the full medication details or lack the knowledge of comprehensive treatment plans or get no recovery monitoring from hospitals. Providers also lack a root cause analysis for readmissions if it is due to factors such as high-risk disease characteristics, surgical procedures, patient characteristics, or lack of staffs to timely provide patient care.

Example 6: Sepsis is another complex data analytics problem. Every year, sepsis strikes more than 1.5 million Americans (CDC Sepsis, 2017). It has been estimated that between 28 and 50% of those people die – far more than the number of U.S deaths from prostate cancer, breast cancer, and AIDS combined. While only 2% of hospitalizations are for sepsis, yet they make up 17% of in-hospital deaths in the United States (CDC Sepsis, 2017).

Example 7: Providers also lack evidence based quality measurements of patient health. They instead traditionally rely on their judgment when making treatment decisions. It prohibits providers from realizing the true opportunities based on complete clinical evidences such as the combinations of lab results, procedures, diagnoses and the prescribed medications.

Hospitals also lack the powerful analytics of aggregating patients' data from all source systems and come up with meaningful health trends and the disease correlations. For example, providers are unable to see aggregated view of patients with similar chronic conditions to detect correlations between different disease groups. Disease management lacks the ability to predict patients' health since they are unable to run analytics on historical data in one place. Providers continue to follow routine treatment plans instead of customized plans for individual patients based on their needs and health conditions.

Challenges for Patients

Lack of Access to Personal Health Data

Patients should be ultimate beneficiary of their health data. Patients, however, continue to lack access to their own health records and as a result are often disengaged to collaborate on care delivery. They also cannot make informed decisions if they lack access to medical history. Providers and nurse practitioners also have to manually manage their care coordination due to lack of a common Internet space that they can share amongst themselves.

Even in instances where a patient portal is available, the portal is often not friendly or does not present a comprehensible language for patients to understand the health information. For example, HbA1c (Glycated Hemoglobin) is a key indicator for diabetes. By measuring HbA1c, providers get an overall picture of average blood sugar level over periods of weeks/months for a patient. Just seeing HbA1c values may not be sufficient for patient enough. Instead, if the portal could show range of HbA1c values, educate the patient on risk likelihood based on medical history, and project health score by comparing to similar patients, those can be powerful analytics that a patient can truly benefit from. These analytics also provide a motivation for patient to actively engage in health improvements.

Ineffective Use of Health Monitoring Data

The biggest gap is in the effective utilization of the rich health data collected by IoT devices. Data simply resides either in local health apps on patient's smartphones or in the data hubs of tracking devices. Many such use cases exist where a provider could make use of this health data and also engage the patient as a partner and collaborator. Currently, healthcare systems lack the architecture and capability to support such use cases for chronic disease management.

Difficult Patient Experience

Both, the long appointment waits and the patient waits at the actual visit, have been long-standing concern in healthcare. It is literally a painful experience for patients to get timely appointments. Providers have busy schedules and are sometimes booked for months. Patients face a continuous dilemma on how to act in those situations. According to a survey, patients wait an average of 29 days nationally to see a dermatologist for a skin exam, 66 days to have a physical in Boston, and

32 days for a heart evaluation by a cardiologist in Washington (Survey of Patient, 2017). This is alarming for chronic disease patients who seek an appointment much sooner such as for heart exam. Sometimes, patients just need a quick consultation but still have to face long waits.

During the appointment itself, the patient reaches on time but then continues to wait for long in the waiting room or in the assigned room until the provider shows up. Then they proceed to wait for diagnosis or lab tests to be done. Patients go through waits throughout the continuum of care, which leaves them dissatisfied and disengaged.

SOLUTIONS AND RECOMMENDATIONS

This section provides the methodology for building an integrated solution architecture to address the various challenges faced by healthcare systems. It also proposes patient-facing analytical solutions and recommendations to positively transform the quality of lives for patients.

Build a Cloud-Based Enterprise Data Warehouse

Figure 2 proposes a cloud-based Enterprise Data Warehouse (EDW). Hosting EDW in cloud provides a scalable cost-effective analytic platform for healthcare system:

1. EDW integrates data from disparate complex source systems (namely Clinical, Claims, Billing, Supply Chain, Operating Room, Scheduling). ETL job runs on an automated schedule to pull data from on-premise sources into the EDW;
2. External data can also be sourced into cloud. Data is stored in EDW if in relational form, or in JSON files in data lake if semi-structured or unstructured;
3. Data lake also contains non-relational data such as imaging files, PDF, Excel and text documents;
4. IoT data is stored in EDW to be used for data analytics. ETL jobs (for example using Apache Spark, Hadoop or Google Cloud Dataflow) are scheduled to ingest and store health-monitoring data on an automated basis;
5. Big Data engine runs against the data from both EDW and data lake to generate insightful benchmarks, patterns and trends. It also sets up the foundation for richer data analytics such as for the recommendation systems and health alerts;
6. Analytical solutions are proposed for both the healthcare team and the patients. The analytics can be delivered through content delivery systems of convenience such as web applications and mobile apps. Analysts can use the integrated data platform to perform complex analytics instantly at their fingertips.

Figure 2. Cloud-based EDW

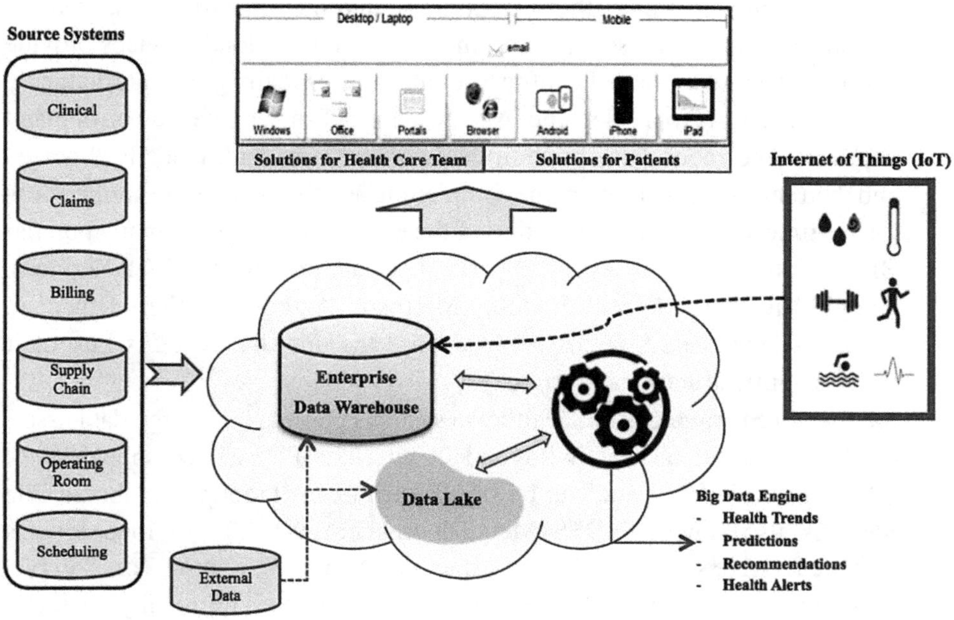

Advantages of Cloud-Based EDW

Hosting the data warehouse in cloud provides a multitude of advantages for healthcare systems:

1. **Significant Infrastructure Cost Savings:** Healthcare systems are under constant pressure to reduce cost of care. Cloud computing rightly fits into this business goal. Firstly, hospitals can avoid huge capital investments. Secondly, cloud solutions give a quick ramp to setup the technical infrastructure. Thirdly, it provides a flexible charge model driven by resource utilization (Zhang, Qi et al., 2017). Depending on the advancements and maturity of healthcare technical infrastructure, they can opt for Infrastructure, Platform or Software as a Service (IaaS, PaaS, SaaS respectively). With SaaS solutions, the responsibility of infrastructure shifts to cloud service provider (Sarkar, Banerjee et al., 2017) and the healthcare systems can turn their focus on building analytical solutions;
2. **Operational Reliability:** Cloud service providers guarantee an extremely high reliability and performance (Leopold, 2015). Healthcare systems can, therefore, focus on building sophisticated analytical solutions instead of bothering about system reliability;

3. **Scalability:** The data can grow large very quickly when a healthcare system is maintaining lives of millions of patients, and they integrate data from all systems along with personal health monitoring data. Cloud services provide scalability options where the infrastructure can elastically grow automatically to adjust or shrink to the demand. Also, when Big Data engines run on demand to process enormous data, the additional resources are automatically allocated and then shrunk back after engine completion, and the charges are applied only for the time and resources used during the execution. There are many popular BDE solutions such as Apache Spark, Big Data on Amazon Web Services, Google BigQuery, Apache Hadoop, Microsoft Azure and others (Imanuel, 2017). A healthcare system should choose the one that best fits into their technical infrastructure and strategy;
4. **Savings in Storage Costs:** Healthcare systems possess very large data assets (multiple source databases, X-Ray and imaging data, PDFs, Excel spreadsheets, XML, genomic files) and can be stored such as in Hadoop Distributed File Systems, also known as HDFS (Matallah, Belalem et al., 2017). Cloud services provide flexible storage options at a fraction of on-premise cost. For example, most frequently accessed data can be kept in reliable and readily available storage tier, whereas, the less frequently used data can be archived in much cheaper storage tiers;
5. **Ease of Access to Centralized Data:** Cloud solutions, by using encryption, provide secured access to data from anywhere on the planet. The study by (Bhardwaj, Subrahmanyam et al., 2016) illustrates cryptographic algorithms for Cloud based applications. Secured access to data is highly important for healthcare model. Cloud solutions can readily empower patients to both own and share their data via cloud data platform. The entire data resides in a centralized warehouse (a single source of truth) that users of various roles (hospital users, providers, care managers, payers, and patients) get to access;
6. **Data Security and Privacy:** Cloud platform is compliant to provide secure and protected access to Protected Health Information (PHI) as per HIPAA regulations. Users, based on their roles and privileges, only get to see what they are allowed to access. For instance, a practice manager will get to see data for the patients in own practice. Provider can see data only for the patients they are responsible for. A patient can only see own medical history. Chief medical officer, on the other hand, can see data of all patients. In case if a user loses the private key, the Cloud solutions can provide an authentication mechanism on the portal for user to generate a new key pair to regain the access;
7. **Ease of Integration with Other Cloud Computing Services:** There are many cloud based offerings and capabilities that healthcare system can integrate with. For instance, Centers for Disease Control and Prevention provides cloud-based

Epi Info ecosystem (Epi Info for Web, 2017) to promote creative healthcare solutions and to provide a highly scalable platform for data collection, analysis and visualization.

While building the data pipeline, the following guiding principles promote the building of the healthcare EDW in a timely fashion and can minimize the cascading impact in end-to-end data flow when something changes upstream in the source systems.

1. **Early Binding of Core Entities:** The common entities that overlap across source systems (e.g. patient, provider, payer, medical coding systems) must be managed upstream in the pipeline for ease of integration. For example, majority of medical data is attached to a patient (receiving the service) and a provider (offering the service). It is essential that a patient record is unique across source systems. If the patient is managed individually by discrete systems (say clinical and billing), then a master data management (MDM) technique must be applied to ensure the duplicate patient records are merged into a single patient record. Patient demographics (name, date of birth, gender, address, phone, email) can be used to identify duplicate patients using a combination of matching algorithms and the business rules. Similar approach must be followed for all overlapping entities such as provider and payer. Code system standardization must also occur upstream so that the coding methodology is consistent across systems. A master reference database should be managed for core entities so that all source systems simply read and write on this shared database. This will naturally enforce a single source of truth and help create a high quality data asset.
2. **Late Binding of Business Rules:** Healthcare business rules are complex. The rules can also be use case specific and may not be a one-size-fits-all set of business rules. Further, due to ever changing medical coding systems, it is important to strategically incorporate volatile business rules later in the pipeline so that rule adjustments has minimal disruption to overall data flow. This concept is also called late binding and ensures that the business rules are aligned and conformed to standards (Dealy, 2016). EDW can still publish the rule-driven enrichments via API calls to any subscriber source systems. For example, the business rule to calculate readmissions using patient encounters or to identify Sepsis cases using diagnosis codes should be implemented in EDW and can then shared with source systems that require it.

EHR is the new standard for medical records and focuses on total care of patient. It brings the digital solution for every aspect of medical practice and can connect to entire service network. According to HealthIT (Electronic Health, 2016), physician's use of EHR/EMR systems increased from 18% in 2001 to 57% in 2011. With EHR, electronic patient records, lab orders, claims, prescriptions, and referral orders can seamlessly travel electronically across hospitals, pharmacies, laboratories, and insurance companies.

(Weldon, 2017) illustrates one such successful story from St. Luke's University which was able to consolidate, master and integrate data from its seven hospitals and more than 270 outpatient sites with the aim of improving population health and reducing the costs.

Integrate With IoT Health Data

IoT allows patients to record and share information in Internet connected world (Dey, Ashour et al., 2017). With patient's consent, healthcare systems can fetch patient's fitness activities into EDW for greater data analytics capabilities (Figure 2). Primary care providers (PCP) can subsequently access this wealth of health data, which was never available before, for the analysis and health tracking. They can study the physical activities, blood pressure, heart rate, sleep pattern, diet and life style, and look at its correlation to patient's existing medical conditions. Integrating patient's real-life health data with electronic health record (EHR) opens many doors of opportunities to streamline patient health analytics and to more accurately predict and assess health conditions.

Many medical systems have already started providing patient portal applications where patient can easily sync their health data from wearable devices, and data then becomes available in EDW. Healthcare facilities can also subscribe to IoT data hubs to get the health data via Representational State Transfer (REST) APIs. REST calls allow to access data using a uniform and predefined set of operations. For example, Fitbit can push the data to hospitals whenever the data is synced or updated during the day from patient's wearable fitness trackers to Fitbit. Fitbit can also send a POST call to notify subscribers that the data has been changed and the subscriber (in this case the healthcare facility) can then invoke REST API calls to capture patient fitness details.

IoT revolutionizes the approach to patient care. Patient medical history and personal health data can be made available to providers prior to patient visit. Having complete patient data allows the provider to study the data beforehand. This way provider can focus on meaningful diagnosis and treatment plan instead of repeatedly asking and learning about medical history and life style details.

Invest into Big Data Analytics

As shown in Figure 2, Big Data analytics are potential powerful solutions to build from cloud-based EDW that combines both EHR and IoT personal health monitoring data (Big Data, 2015). This powerful combined data asset can uncover meaningful analytics and trends, never seen otherwise (Kamal, Ripon et al., 2016). The chapter subsequently proposes analytical solutions that can serve as the backbone of rich cloud-based data platform.

Generate Similar Patient Groups

Patient similarity evaluation can predict clinical affinity and diagnostic proximity of patients (Wang, 2015). Following algorithm proposes to detect similar patients based on their age group, gender and chronic conditions (Figure 3). Here the total number of groups is limited by finite data elements (few age groups, two genders, finite chronic conditions). Algorithm is further scalable to add other influential variables such as race and ethnicity. A sicker patient with multiple chronic conditions will belong to multiple patient groups, and therefore, indicating a higher risk.

Mining data by patient groups produces more relevant analytics for patients (Jensen, Jensen & Brunak, 2012). A patient can readily see his relative health performance to see where he falls in the health scale compared to the benchmark. Data can also be mined to see which patients are doing better on reducing chronic

Figure 3. Similar Patient Algorithm

Age Group
0-18
18-30
30-40
40-50
...

Chronic Condition
Diabetes
Health Disease
Hypertension
Asthma
...

Gender
Male
Female

Patient Group	Combination
Patient Group 1	Diabetes, Male, 0-18
Patient Group 2	Diabetes, Female, 0-18
Patient Group 3	Diabetes, Male, 18-30
..	
...	

illness and how it correlates to their health monitoring data (fitness activity, stress, sleep, diet). The concept of similar patients also creates a natural and unique emotional connection when patients realize that they are being compared to similar patients in a community setting.

Mine Data to Discover Performance Benchmarks

Healthcare performance is measured in three aspects, firstly the imperative to control high cost of care, secondly to meet the desired goals to improve the quality of care, and thirdly to improve patient satisfaction.

Discovering meaningful insights require rich benchmarks. Benchmarking is a well-defined management concept and is widely used in business industries to set up goals for best practices at best cost. The key feature of benchmarking is its integration within a comprehensive and participatory policy of continuous quality improvement (CQI) (Ettorchi-Tardy, 2012).

Healthcare benchmarks can help lead the quality improvement efforts. By creating performance benchmarks, hospitals can compare two or more peers to accurately detect the improvement opportunities. Healthcare leaders can define strategic goals to assess peer performances such as by departments, providers, medical procedures or diagnosis. Medical leaders can establish Key Performance Indicators (KPI) metrics, such as reduction in readmission rate or percentage of quality measures meeting CMS criteria, to better manage quality of care and track it against pay-for-performance results. Having a clean and standardized data asset is essential to produce high quality benchmarks and EDW helps fill in this gap.

Below are some examples of benchmarks that can be generated through the Big Data engine and subsequently be incorporated in hospital settings. Apache Spark is one such powerful engine with ability to develop applications in multiple languages (such as Java, Scala, Python, R) to read and process large volumes of data. Spark also provides SQL and DataFrames capabilities to easily embed within same applications. Data from EDW can be loaded into Spark memory to process and generate insightful benchmarks. Hadoop and Google BigQuery can be some of the other options.

Internal Quality Benchmarks

Internal benchmarks are generated by using medical data within same hospital systems. It can also be rapidly rolled out across all healthcare facilities to evaluate and monitor the performance of peers.

Example 1: Benchmarks against various metrics such as length of stay, inpatient and outpatient costs, direct versus indirect costs, total CDM costs and charges, risk of mortality level, severity level, complication of patient care are generated from EDW. These comparative metrics provide medical specificity in scorecards to indicate where the quality of care and outcomes can be improved.

Example 2: Patient readmissions is a key focus area where benchmarking can play a huge role. Big Data engines can run algorithms on entire patient encounter database to create a comparable readmission rate based on patients admitted with similar risks. Similar patient algorithm (Figure 3) can be used to accurately measure the readmissions based on chronic diseases. Actual readmission rate when compared with benchmarks gives transparency into performance. It also gives projections into potential financial loss due to excess readmissions that could have been avoided otherwise. Providers can dig into specific patient groups or medical conditions if it has correlations with readmission rate. Providers can look at different service lines, for example Cardiac Services, General Surgery, Neurosurgery, and Obstetrics, and analyze if low quality of service in a particular service line is resulting in majority of patient readmissions.

Example 3: Benchmarks open up endless opportunities to look at complex and expensive health issues such as opioid crisis. Opioid was recently declared a national epidemic in the United States. Hospitals can study opioid cases based on performed diagnosis and prescribed drugs and also can identify providers who have higher Opioid cases compared to average.

Example 4: Sepsis benchmarks can help analyst to target specific medical conditions where patients are being infected more than average and which hospital settings require in-depth analysis. For example, benchmarks can help discover if hospital is non-compliant on protocols resulting in higher septicemia cases. Timely diagnosis and tests is critical for sepsis cases. Benchmarks could detect such bottlenecks, for example data could show if Symptomatic Inflammatory Response Symptom (SIRS) criteria were being followed regarding potential infections.

External Quality Benchmarks

Using the Agency for Healthcare Research and Quality (AHRQ) logic, these benchmarks provide an analysis of patient safety and quality measures well in advance of any public reporting (Healthcare Performance, 2017).

Patient Satisfaction Scores

This comparative analysis combines patient encounters with data from patient satisfaction survey to link patient experience to overall outcome and cost of care.

Supply Chain Benchmarks

Healthcare supply chain consumes nearly one third of hospital operating budget and continues to be under tremendous pressure to cut costs. It is another complex domain in which tens of thousand of vendors, distributors and manufacturers participate are selling millions of products to the health systems. Procurement department need insights into item prices to avoid overpayments. Using EDW, health systems can generate price benchmarks and proactively use it to identify pricing opportunities and subsequently to negotiate with vendors for price reductions.

Generate Medical Recommendations

The entire healthcare data is bucketed by similar patient groups to generate recommendations for individual patients who are suffering from chronic illness. For instance, it can detect missing diagnoses or lab tests by comparing patient medical history with similar patients. This is essential for patients who are suffering from chronic conditions and require more frequent check ups and health monitoring. Data within similar patient groups (for instance, same age group and gender but different chronic conditions) can be mined for correlations among multiple chronic diseases to proactively inform patients of likelihood of obtaining other chronic illness and to help them prevent or mitigate by taking timely health actions.

Healthcare system can also publish anonymized aggregated data to public entities such as Government agencies, which accumulate data from all healthcare systems, mine and broadcast wider population health trends and chronic disease situations.

Example 5: Chronic Obstructive Pulmonary Disease (COPD) is a progressively life-threatening lung disease that causes breathlessness (initially with exertion) and predisposes to exacerbations and serious illness (Chronic obstructive, 2016). Respiratory issues are caused by exposure to tobacco smoke (active or passive), by indoor and outdoor air pollution, and by dusts and fumes. COPD is also the third leading cause of death in the United States and is an extremely costly disease, both in its actual medical costs and the impairments that limit patients with this illness (Ni & Xu, 2016) Many IoT solutions provide GPS-enabled inhalers such as ("Asthmapolis Relaunches", 2013). Various data elements about inhaler such as its usage, symptoms and triggers, time and location of use are shared with healthcare systems. This data, when combined with patient medical history, can be powerful for risk analytics. Providers and care managers can more accurately predict asthma level of COPD patients. For example, based on real time pollution and seasonal allergy data, providers can provide recommendations for proper care.

PROVIDE ANALYTICAL SOLUTIONS FOR PATIENTS

Provide Access to Individual Medical Data

Healthcare organizations must engage patients to work together for better quality of care. To engage the patients, a healthcare system must be able to firstly provide patients full access to their data. The content delivery system can be a web portal or a mobile app. A patient can look at medical history, can look at risk scores based on own chronic diseases, and can be motivated to advices such as on drug adherence and life style changes. They are also less likely to miss on appointments. Patients can also monitor trends over time of their quality of health. For instance, if a diabetic patient sees a continuous rise in HbA1c over last six months, this can raise a health alarm for him to consciously take corrective actions before it gets worse.

Provide Personalized Care

Another usage of this data is that patients can receive personalized messages and videos based on their medical conditions and can interact with providers using the portal. System can also share population health trends with patients via emails and text messages. PCPs can advise patients to track against specific health goals where IoT devices help monitor against those targets. Based on best practices, advisory solutions can even be built to automatically alert patients based on real time events and the life style.

Example 6: If a patient is diabetic with high HbA1c level, he or she could be provided a risk assessment by comparing with diabetic population. Providers can proactively reach these risky patients to have them take preventive measures. Patient can further receive automated custom messages based on HbA1c lab results with proper directions on what actions to take.

Example 7: Providers can view food diary from patient fitness trackers and provide suggestions such as good versus bad food based on patient health conditions. They can notify such as by giving thumbs up for healthy choices and thumbs down for unhealthy choices. Here, the analytics depends on patient's ability to maintain an accurate food diary.

Integrate With Remote Patient Monitoring

Medical visits continue to be a painful and lengthy (besides being a costly process). Instead in smart cities, patients can share health-monitoring data through IoT solutions (Jain & Bhatnagar, 2017). IoT solutions store the data in cloud. Data is then accessed by healthcare systems. A patient suffering from a chronic disease is less likely to develop complications if provider can detect risks and recommend early diagnosis or offers timely treatment plans. IoT also enables a preventive approach to continuum of care by forging a strong provider-patient partnership.

Timely Response to Emergency

IoT can help benefit both sick as well as healthy patients. For example, fall risk in the elderly is a common chronic disease and is a frequent source of injury. It is projected that 20% of United States population will be over the age of 65 by 2030 (Fueled by, 2014). Elderly are eager to continue maintaining their independent lifestyle. However, due to age and health, at least one in three falls every year. Fall injury also imposes sizable expenses (Smits & Ryter, 2015). Using patient's health data, providers can proactively offer caution of chronic falls risk. Big Data can uncover hidden trends such as fall risk based on medical conditions and lifestyle (for instance, a senior citizen staying alone, some chronic conditions, particular medications). If a fall incident occurs, the patient can immediately report it to the emergency and to the PCP. PCP can instantly look at entire medical history to recommend appropriate actions.

Another use case can be PCP's strategic response to anomalies in health data. Sometimes patient might feel normal but underlying health data might raise alarms. For example, if the tracker reports patient heart rate at unusually low rate (Bradycardia) or abnormally raised heart rate (Tachycardia), provider can promptly initiate a communication via call or text message or telehealth to ensure it is not life threatening. Similarly, if patient's blood pressure increases alarmingly, the tracker device can send a real time alert to the provider who can then take appropriate action to reach out and administer the patient to bring down the blood pressure. This data is also logged in EDW for subsequent analytics and trends.

Telehealth

In smart cities enabled with modern technologies, providers can also offer telehealth visits over in-person appointments for minor check-ups and follow-ups. This can bring a huge efficiency by saving time for both provider and patient. For frequent consultations, patients need not visit the medical office but instead can consult the provider and hospital staffs over video talk, chat or phone. With telehealth, patients can easily interact with providers on health alerts, advices and medication questions. Video talks also provide audio- visual experience similar to the office setting. Telehealth options can also be immensely helpful for patients who are just getting discharged from hospitalization and thus need a close monitoring at home.

Predictable Scheduling and Waits

- **Scheduling Appointments:** Providers can offer empty appointment slots online for patients to book appointments at their convenience. This will provide a stress-free appointment experience for patients. It can also result in fewer open spots for medical practice resulting in better revenue. If a patient with relatively routine chronic disease treatments cannot see the provider on time, it only results in more expensive treatments otherwise;
- **Predictable Wait Time:** With Big Data solutions, healthcare organization can easily track wait times. Historical wait times help rate the providers on a scale of patient experience. Furthermore, a predictable wait time relaxes the sick patient who otherwise is eager to see the provider. Historical visits can also be mined to discover hidden trends such as timeslots with high wait times. Also, if appointment for specific patient age group or a chronic disease typically takes longer, then predictive wait time can factor in these variables for higher accuracy. Predictive wait times can further help chief nursing officer in resource planning to ensure that enough nursing staffs are available during busier time slots. Hospital system at Assistance Publique-Hôpitaux de Paris is one such success story, which used data from various sources to come up with daily and hourly predictions of how many patients are expected at each hospital and accordingly plan for optimal staffing (Nine Examples, 2017).

Create a Feedback Loop

Patients can provide real time feedback to providers about the quality of care for a specific visit. Feedback system can be extensively designed to take multiple variables such as office cleanliness and comfort, staff courteousness, wait times, level of trust in provider's decisions, provider's attention to details and empathy towards patient. Patient can also provide comments for improvements. This feedback data loops into EDW and generates real time analytics about hospital quality of care and patient satisfaction. Higher the patient satisfaction score, the more engagement and the continuation of business from the patient are expected.

IMPLEMENTATION STRATEGY

Due to complexity in data, multiple disparate systems, and number of parties involved, it can be a daunting project for healthcare system to implement an end-to-end cloud EDW solution. It is important to layout a strategic roadmap to accomplish the entire vision in multiple phases using agile software development methodology. The phased approach provides a continued success watermark and also ensures that healthcare model is incrementally scaled to add new data feeds (IoT, third party data such as external benchmarks) over time. By adopting agile principles, healthcare system can make rapid iterative progress and be flexible to design changes without needing to build time-consuming lengthy requirements and entire solution design in advance.

The following implementation approach provides the details on implementing a cloud-based data platform using IoT:

1. Setup a cloud-based data platform that supports HIPAA compliance (such as in Amazon AWS, Microsoft Azure, Google Cloud);
2. Define specifications for early binding of core entities and implement them upstream prior to EDW. Implement business rules after integrating data in EDW;
3. Implement first version of cloud-based EDW (and continue over phases to expose more data from source systems);
4. Set up data security framework: This ensures that users will access authorized data from EDW in a secured and protected manner;
5. Open up EDW for both the healthcare team and the patients (via web portal and mobile apps);
6. Start collecting IoT data via APIs (fitness data, sleep cycle, health monitoring data);
7. Build Big Data engine to start generating analytic capabilities for hospitals (such as physician performance, cost optimization opportunities, quality improvement opportunities);

8. Build advanced predictive analytics using Big Data engine (such as using Apache Spark or Hadoop):
    a. Identify similar patients;
    b. Detect chronic disease trends;
    c. Generate performance benchmarks;
9. Expose advanced analytics to healthcare team and the patients;
10. Build APIs to publish anonymized healthcare trends to external parties such as payers or CDC to enable them to discover long-term health trends across entire population. Before sharing publicly, data should be anonymized to remove Protected Health Information (PHI) of individual patients.

FUTURE RESEARCH DIRECTIONS

The authors plan to further explore the IoT integration into cloud-based enterprise data warehouse for healthcare systems. IoT can generate data at an astonishing rate that IoT companies need to account for. Healthcare systems should also look for scalable design when integrating IoT data with pre-existing large healthcare data. If different IoT devices define different data standards, the data can stay locked within the device in siloes. Instead, it is important for various IoT devices to store data using common standards so it can be seamlessly shared and integrated with existing medical records. With multiple platforms, different data structures, and different API protocols, it requires a standardization to integrate medical data with monitoring data. Also, patient privacy is of utmost important, and therefore all IoT devices, cheap or expensive, need to comply with HIPAA regulations to be able to share data with third parties. Compliance agreements need to be properly established as well so healthcare systems and devices have trustworthy relationship to share data with each other.

CONCLUSION

In Value-based care model, healthcare systems are under tremendous pressure to reduce the cost of care and to improve the quality of patient outcomes. They continue to battle due to lack of insights into comprehensive view of patient health. With billions of smart devices already in use, the Internet of Things has seen a rapid acceleration. IoT is also proving to play a pivotal role in health monitoring due to its low-cost infrastructure, Internet connectivity, and by allowing the ease of access to health data from anywhere. The chapter has sought to first highlight the complexity and challenges in healthcare systems. The chapter then proposed cloud-based solutions to build an enterprise data platform that integrates with IoT devices. Finally, Big Data solutions and smart applications are presented for both healthcare systems and for patients. The chapter concludes by discussing great potentials of IoT. It also illustrates new research challenges faced by rapid growth of multitude of smart devices and the potentials for Big Data analytics with potential integration with healthcare data.

REFERENCES

Ali, A. H., Atia, A., & Mostafa, M. S. M. (2017). Recognizing driving behavior and road anomaly using smartphone sensors. *International Journal of Ambient Computing and Intelligence*, *8*(3), 22–37. doi:10.4018/IJACI.2017070102

Asthmapolis. (2013, September). Asthmapolis Relaunches as Propeller Health to Advance Broader Respiratory Mission [Press Release]. Retrieved from https://www.propellerhealth.com/2013/09/10/asthmapolis-relaunches-as-propeller-health-to-advance-broader-respiratory-mission/

Bhardwaj, A., Subrahmanyam, G. V. B., Avasthi, V., & Sastry, H. (2016). Security algorithms for cloud computing. *Procedia Computer Science*, *85*, 535–542. doi:10.1016/j.procs.2016.05.215

Bhatt, C., Dey, N., & Ashour, A. S. (Eds.). (2017). Internet of things and big data technologies for next generation healthcare. Springer. doi:10.1007/978-3-319-49736-5

HIMSS Asia Pacific. (n.d.). Big Data Improves Chronic Disease Management. Retrieved from http://www.himssasiapac.org/content-library/exclusive-articles/big-data-improves-chronic-disease-management

Boccuti, C., & Casillas, G. (2017, March). Aiming for fewer hospital U-turns: The medicare hospital readmission reduction program. KFF. Retrieved from https://www.kff.org/medicare/issue-brief/aiming-for-fewer-hospital-u-turns-the-medicare-hospital-readmission-reduction-program/

Center for Disease Control and Prevention (CDC). (2014). Morbidity and Mortality Weekly Report. Retrieved from https://www.cdc.gov/mmwr/preview/mmwrhtml/su6304a2.htm

Center for Disease Control and Prevention (CDC). (2017). Sepsis: Data & Reports. Retrieved from https://www.cdc.gov/sepsis/datareports/index.html

Dealy, G. (2016). The importance of business rules in actionable healthcare data. *Health IT Outcomes*. Retrieved from https://www.healthitoutcomes.com/doc/the-importance-of-business-rules-actionable-healthcare-data-0001

Dey, N., Ashour, A. S., & Bhatt, C. (2017). Internet of things driven connected healthcare. In *Internet of things and big data technologies for next generation healthcare* (pp. 3–12). Cham: Springer. doi:10.1007/978-3-319-49736-5_1

Dey, N., Hassanien, A. E., Bhatt, C., Ashour, A., & Satapathy, S. C. (Eds.). (2018). *Internet of Things and big data analytics toward next-generation intelligence*. Springer. doi:10.1007/978-3-319-60435-0

Dickson, B. (2015). Why IoT security is so critical. *Techcrunch*. Retrieved from https://techcrunch.com/2015/10/24/why-iot-security-is-so-critical/

HealthIT.Gov. (2016). Electronic Health Records Infographic. Retrieved from https://www.healthit.gov/patients-families/electronic-health-records-infographic

Elhayatmy, G., Dey, N., & Ashour, A. S. (2018). Internet of Things based wireless body area network in healthcare. In *Internet of Things and Big Data Analytics Toward Next-Generation Intelligence* (pp. 3–20). Cham: Springer. doi:10.1007/978-3-319-60435-0_1

Elliott, T. E., Holmes, J. H., Davidson, A. J., La Chance, P.-A., Nelson, A. F., & Steiner, J. F. (2013). Data warehouse governance programs in healthcare settings: A literature review and a call to action. *eGEMs. The Journal for Electronic Health Data and Methods*, *1*(1), 1010. PMID:25848561

Centers for Disease Control and Prevention (CDC). (2017). Epi Info for Web & Cloud. Retrieved from https://www.cdc.gov/epiinfo/cloud.html

Ettorchi-Tardy, A., Levif, M., & Michel, P. (2012). Benchmarking: A method for continuous quality improvement in health. *Healthcare Policy*, *7*(4), 101–119. PMID:23634166

Evans, D. (2011). The Internet of Things: how the next evolution of the Internet is changing everything. *Cisco Systems*. Retrieved from https://www.cisco.com/c/dam/en_us/about/ac79/docs/innov/IoT_IBSG_0411FINAL.pdf

Evans, R. S., Lloyd, J. F., & Pierce, L. A. (2012). Clinical use of an enterprise data warehouse. *AMIA ... Annual Symposium Proceedings* (pp. 189–198). PMID:23304288

Farooq, M. U., Waseem, M., Khairi, A., & Mazhar, S. (2015). A critical analysis on the security concerns of Internet of Things (IoT). *International Journal of Computers and Applications*, *111*(7), 1–6. doi:10.5120/19547-1280

Rios, M. (2014Fueled by Aging Baby Boomers, Nation's Older Population to Nearly Double in the Next 20 Years. *Census Bureau Reports*. Retrieved from https://www.census.gov/newsroom/press-releases/2014/cb14-84.html

Gale, L., Burdick, J., Wagstaff, P., Flowers, S., Plott, P. C., Furlong, D., & Daniels, M. (2013). The enterprise data warehouse: Moving from retrospective to proactive analytics. *Journal of Healthcare Information Management*, *27*(3), 38–44.

Gandomi, A., & Haider, M. (2015). Beyond the hype: Big data concepts, methods, and analytics. *International Journal of Information Management*, *35*(2), 137–144. doi:10.1016/j.ijinfomgt.2014.10.007

Hay, S. I., Abajobir, A. A., Abate, K. H., Abbafati, C., Abbas, K. M., Abd-Allah, F., ... Aboyans, V. (2017). Global, regional, and national disability-adjusted life-years (DALYs) for 333 diseases and injuries and healthy life expectancy (HALE) for 195 countries and territories, 1990–2016: A systematic analysis for the Global Burden of Disease Study 2016. *Lancet*, *390*(10100), 1260–1344.

Health Expenditures, C. D. C. (2017). *Center for Disease Control and Prevention*. Retrieved from https://www.cdc.gov/nchs/fastats/health-expenditures.htm

Healthcare Performance Benchmarks. (2017). *McKesson*. Retrieved from http://www.mckesson.com/providers/health-systems/enterprise-intelligence/healthcare-performance-benchmarks/

High Blood Pressure Fact SheetC. D. C. (2016). Retrieved from https://www.cdc.gov/dhdsp/data_statistics/fact_sheets/fs_bloodpressure.htm

Imanuel. (2017). Top 50 Big Data Platforms and Big Data Analytics Software. *Predictive Analytics Today*. Retrieved from https://www.predictiveanalyticstoday.com/bigdata-platforms-bigdata-analytics-software/

Islam, S. M. R., Kwak, D., Kabir, M. H., Hossain, M., & Kwak, K. S. (2015). The Internet of Things for health care: A Comprehensive Survey. *IEEE Access: Practical Innovations, Open Solutions*, *3*, 678–708. doi:10.1109/ACCESS.2015.2437951

Jain, A., & Bhatnagar, V. (2017). Concoction of ambient intelligence and big data for better patient ministration services. *International Journal of Ambient Computing and Intelligence*, *8*(4), 19–30. doi:10.4018/IJACI.2017100102

Jensen, P., Jensen, R., & Brunak, S. (2012). Mining electronic health records: Towards better research applications and clinical care. *Nature Reviews. Genetics*, *13*(6), 395–405. doi:10.1038/nrg3208 PMID:22549152

Jesdanun, A. (2015). Challenges for doctors using fitness trackers and apps. *Science*, *10*. Retrieved from https://phys.org/news/2015-02-doctors-trackers-apps.html

Kamal, M. S., Dey, N., & Ashour, A. S. (2017). Large scale medical data mining for accurate diagnosis: a blueprint. In *Handbook of Large-Scale Distributed Computing in Smart Healthcare* (pp. 157–176). Cham: Springer. doi:10.1007/978-3-319-58280-1_7

Kamal, S., Ripon, S. H., Dey, N., Ashour, A. S., & Santhi, V. (2016). A MapReduce approach to diminish imbalance parameters for big deoxyribonucleic acid dataset. *Computer Methods and Programs in Biomedicine*, *131*, 191–206. doi:10.1016/j.cmpb.2016.04.005 PMID:27265059

Kayyali, B., Knott, D., & Kuiken, S. V. (2013). The big-data revolution in US health care: Accelerating value and innovation. *McKinsey & Company*. Retrieved from https://www.mckinsey.com/industries/healthcare-systems-and-services/our-insights/the-big-data-revolution-in-us-health-care

Kilbridge, P. (2015). Introducing the acute care patient portal. *Advisory Board Company*. Retrieved from https://www.advisory.com/research/health-care-it-advisor/research-notes/2015/introducing-the-acute-care-patient-portal

Kohler, J., Lorenz, R., Gumbel, M., Specht, T., & Simov, K. (2017). A security by distribution approach to manage big data in a federation of untrustworthy clouds. In Privacy and Security Policies in Big Data. Hershey, PA: IGI-Global.

Kohler, J., Simov, K., & Specht, T. (2015). Analysis of the join performance in vertically distributed cloud databases. *International Journal of Adaptive, Resilient and Autonomic Systems*, *6*(2), 65–87. doi:10.4018/IJARAS.2015070104

Kohler, J., & Specht, T. (2014, November). Vertical query-join benchmark in a cloud database environment. In *2014 Second World Conference on Complex Systems (WCCS)* (pp. 581-586). IEEE. 10.1109/ICoCS.2014.7060950

Krebs, P., & Duncan, D. T. (2015). Health App Use Among US Mobile Phone Owners: A National Survey. *JMIR mHealth and uHealth*, *3*(4), e101. doi:10.2196/mhealth.4924 PMID:26537656

Landi, H. (2016). The business case for increasing patient portal adoption. *Healthcare Informatics*.

Laney, D. (2001). 3D Data Management: Controlling Data Volume, Velocity, and Variety. *Gartner*. Retrieved from http://blogs.gartner.com/doug-laney/files/2012/01/ad949-3D-Data-Management-Controlling-Data-Volume-Velocity-and-Variety.pdf

Lee, C. H., & Yoon, H.-J. (2017). Medical big data: Promise and challenges. *Kidney Research and Clinical Practice*, *36*(1), 3–11. doi:10.23876/j.krcp.2017.36.1.3 PMID:28392994

Leitheiser, R. L. (2001). Data quality in health care data warehouse environments. In *Proceedings of the 34th Annual Hawaii International Conference on System Sciences*, Maui, HI (pp. 1-10). 10.1109/HICSS.2001.926576

Leopold, G. (2015). AWS rates highest on cloud reliability. *Enterprise Tech*. Retrieved from https://www.enterprisetech.com/2015/01/06/aws-rates-highest-cloud-reliability/

Manyika, J., & Chui, M. (2015). By 2025, Internet of things applications could have $11 trillion impact. *McKinsey & Company*. Retrieved from https://www.mckinsey.com/mgi/overview/in-the-news/by-2025-internet-of-things-applications-could-have-11-trillion-impact

Matallah, H., Belalem, G., & Bouamrane, K. (2017). Towards a new model of storage and access to data in big data and cloud computing. *International Journal of Ambient Computing and Intelligence*, *8*(4), 31–44. doi:10.4018/IJACI.2017100103

Mayer-Schönberger, V., & Cukier, K. (2013). Big Data: A revolution that will transform how we live, work, and think. *American Journal of Epidemiology*, *179*(9), 1143–1144.

Medicare hits hospitals with readmission penalties; does the regulation need to be tweaked? (2014). *Real Time Medicare Data*. Retrieved from http://rtmd.org/topics/2014/11/medicare-hits-hospitals-with-readmission-penalties-does-the-regulation-need-to-be-tweaked/#.WfQixxNSyqB

Misra, S. M. D. (2015). New report finds more than 165,000 mobile health apps now available, takes close look at characteristics & use. *iMedicalApps*. Retrieved from https://www.imedicalapps.com/2015/09/ims-health-apps-report/

Mohamed, K., & Zabani, I. (2016). Utilizing health analytics in improving the performance of healthcare services: A case study on a tertiary care hospital. *Journal of Infection and Public Health*, *9*(6), 757–765. doi:10.1016/j.jiph.2016.08.016 PMID:27663517

Murdoch, T. B., & Detsky, A. S. (2013). The inevitable application of big data to health care. *Journal of the American Medical Association, 309*(13), 1351–1352. doi:10.1001/jama.2013.393 PMID:23549579

National Center for Chronic Disease Prevention and Health Promotion. (2017). Retrieved from https://www.cdc.gov/chronicdisease/overview/index.htm

Ni, H., & Xu, J. (2016). COPD-related Mortality by Sex and Race Among Adults Aged 25 and Over: United States, 2000–2014. *NCHS Data Brief, 256*, 1–8. PMID:27648679

Nine Examples of Big Data Analytics in Healthcare That Can Save People. (2017). *Elogic Square*. Retrieved from http://elogicsquare.com/blog/9-examples-of-big-data-analytics-in-healthcare-that-can-save-people/

Pennic, F. (2015). Novant Health To Integrate Epic MyChart with FitBit & HealthKit. *HIT Consultant*. Retrieved from http://hitconsultant.net/2015/08/21/novant-health-to-integrate-epic-mychart-with-fitbit-healthkit/

Press, G. (2014). A Very Short History Of The Internet Of Things. *Forbes*. Retrieved from https://www.forbes.com/sites/gilpress/2014/06/18/a-very-short-history-of-the-internet-of-things/3/#52cb61da350a

Raghupathi W, & Raghupathi V. (2014). Big data analytics in healthcare: promise and potential. *Health Information Science and Systems, 2*(3).

Rahman, M., Carbunar, B., & Topkara, U. (2016). Secure management of low power fitness trackers. *IEEE Transactions on Mobile Computing, 15*(2), 447–459. doi:10.1109/TMC.2015.2418774

Rao, R. V., & Selvamani, K. (2015). Data security challenges and its solutions in cloud computing. *Procedia Computer Science, 48*, 204–209. doi:10.1016/j.procs.2015.04.171

Ren, K., Wang, C., & Wang, Q. (2012). Security challenges for the public cloud. *IEEE Internet Computing, 16*(1), 69–73. doi:10.1109/MIC.2012.14

Rodrigues, J. (2010). Health Information Systems: Concepts, Methodologies, Tools, and Applications (Vol. 1). Hershey, PA: IGI Global. doi:10.4018/978-1-60566-988-5

Rodrigues, J., de la Torre, I., Fernández, G., & López-Coronado, M. (2013). Analysis of the security and privacy requirements of cloud-based electronic health records systems. *Journal of Medical Internet Research, 15*(8), e186. doi:10.2196/jmir.2494 PMID:23965254

Rosenthal, E. (2014). The Health Care Waiting Game. Long Waits for Doctors' Appointments Have Become the Norm. *The New York Times*. Retrieved from https://www.nytimes.com/2014/07/06/sunday-review/long-waits-for-doctors-appointments-have-become-the-norm.html

Rumsfeld, J. S., Joynt, K. E., & Maddox, T. M. (2016). Big data analytics to improve cardiovascular care: Promise and challenges. *Nature Reviews. Cardiology*, *13*(6), 350–359. doi:10.1038/nrcardio.2016.42 PMID:27009423

Sarkar, M., Banerjee, S., Badr, Y., & Sangaiah, A. K. (2017). Configuring a trusted cloud service model for smart city exploration using hybrid intelligence. *International Journal of Ambient Computing and Intelligence*, *8*(3), 1–21. doi:10.4018/IJACI.2017070101

Schneeweiss, S. (2014). Learning from big health care data. *The New England Journal of Medicine*, *370*(23), 2161–2163. doi:10.1056/NEJMp1401111 PMID:24897079

Settles, C. (2014). Integrating Patient Devices into Health IT Systems. *Medical Design Technology*. Retrieved from https://www.mdtmag.com/article/2014/12/integrating-patient-devices-health-it-systems

Shcherbina, A., Mattsson, C. M., Waggott, D., Salisbury, H., Christle, J. W., Hastie, T., ... Ashley, E. A. (2017). Accuracy in wrist-worn, sensor-based measurements of heart rate and energy expenditure in a diverse Cohort. *Journal of Personalized Medicine*, *7*(2), 3. doi:10.3390/jpm7020003 PMID:28538708

Smits, T., & Ryter, A. (2015). Chronic Conditions and Fall Risk. *Philips*. Retrieved from https://www.lifeline.philips.com/resources/blog/2015/09/report-chronic-conditions-and-the-high-risk-of-falling.html

Solanki, V. K., Katiyar, S., BhashkarSemwal, V., Dewan, P., Venkatasen, M., & Dey, N. (2016). Advanced automated module for smart and secure city. *Procedia Computer Science*, *78*, 367–374. doi:10.1016/j.procs.2016.02.076

Survey of Patient Appointment Wait Times. (2017). *Merritt Hawkins*. Retrieved from https://www.merritthawkins.com/2014-survey/patientwaittime.aspx

The impact of medical coding on a hospital's bottom line. (2014). *Real Time Medicare Data*. Retrieved from http://rtmd.org/topics/2014/07/the-impact-of-a-wrong-medical-coding-on-a-hospitals-bottom-line/#.WfQdGxNSyqB

Vos, T., Barber, R. M., Bell, B., Bertozzi-Villa, A., Biryukov, S., Bolliger, I., ... Murray, C. J. L. (2015). Global, regional, and national incidence, prevalence, and years lived with disability for 301 acute and chronic diseases and injuries in 188 countries, 1990–2013: A systematic analysis for the Global Burden of Disease Study 2013. *Lancet*, *386*(9995), 743–800. doi:10.1016/S0140-6736(15)60692-4 PMID:26063472

Wang, F. (2015). Adaptive semi-supervised recursive tree partitioning: The ART towards large scale patient indexing in personalized healthcare. *Journal of Biomedical Informatics*, *55*, 41–54. doi:10.1016/j.jbi.2015.01.009 PMID:25656756

Wang, H., Liu, W., & Soyata, T. (2014). Accessing big data in the cloud using mobile devices. In Handbook of Research on Cloud Infrastructures for Big Data Analytics (pp. 444-470). Hershey, PA: IGI Global.

Weldon, D. (2017). St. Luke's University uses data warehouse to improve performance. *Health Data Management*. Retrieved from https://www.healthdatamanagement.com/news/st-lukes-university-uses-data-warehouse-to-improve-performance

Wikipedia. (n.d.). Health Level 7. Retrieved from https://en.wikipedia.org/wiki/Health_Level_7

World Health Organization (WHO). (2016). Chronic obstructive pulmonary disease (COPD) Retrieved from http://www.who.int/mediacentre/factsheets/fs315/en/

Yang, W., Wang, X., Song, X., Yang, Y., & Patnaik, S. (2018). Design of intelligent transportation system supported by new generation wireless communication technology. *International Journal of Ambient Computing and Intelligence*, *9*(1), 78–94. doi:10.4018/IJACI.2018010105

Yu, S., Wang, C., Ren, K., & Lou, W. (2010). Achieving Secure, Scalable, and Fine-grained Data Access Control in Cloud Computing. In *Proceedings of IEEE International Conference on Computer Communications*, San Diego, CA (pp. 1-9). 10.1109/INFCOM.2010.5462174

Zhang, W., Qi, Q., & Deng, J. (2017). Building Intelligent Transportation Cloud Data Center Based on SOA. *International Journal of Ambient Computing and Intelligence*, *8*(2), 1–11. doi:10.4018/IJACI.2017040101

KEY TERMS AND DEFINITIONS

CDC: Centers for Disease Control and Prevention; a federal agency to conduct and support disease prevention and population health improvements.

Chronic Disease: A disease that persists for a long time and cannot simply be cured with vaccines or medications such as diabetes, arthritis, hypertension.

Cloud Computing: Delivery mechanism of computing services (server, storage, database, networking, software) over the Internet.

EDW: A central repository of integrated data used for reporting, analytics and machine learning.

EHR: Maintains electronic version of patient medical history and makes it easy to share across healthcare entities.

Patient Encounter: An interaction between patient and healthcare providers to receive the care and treatments.

Personalized Care: Custom medical treatment approach depending on individual patient's health and medical history.

Population Health: Emphasis on improving health outcomes of a group of patients such as diabetic or cardiovascular patients.

Quality of Care: The degree to which medical treatments and care services provided to patient yields to the increase in likelihood of positive results.

Wearable Device: Technology gadget that a consumer can wear to monitor the fitness and health.

Chapter 10
Smart Waste Management System

S. Umamaheswari
Dr. G.R. Damodaran College of Science, India

ABSTRACT

Smart connected devices are emerging from the internet today. These devices are embedded with intelligence due to the development in technology. These intelligent (smart) devices are capable of interacting with humans and other smart devices, which has led to the development of "Internet of Things" (IoT). The environment can be made smart and self-aware by the direct communication between more and more devices that are part of the IoT. The IoT concept can be applied to design a smart waste management system that makes the environment clean. A smart cities project has been initiated in most developing countries. Waste management is one of the prime components of the smart city. Garbage bins can be made intelligent, so that the fill level of the bins can be sensed, and the waste can be collected. This system will minimize the time and the cost of collecting waste.

PREFACE

This chapter gives an overview of the Smart Waste Management system. This system is modeled by integrating the IoT, Cloud Computing and Big Data Analytics. Cloud Database maintains the data received from the sensors that can be used for analysis. Big Data Analytics can be applied for analyzing the data for decision-making. Keeping the environment clean is the motto of Smart Waste Management System. At present, smart waste disposal methods are not fully implemented. The ideas and the concepts identified through this system would benefit the public in providing the lowest cost

DOI: 10.4018/978-1-5225-6207-8.ch010

service. Smart environment makes the country clean and green. This System aims at creating such an environment by having smart waste bins in the public places that leads to the healthy life of the people. A Smart Waste Management Framework for managing the waste disposal has been presented in this chapter.

INTRODUCTION

Nowadays population is increasing rapidly, which results in lack of public awareness and people invest less money in programs related to the waste management. This has created a huge health issue all over the world. Solid waste management is a big challenge in urban areas for most of the countries throughout the world.

A rapid growth of large and dense residential area and a critical requirement for urban environmental protection lead to the need for the implementation of an efficient waste management framework. Proper management of waste materials is important to maintain a healthy and hygienic environment to live. As per the research of CPHEEO (Central Public Health and Environmental Engineering Organization) the total amount of waste generated in India is approximately 1.3 pounds per person every single day, which is comparatively less compared to 4.6 pounds of waste generated per person every day in the United States (U.S.). But the U.S. population was approximately 307 million in July 2009, whereas India's population was 1.2 billion. These statistics show that India is generating almost 27 million more tons of waste than the U.S. every year.

A better Waste Management system is a primary perquisite for a Smart City for maintaining a clean environment as the different kinds of waste disposal are on the rise. Many technologies are adopted for waste collection as well as for well managed recycling. Problems due to pollution can be controlled by implementing Smart Waste Management.

BACKGROUND

Smart City

A Smart City is a one, which provides the needs of the citizens of developing the entire urban eco-system that is characterized by institutional, physical, social and economic infrastructure. The use of Information and Communication Technology (ICT) changed the quality of life, mobility of the people and the economy (Jeff McIntire-Strasburg, 2015). In a Smart City project, all the processes will be automated with IoT devices. Smart Healthcare, Smart Education, Smart Waste Management,

Smart Transportation and Smart Governance are some of the features of a Smart City. The main objective of a Smart City is to provide a good quality lifestyle, cost effective services and easy access to the services and infrastructure to the citizens of the country.

The Internet of Things (IoT)

The Internet is emerging with a whole lot of smart connected devices today. There is an exponential growth in the number of smart devices connected to the mobile internet. The devices are embedded with high intelligence due to the development of technology. These intelligent (smart) devices are capable of interacting with humans and other smart devices, as well, which led to the development of the "Internet of Things" (IoT). Kevin Ashton has coined the term "Internet of Things" in the context of supply chain management in the year 1999 that describes a technology of the future based on the Internet and involves sharing of information (K. Ashton, 2009). The People use the internet through the World Wide Web through the interconnected autonomous devices called as the Internet of Things (IoT), which shares the information amongst each other over the Internet. There are no machine equivalent web browsers for these devices and the Development of new tools/services are in progress by the standardization bodies. The features of the Smart City can be implemented with the realization of IoT.

Big Data Analytics

In a Smart City project, IoT devices will automate all the processes with a number of sensors. These sensors generate a huge amount of data used for some decision-making process. Sensor generated data cannot give the exact solution because it is a noisy data. It has to be used for some process like cleaning to avoid redundancy, unwanted data, etc. to get the exact data. Message Queue Telemetry Transport (MQTT) and Data Distribution service are the two protocols, which help IoT device to connect with the M2M network. MQTT protocol collects the data from various devices and puts those data over the internet. There is a need for strong, reliable, and scalable IT infrastructure for high complex process and securing those data.

Big Data refers to the huge volume of data. They are generated from the sensors, devices, video/audio, networks, log files, transactional applications, web, and social media in a real time environment on a very large scale. Volume – the quantity of data, Variety – type and nature of data, Velocity – the data generation speed, Variability –data inconsistency and Veracity – the quality of data that affects the analysis accuracy are the main characteristics of the Big Data.

Big Data Analytics refers to making use of advanced analytic techniques such as machine learning, data mining, natural language processing and predictive analysis against very large, structured, semi-structured and unstructured data that results in better and faster decision-making in businesses. The Hadoop tool helps in collecting and capturing data from the sensor, users, card readers and some other sources.

Cloud Computing

Cloud computing provides various services that include infrastructure, applications and/or storage space. Cloud computing can be defined as a pay-per-use model for facilitating on-demand access to the resources the user is in need of. The cloud service provider offers the services that include servers, application programs, operating systems, storage, etc. If the organization is planning to implement big data strategy for decision-making, usage of the cloud environment will be a better solution. The data is stored in the cloud storage system can be retrieved from any location that has Internet access. Amazon Web Services, Microsoft Azure, IBM Cloud and Google Cloud are offering different cloud services for the organizations.

RELATED WORK

The smart city can be realized only when the environment is fully automated. The literature states that various authors have proposed or implemented new approaches for making the city smart. Big Data, Cloud Computing, Embedded Systems and Wireless networks are the key technologies adopted for different smart city applications.

The complicated procedures are a major issue of the local municipal authorities due to problems in collecting, transporting and processing the solid waste disposal. A solution that saves time, money, and environment, which use sensors for measuring the fill level of the dustbins. This system generates collections plans with efficient schedules and routes (Twinkle Sinha et.al., 2015).

Smart garbage system with battery-based smart garbage bins is proposed by Insung Hong et al. (2014). The smart bins exchange information using wireless mesh networks and a router. The battery lifetime is increased, and user convenience is considered with various IoT technologies.

A garbage framework is implemented which has a garbage collection point with a camera and trash cans with load cell sensors. A snapshot of the garbage can and the load sensor is compared with the threshold value. This will decide the level of garbage and information is sent accordingly for collecting the bin (Peter Barry, Patric Crowley)

Some garbage collection systems check the weight of the bins to identify whether the bin is full or not. Weight sensors can be embedded in the bins. As these sensors won't provide the information about the level of garbage in the bins, Infrared (IR) sensors are used. Global System for Mobile (GSM), IR Sensor and Graphical User Interface (GUI) are the technologies used for efficient waste disposal when the garbage level reaches its maximum (Narendra et al., 2014).

The use of digital technologies is on the rise in the development of a Smart City for enhancing the quality of life of the people. The resources of the smart city include Electrical energy, Traffic signals, sensors deployed in different places etc. All these resources have to be maintained and secured in order to stop the draining and improve the standard of living. An implementation of power anti-theft module which is a camera-based surveillance and real-time power monitoring subsystem to ensure the security of people in public places. The security framework provides security for the electricity, public and traffic control system (Solanki et al., 2016).

With the increase in IoT devices, many processes are become automated in the applications like energy, agriculture, transportation, healthcare and e-governance. The data emitted by the sensor devices that are part of the IoT network has to be analyzed and visualized for a predictive analysis and decision making. There is a need for a better agile infrastructure, the big data for storing the data generated from the IoT devices. The visualization tools are applied for predicting the behaviors and statistical trends of the data accurately to make a better decision making. Big Data Analytics, which deals with the mining, analyzing and representing in different forms can be integrated with IoT. The FLUME and HIVE tools for data analysis are applied for big data analysis (Dey et al., 2018).

Wireless Body Area Network (WBAN) is a system that consists of multiple sensor nodes implanted within the human body. These sensors are capable of capturing vital signs of the body such as blood pressure, temperature, sugar level, humidity and heart pulse and transmitting them to the doctors. The integration of the IoT and WBAN connects the doctors, patients and nurses for better healthcare. The IoT based WBAN has been proposed by considering the challenges in realizing WBAN on the IoT (Elhayatmy et al., 2018).

The convergence of information technologies and medicine transforms healthcare more accurate and efficient. The integration of IoT with Health care applications leads to the smart devices to make user-machine interaction Dey et al., 2017).

Machine learning is mostly used as a decision support mechanism that uses a large set of data. Machine learning techniques assist the doctors in medical diagnosis process and decision making. Bio inspired algorithms and advanced data mining techniques are used for analyzing the large-scale data sets (Kamal et al., 2016).

The processing of the large data sets to retrieve knowledge is a critical task. K-Nearest Neighbor (*K*-NN) classification approach has been used to design a contemporary distributed clustering methodology for imbalance data reduction. A MapReduce-oriented framework is implemented using different clusters of automated contents that consist of multiple algorithmic approaches (Kamal et al., 2017).

The design of Intelligent Transportation System (ITS) will changeover into a next level development of variously new requirements coupled with the new generation wireless communication technology. The new generation of wireless communication technology is adopted to establish ITS (Yang et al., 2018).

Service-oriented Architecture (SOA) is an architectural model that applies the component's services for the use of standardized communication protocols by the other users. A distributed traffic cloud data center based on SOA and application of a network-aware energy conservation scheduling algorithms on cloud data center makes the full utilization of all the resources in the cloud data center (Zhang et al., 2017).

In general, smart-city services are provided on on-demand process in single domains like Internet service, building management, transportation, and health care. The cloud computing concept is put in to practice by means of virtualized on on-demand network services. There are many security challenges faced by the Cloud users when the sharable resources are used. A trusted cloud service model is proposed for the Smart city users. The Multivariate Normal Distribution selects various attributes of security entities (Sarkar et al., 2017).

The transport system embedded with a recognition method for driving event has been proposed by Ali et. al. (2017). K-NN algorithm and Dynamic Time Warping algorithms were applied for evaluation.

Ambient Intelligence (AmI) is closely associated with an intelligent system in which technologies, capable enough to automate a platform at same time embedding the other necessary services such as anticipatory, personalized, adaptive etc. AmI makes the automation of the environment that is responsive and sensitive. The integration of AmI and Big Data in a smart health care environment provides the better treatment for the patients at home (Jain & Bhatnagar, 2017).

The massive increase in the data leads to the development of new techniques for data access. Cloud and Big Data are the emerging technologies for data analysis. The implementation of the Cloud and Big Data storage is certainly the Hadoop Distributed File System (HDFS). A new storage and data access model is proposed to improve the service metadata for Hadoop that maintains consistency without affecting the performance and scalability of metadata (Matallah et al., 2017).

TECHNICAL REQUIREMENTS

Wyzbee IoT Platform

The WyzBeeIoT platform provides support for the developers to create new IoT devices. This platform comprises of hardware boards, cloud software and services framework and application development environment along with a product synthesis solution for synthesizing the final product. The WyzBeeIoT platform considerably lessens the time taken for creating IoT devices by offering support to sense, compute, communicate and develop applications. This board consists of Redpine's Wireless Secure MCU (WiSeMCU™) with a multi-protocol wireless module providing Wi-Fi, Bluetooth 4.1, and ZigBee connectivity, six-axis inertial sensors, an infrared receiver, a debug port, LEDs, push-buttons, USB ports, and WyzBee THING™ expansion connector.

The WiSeMCU module can be operated on embedded TCP/IP networking stack with SSL/TLS/HTTPS security and also on complete Wi-Fi, Bluetooth 4.1 and ZigBee stacks. The WyzBee THING expansion header has the provision of adding additional peripherals such as audio, GSM, GPS, capacitive touch display and rechargeable battery provided by Redpine provides additional peripherals which are known as 'Things'. The developers can choose the application development environment, which includes IAR, Keil and free CoIDE from CooCox based on their requirements. The application development environment comprises of integrated stacks (MQTT, CoAP, 6LoWPAN, Thread, etc.), analytics engines (big data, streaming, machine learning etc.) and tools for data visualization and data analytics. PUF-based hardware security block is also integrated with this board for individual authentication of IoT devices.

Figure 1. The WyzBeeIoT Platform

Ultrasonic Sensor

Ultrasonic sensors are used for stable detection of liquids, clear objects and uneven surfaces in dirty environments. The fill level of the bin can be measured from the upper point of the bin.

GPS Device

A GPS device calculates the geographical position of the device where it is attached. The GPS device is embedded in the garbage bins and the collection truck.

GSM Module

The GSM module is used to send messages to the Garbage truck when the bin reaches the threshold level. This module also sends short text messages (SMS) to the respective local authorities.

Keil IDE

Keil® Microcontroller Development Kit (MDK) is the most comprehensive software development solution for Arm®-based microcontrollers and comprises of all components that are needed to create, build, and debug embedded applications.

Embedded C

Embedded C is a programming language used to develop programs for embedded devices.

MySQL

MySQL is the powerful Open Source Relational SQL database management system which is used for storing data used by the web-based software applications.

Cloud Service Provider

WyzBeeIoT Cloud provides flexible and customizable connectivity, analytics and user interfaces. The developer can also make use of Amazon Web Services (AWS), Microsoft Azure, Google Cloud or IBM Cloud for data storage.

WyzBee API

WyzBee API is a set of application programming interfaces (APIs) that can be used for integrating the devices and the applications. This WyzBee API set includes the API Library for Application Protocols, Board peripherals, MCU peripherals, Things and Wireless.

Data Analytics Tools

The data analytics and visualization are done for effective decision making. The following table provides some of the tools thatcan be used based on the type of the data storage and the requirements of the users.

SMART WASTE MANAGEMENT SYSTEM FRAMEWORK

The Smart Waste Management system framework is designed with the support IoT. This framework deals with the design and development of a Smart Waste management system. The primary objectives of this system are:

Table 1. Tools for data analytics and visualization

Big Data Analytic tools	Apache Hadoop HDInsight Hive NoSQL Sqoop Bigdata in Excel
Open Source Data tools	R-Programming Orange Rapid miner Weka Pentaho
Data Visualization tools	Data Wrapper Solver Qlick Google fusion Table Infogram
Cloud Analytics tools	KVM Delta Cloud Eucalyptus Open Nebulla Open Stack Synnefo Open QRM Cloud foundary

- To keep the city clean and green;
- To reduce the cost of the waste collection process;
- To minimize the time taken for garbage collection;
- To design a web-based solution to monitor the fill level of the garbage bins;
- To schedule the collection process and to inform to the vehicle driver;

The methodology of this system is as follows:

- Initially, smart garbage bins that are embedded with the level sensors are deployed in the public places;
- These smart bins will send the fill level of the IoT cloud;
- The fill level data received from the bins are analyzed to generate schedules and optimized routes for collection;
- A Web Application is developed for data analysis and report generation;
- Reports can be sent to the authorities through email/SMS;
- A mobile-based web service will send the information to the vehicle drivers who collect the garbage.

The following figure depicts the system architecture of the Smart Waste Management System.

Figure 2. System Architecture of Smart Waste Management System

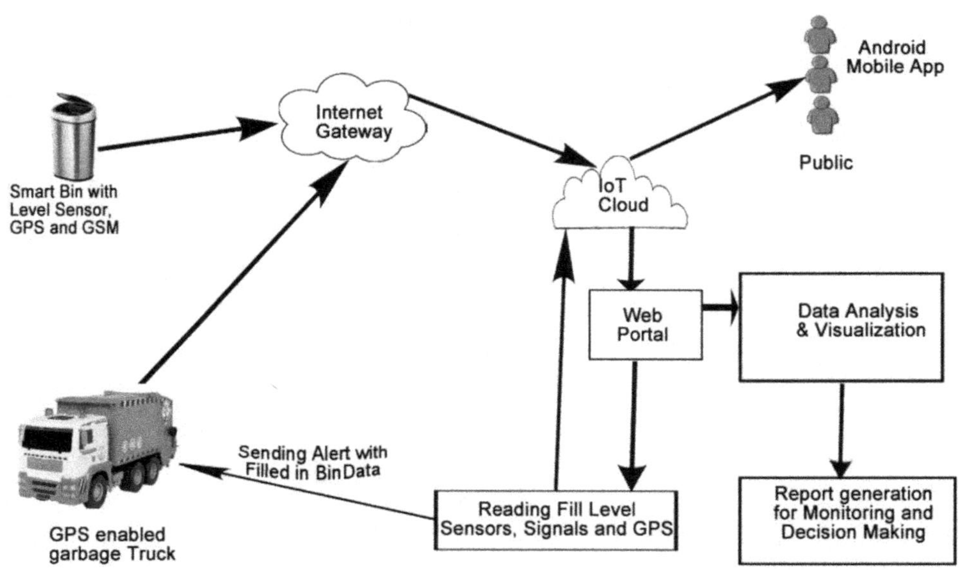

The garbage bins are embedded with the Ultrasonic sensor, GSM module and a GPS system. Wyzbee board is used to fit the sensor. The fill level is sensed by the Ultrasonic sensor. The GSM module communicates information on the fill level of the garbage bins to the IoT Cloud. The GPS feature tracks the location of the bin so that the communication can be sent to the nearest truck driver. The garbage collecting truck is also enabled with the GPS feature.

A Web portal is developed to retrieve data from the cloud and to send alter to the Garbage Vehicle driver. Google map is embedded in the portal so that the location of the garbage bins and the garbage collecting vehicles can be tracked. MQTT and CoAP protocols are used for message exchange from the sensors and the Web Portal. The cloud database will maintain the details of the filled and unfilled bins. The information about the bins will be provided to the authorized users who are sending the request.

The Android mobile application can be used by both the public and the vehicle driver. The public will get the information about the nearest unfilled bins. The garbage vehicle drivers will get the information about the filled in bins.

The public will put the waste into the smart bins. These bins will check the threshold fill level and send the status along with the coordinates to the Cloud Database once it reaches the level. The Web portal will read these data and checks for the nearest garbage vehicle to empty the waste. A notification is sent to the local authorities once the bin is emptied.

The information about the waste production can be analyzed for the following reasons:

- To find the amount of waste produced every year;
- To analyze the economic calculations spent for the waste disposal;
- To identify the locations where the bins are filled very frequently, so that additional bins can be deployed;
- To reduce the transportation cost by Route Optimization.

FUTURE DIRECTIONS

A framework for Smart Waste Management System that uses the MQTT and CoAP protocol has been presented in this chapter. The category of waste is not considered in this model. Suitable sensors can be deployed in the bins to sense the category of the waste. The waste that can be recycled can be segregated. As the sensors might be damaged due to improper usage, proper awareness about the usage of the smart bins and the mobile app should be created among the public before it is implemented on a large scale.

SUMMARY

This chapter presented the design and development of Smart Waste Management System that is one of the primary components of a smart city. The Web Portal application that applies the MQTT and CoAP protocol for communicating data from the bins is presented. The usage of IoT Cloud, Data Analytics and Visualization makes the decision making effective. This system avoids human intervention, saves human time and effort and provides a smart technology for waste management that leads to a healthy and pollution free environment. This system can be implemented in a Smart City, where the public can be benefitted.

REFERENCES

Ali, A. H., Atia, A., & Mostafa, M. S. M. (2017). Recognizing Driving Behavior and Road Anomaly using Smartphone Sensors. *International Journal of Ambient Computing and Intelligence*, *8*(3), 22–37. doi:10.4018/IJACI.2017070102

Ashton, K. (2009). That 'internet of things' thing. *RFID journal, 22*(7), 97-114.

Barry, P., & Crowley, P. (2012). *Modern embedded computing: designing connected, pervasive, media-rich systems*. Elsevier.

Dey, N., Ashour, A. S., & Bhatt, C. (2017). Internet of things driven connected healthcare. In *Internet of things and big data technologies for next generation healthcare* (pp. 3–12). Cham: Springer. doi:10.1007/978-3-319-49736-5_1

Elhayatmy, G., Dey, N., & Ashour, A. S. (2018). Internet of Things Based Wireless Body Area Network in Healthcare. In *Internet of Things and Big Data Analytics Toward Next-Generation Intelligence* (pp. 3–20). Cham: Springer. doi:10.1007/978-3-319-60435-0_1

Hong, I., Park, S., Lee, B., Lee, J., Jeong, D., & Park, S. (2014). IoT-Based Smart Garbage System for Efficient Food Waste Management. *The Scientific World Journal*, *2014*, 646953. doi:10.1155/2014/646953 PMID:25258730

Jain, A., & Bhatnagar, V. (2017). Concoction of Ambient Intelligence and Big Data for Better Patient Ministration Services. *International Journal of Ambient Computing and Intelligence*, *8*(4), 19–30. doi:10.4018/IJACI.2017100102

Kamal, M. S., Dey, N., & Ashour, A. S. (2017). Large Scale Medical Data Mining for Accurate Diagnosis: A Blueprint. In *Handbook of Large-Scale Distributed Computing in Smart Healthcare* (pp. 157–176). Cham: Springer. doi:10.1007/978-3-319-58280-1_7

Kamal, S., Ripon, S. H., Dey, N., Ashour, A. S., & Santhi, V. (2016). A MapReduce approach to diminish imbalance parameters for big deoxyribonucleic acid dataset. *Computer Methods and Programs in Biomedicine*, *131*, 191–206. doi:10.1016/j.cmpb.2016.04.005 PMID:27265059

Kumar, N., Swami, G. C., & Nagadarshini, K. N. (2014). Efficient Garbage Disposal Management in Metropolitan Cities Using VANETs. *Journal of Clean Energy Technologies*, *2*(3).

Kunzmann, K. R. (2014). *Smart Cities: A New Padigm of Urban Development* (pp. 9–20). Crios; doi:10.7373/77140

Matallah, H., Belalem, G., & Bouamrane, K. (2017). Towards a New Model of Storage and Access to Data in Big Data and Cloud Computing. *International Journal of Ambient Computing and Intelligence*, *8*(4), 31–44. doi:10.4018/IJACI.2017100103

McIntire-Strasburg, J. (2015). An "Internet of Things" Approach to cleaner, greener public space.

Sarkar, M., Banerjee, S., Badr, Y., & Sangaiah, A. K. (2017). Configuring a trusted cloud service model for smart city exploration using hybrid intelligence. *International Journal of Ambient Computing and Intelligence*, *8*(3), 1–21. doi:10.4018/IJACI.2017070101

Sinha, T., Kumar, K. M., & Saisharan, P. (2015). smart Dustbin. *International Journal of Industrial Electronics and Electrical Engineering*, *3*(5), 101–104.

Solanki, V. K., Katiyar, S., BhashkarSemwal, V., Dewan, P., Venkatasen, M., & Dey, N. (2016). Advanced automated module for smart and secure city. *Procedia Computer Science*, *78*, 367–374. doi:10.1016/j.procs.2016.02.076

Yang, W., Wang, X., Song, X., Yang, Y., & Patnaik, S. (2018). Design of Intelligent Transportation System Supported by New Generation Wireless Communication Technology. *International Journal of Ambient Computing and Intelligence*, *9*(1), 78–94. doi:10.4018/IJACI.2018010105

Zhang, W., Qi, Q., & Deng, J. (2017). Building Intelligent Transportation Cloud Data Center Based on SOA. *International Journal of Ambient Computing and Intelligence*, *8*(2), 1–11. doi:10.4018/IJACI.2017040101

KEY TERMS AND DEFINITIONS

6LoWPAN: IPv6 over Low Power Wireless Personal Area Networks: 6LoWPAN acts as an adaptation layer to transport IPv6 packets over 802.15.4 links.

CoAP: Constrained Application Protocol: CoAP is a specifically designed application layer protocol for use with constrained devices and constrained networks in the Internet of Things.

Data Analytics: The science of studying raw data with pattern matching and finding conclusions about the information.

GSM: Global System for Mobile used for communicating information between the devices in different locations.

IoT: Internet of Things: IoT refers to the interconnection of the IP smart objects, such as sensors and actuators.

MQTT: Message Queuing Telemetry Transport: MQTT is a publish/subscribe messaging protocol designed for lightweight M2M communications.

Section 5

The Case Studies:

Urban Areas in the Conext of Smart City Infrastructures

Chapter 11
Big Data Analytics and Internet of Things for Urban Transportation:
A Case of Pune City, Maharashtra, India

Jyoti Chandiramani
Symbiosis International University (Deemed), India

Sushma Nayak
Symbiosis International University (Deemed), India

ABSTRACT

The idea of smart city has assumed popularity in numerous countries across the globe. In 2015, the Government of India embarked on a mission of creating 100 smart cities to sustain the burgeoning urban population. While a wide-ranging set of fundamentals has a key role in enhancing the quality of life of citizens, the chapter revolves around transportation issues and traffic management concerns in one of India's smart cities, Pune. Transport is one of the few areas where Pune lags behind compared to its urban counterparts in the country. Public transportation in the city has been ineffectual, and auto rickshaws have been unyielding and pricey, thus making it imperative to possess personal vehicles or resort to app-based cab services. A palpable outcome of this has been traffic congestion that leads to slower travelling speeds, extended trip times, and amplified vehicular queuing. Big data and IoT can make a considerable impact in realizing the smart city objectives for efficient transportation in Pune by serving as complementary measures to supply-side policies.

DOI: 10.4018/978-1-5225-6207-8.ch011

INTRODUCTION

The idea of smart city – an urban region where information and communications technology (ICT) is the fundamental infrastructure for offering essential services to citizens – has assumed popularity in numerous countries across the globe. The smart city transformation is fueled by technology advancements. A smart city is known to efficiently leverage technology, transportation, communications, amenities, government policies and citizen involvement for designing an urban setting that reinforces modernism, progress, productivity, and sustainability. It is a concept that primarily considers the needs of citizens to draw up plans for meeting their requirements in real time. A smart city makes us envision a host of conveniences. For instance, streetlights would turn up or turn off depending upon the movement of people and vehicles. Trash would be cleared as soon the garbage container gets crammed with the waste. Notification on traffic and air pollution could be received on mobile phones with respect to the intended travel destination. Availability of parking spaces could be learnt through mobile apps. Eateries and cafés would put forward menus suiting popular tastes and preferences. Billboards and hoardings would flash advertisements in conformity with the bystanders' latest purchase patterns. Such services are deemed to improve livability for city dwellers.

A smart city, although, is not just about infusion of technology into the life of people, but a lot more. It is also about smart governance and smart citizenry (Khanna, 2015), wherein the benefactors and the beneficiaries are expected to be committed towards fulfillment of their obligations for the realization of smart city initiatives. Smart cities are "places where information technology is combined with infrastructure, architecture, everyday objects, and our own bodies to address social, economic, and environmental problems" (Townsend, 2014).

The concept of smart city has not originated today but has prevailed since several years in different epithets and forms. The introduction of programmed traffic lights in Houston, Texas, in the early 1920s, was categorically a smart initiative (Poole, 2014). However, the model that typifies the present-day smart cities and those likely to emerge in the future has undergone massive transformation in recent decades to reflect sizeable concepts upheld by advocates and stakeholders. Likewise, the significance of various terminologies has altered over time, contingent upon the ideas forwarded by academia, entrepreneurs, political groups and civil society (Eremia, Toma & Sanduleac, 2017). A major development in this regard happened in 2008 when the American multinational IBM commenced working for a 'Smarter Planet' vision to come up with intelligent systems that were far more advanced and user-oriented than ever before (Puri, 2014).

One of the contemporary ways having enormous potential to advance smart city services is big data analytics (Al Nuaimi, Al Neyadi, Mohamed & Al-Jaroodi,

2015). Big data implies extensive datasets which were not within the realm of perception, acquisition, organization, and processing of conventional information technology and software/hardware tools within a suitable time dimension (Chen, Mao, & Liu, 2014). The vastness of big data is evident from five Vs – Volume, Veracity, Variety, Value and Velocity (Yin & Kaynak, 2015). The progression of big data and the development of Internet of Things (IoT) technologies have played a crucial role in the viability of smart city initiatives. Big data extends the potential for cities to get hold of meaningful insights from copious datasets gathered through a range of sources; likewise, IoT permits the adaptation of sensors, Radio-Frequency Identification (RFID), and Bluetooth in the real-world setting by means of highly networked services (Hashem et al., 2016). With the technical support from big data and IoT, a smart city thus leans on a triad – being instrumented (mechanized), being interconnected (unified) and being intelligent (smart, automatic and self-regulating).

A concept, popularized by IBM and pulled off by numerous nations across the world, caught India's attention in 2015 when the government embarked on a mission of creating 100 smart cities to sustain the burgeoning urban population. Concomitantly, the Ministry of Housing and Urban Affairs, Government of India, stated that there is no time-honored general definition of smart city as the same varies for different cities and distinct nations, depending on the extent of development, availability of resources, readiness to change and aspirations of the city dwellers. Nevertheless, some of the principal infrastructure rudiments in a smart city include sufficient water supply; regular distribution of electrical energy; hygiene and sanitation, inclusive of solid waste management; well-organized means of transport and urban mobility; housing within means for the deprived; robust IT connectivity; effectual governance, principally e-governance; citizen engagement; sustainable environment; safety and protection of general public; and provision of efficient healthcare and education facilities ("What is Smart City," 2017). The Indian government, since 2015, has introduced a string of smart city initiatives for the furtherance of urban development.

Even as an all-encompassing set of fundamentals has a key role in enhancing the quality of life of citizens in a smart city, the present study calls attention to the significance of transportation in the urban setting. Urbanization brings with it a series of challenges owing to the increasing numbers of residents in cities. As the cities develop and expand, there emerges a need for improved transportation services to facilitate enhanced mobility. This leads to a rise in the purchase and ownership of private vehicles (Reddy & Balachandra, 2012). According to a study by Dargay, Gately, & Sommer (2007), vehicle ownership across different countries in the world exhibits a unique S-shaped pattern in conformity with the Gompertz function of per-capita income; thus, countries with low levels of per-capita income in their initial development stage show evidence of less vehicle ownership, which increases eventually with the increase in per-capita income, but then declines at exceptionally high income levels to come to a point of saturation. Developing countries, particularly in Asia, are presently experiencing a phase of gradual increments in income complemented with a rise in ownership of private vehicles – a trend that is likely to continue for one more decade (Dargay et al., 2007).

Although increase in vehicles is an evident and palpable outcome of urbanization, there are several implications of this inexorable phenomenon. One of these implications includes traffic congestion that leads to slower travelling speeds, extended trip times, and amplified vehicular queuing. Insufficient capacity of public transportation with respect to reliability and security, pathetic road conditions, congested and jammed road network, substandard traffic management, parking woes, and dearth of modal alternatives together with pedestrian walkways, linger as critical issues in most of the cities (Prabhakar, Gupta & Mehrotra, 2015). Additionally, a large number of cities also fall short of integrated transportation plans resulting in considerable demand-supply mismatch and ineffectual transportation network. Thus, improving mobility and lessening traffic congestion constitute some of the major challenges confronting smart cities today.

The present chapter revolves around transportation issues and traffic management concerns in one of India's smart cities – Pune. Pune is the ninth most densely inhabited city in India ("Cities Profile of 20 Smart Cities," 2017) and the second largest in Maharashtra (Krishnamurthy, Mishra & Desouza, 2016), after Mumbai (*Smart Pune's Step Towards Sustainable Transportation*, 2016). A report released in 2017 by World Economic Forum on the migratory trends in cities bears out the fact that while India witnessed a twofold increase in interstate migration during 2001-2011 in contrast to the preceding decade, Pune stood as one of the two foremost influenced cities in India and among four most influenced in Asia, with respect to increase in inflow of migrants ("Interstate migration in India doubled between 2001-2011: WEF," 2017). The report further stated that interstate migration is escalating at 4.5 percent per annum. With respect to Pune, it was observed that population of urban agglomeration expanded by 34% for the decade 2001-2011, not as an upshot of natural growth but due to migration in vast numbers (Umbrajkar, 2012). Pune's vehicular density trends since 1999-2000, as exhibited in table 1, clearly point the rationale for the extent of traffic congestion the city has been experiencing during this period. A Google Maps impression of Pune at 7:30 pm on weekdays further corroborates the magnitude of the problem, with 70 percent of prominent roads marked "deep red," signifying acute traffic congestion ("Reimagining Pune: Mission Smart City," n.d.).

With the launch of smart city initiatives in Pune in the year 2016, several projects specifically relating to transportation were outlined. While the city has successfully

Table 1. Vehicles registered with Regional Traffic Office (RTO), Pune (1999-00 to 2013-14) Source: Regional Transport Office (RTO), Pune

Year	Two Wheelers	Four Wheelers	Number of Vehicles
1999-00	54463	10807	72037
2000-01	59436	10803	78605
2001-02	62409	12823	82643
2002-03	69426	12766	90308
2003-04	77382	15030	100269
2004-05	84186	19711	113279
2005-06	95073	23561	129673
2006-07	104406	28608	148403
2007-08	87499	27893	129899
2008-09	94346	28266	131756
2009-10	102620	30962	143116
2010-11	127960	40499	181604
2011-12	155540	49194	219014
2012-13	146240	47374	202556
2013-14	145794	41507	197028

cracked a few, one of which is the linking of e-challan system with the CCTV camera to check unbridled traffic outlawry (Banerjee, 2017), numerous others such as delineation of zones for striders, bicycles and motors vehicles; traffic mobility plan for the city; development of mobile applications for obtaining details on the arrival and departure time of buses; etc (Bari, 2017) are yet to be accomplished. It is further observed that for a number of projects, the targets have not been realized within the preset deadlines: only 10% of the total projects have actually commenced, some are pending in anticipation of approval, and few have been written off altogether on account of controversies and debates surrounding them (Kulkarni, 2017).

This chapter shall closely examine the role of big data analytics and IoT in actualizing the smart city initiatives for Pune city from the standpoint of transportation and traffic management; the focus will be on objectives identified in the smart city plan, those attained so far, those waiting to see the light of day, and the way ahead for further action. The outcomes of this chapter can go a long way in realizing the objectives of citizen-friendly transportation for smart city Pune, and also serve as a cue to accomplish the same in other Indian cities having similar concerns. This is further important considering the fact that India reports the highest number of fatalities in road accidents across the globe, and one in every ten fatality occurring is reported from India (Dash, 2017).

BACKGROUND

While the idea of smart city has assumed recognition today, it can be modeled in distinct ways, from the exhaustive use of ICT in urban milieu to the initiation of ground-breaking services, managerial capacities, and physical infrastructure (Rodríguez Bolívar, 2015). Technology investiture is one of the prominent ways in which a smart city attains 'smartness.' With respect to transportation, technological enhancements supported by big data analytics can assist in augmenting reliability and stability of public transportation network by imparting information at fingertips on arrivals/departures/route information to travelers for trouble-free journey, managing traffic, locating parking spaces, facilitating easy payments, etc. However, a city needs to meet critical supply requirements with respect to transportation for its residents, before heading for technological enhancements. As an instance, Pune's transport system, akin to that in other Indian cities, has for long been reeling under the demands of growing populace, insufficient bus fleets and derisory resource availability, mounting competition from personal modes of transport, and disregard from the political and executive machinery (Sekhsaria, 2004). The fleet is old and tottering; the buses are in a dismal state; few buses lack window panes; few others are either short of seats or have broken seats; frequent breakdowns are reported; there is paucity of qualified people to guide passengers; there are inadequate bus depots and washing facilities; hygiene is abysmally poor; bus stops and bus stands are pathetic. For years together, the scenario has been so appalling that the administration lacks funds to complete even preventive maintenance that is required to keep the buses running and forestall accidents. Even if the funds are available, there is problem with regard to judicious utilization of funds. With supply-side bottlenecks in public transport system, the foremost priority for the administration is to strengthen the city's public transport in terms of availability, regularity and frequency. Only when the basic supply requirements are met adequately, the potential of big data analytics can be realized for smart transportation in the city, which the present chapter draws attention to.

URBANIZATION AND SMART CITIES

Rationale for Smart City Initiative in India

India, second most populous country in the world after China, has witnessed sweeping internal migration on account of regional disparities prevalent in the country. The past decade has seen a phenomenal rate of migration within the country wherein people have moved from rural areas and hinterlands to cities in search of employment opportunities. The Census 2011 reports an increase in urban residents by 9 million from what was reported in the earlier Census (Joshi, 2016). Cities have customarily been the focal point of financial influence of a nation. In the case of India, they occupy roughly 31% of existing populace (refer to Figure 1) and put in 63% of GDP ("Census 2011 - Census of India", n.d.). The present-day cities persist with this drive, befitting economic hubs, both at a national and global level, because of the economies of scale they command; they exert a pull on foreign investment, draw in business, and attract first-rate intellectuals from around the world (Khanna, 2015). If this trend continues, the cities shall be required to equip themselves to accommodate such a growing population, provide them housing, work, civic amenities, effective governance, speedy means of communication, swift transportation, continual access to power, water supply and comfortable living. The smart city initiative in India was mainly instated to enable the cities offer a decent quality of living to their inhabitants, considering the fact that Indian cities have been performing poorly in global rankings with respect to livability and workability, as reflected by Mercer's Quality of Living Index (Mathur, 2017).

Smart City Mission in India: Execution, Features and Strategies

The execution of smart city initiative in India is brought into effect through a host of selected consulting firms entrusted with the task of supporting the cities to partake in a contest. A total of 100 cities, evenly strewn among the States and Union Territories, shall be chosen over the span of 5 years ranging from financial year 2015-2016 to financial year 2019-2020 ("Coverage & Duration: Smart Cities Mission, Government of India," 2017).

The selection process shall be elaborate and meticulous, and run through a series of well-defined steps. Each aspiring city shall be in the race known as a 'challenge.' The first part of challenge for a city would be to compete with its counterparts within the State/Union Territory; those emerging successful at this stage shall move on to the next stage whereby the subsequent challenge will be to create proposals reflecting incorporation of the strategic elements (discussed later in this segment).

Figure 1. Trends in Indian urban population (1951-2011)
Source: Census 2011 - Census of India

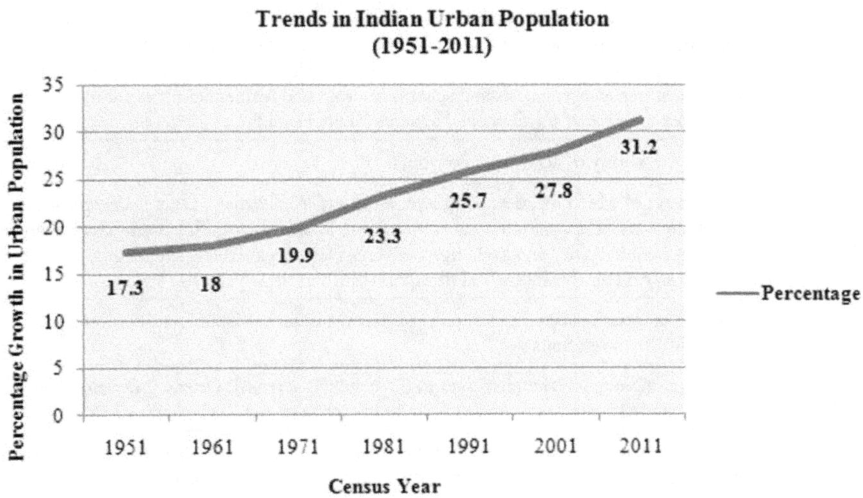

These proposals, made with due consideration to estimations held by city dwellers and stakeholders, shall be communicated to the Ministry of Housing and Urban Affairs, Government of India. A panel of connoisseurs and experts from India and overseas, alongside a group of recognized organizations and institutions, shall then scrutinize these proposals to arrive at the list of winning cities.

The Smart City Mission in India aims at all-inclusive development in the selected cities, as listed in Table 2.

According to the Ministry of Housing and Urban Affairs, Government of India, the strategies outlined for the fulfillment of the smart city mission in India have been presented in Figure 2.

The strategies are further discussed as follows:

1. **Retrofitting:** This strategy mainly emphasizes development of a prevailing inner-city area to make it more resourceful, ecologically aware and livable. For this, the principle of citizen engagement shall be brought to the fore for selection of area that exceeds 500 acres. Considering the existing state of infrastructure and intent of local residents, a plan shall be designed for initiation of smart initiatives within the identified area. The current structures would mostly stay untouched in this paradigm; nevertheless, additional comprehensive infrastructure and numerous smart applications would be integrated and fitted into the retrofitted smart city. The premeditated objectives of this plan would be attained over a shorter duration in order to simulate similar strategies in other parts of the city.

Table 2. Features of India's smart city mission

Dimension	Explication
Mixed land use	Encourage effective use of land in target areas that supports a host of interconnected activities: positioning distinct kinds of activities in close proximity (contiguous); for instance, stores, supermarkets and schools in or around residential neighborhoods that moderates the time of travel needed to access daily errands
Housing	Ensure availability of affordable housing to all
Walkability	Develop zones which make walking a pleasing experience and warrant safe pedestrian movement – moderate jamming and congestion, lessen air pollution, conserve resources, encourage connectivity and exchanges, ensure safety; create roads supporting movement of vehicles as well as non-motorized transport – striders and cyclists
Open spaces	Building up open spaces – parks, playing fields, and recreational areas to offer a better quality of life to residents
Transport	Provide numerous travel alternatives to the residents - Transit Oriented Development (TOD), an efficient mass transport system that includes regular bus service, Metro, etc.
Governance	Offer governance that suits the comfort and convenience of citizens with respect to cost of delivery, ease of access, answerability, intelligibility and transparency; recourse to electronic modes – internet and mobile – for reaching out citizens, obtaining their views and feedback; bring into play online supervision of programs, schemes and actions with the assistance of cyber journey of worksites
Distinctiveness	Imparting distinctiveness and exclusivity to the city based on economic pursuits such as native cuisine, wellbeing, edification, civilization, fine art, ethos, ethnicity, sports merchandise, furnishings, fabric, dairy, etc.
Infrastructure	Application of Smart Solutions to create disaster-free zones, deployment of limited resources, providing economical services

Source: Smart City Features: Smart Cities Mission, Ministry of Housing and Urban Affairs, Government of India, 2017

Figure 2. Strategic outline for India's smart city mission
Source: Strategy: Smart Cities Mission, Ministry of Housing and Urban Affairs, Government of India, 2017

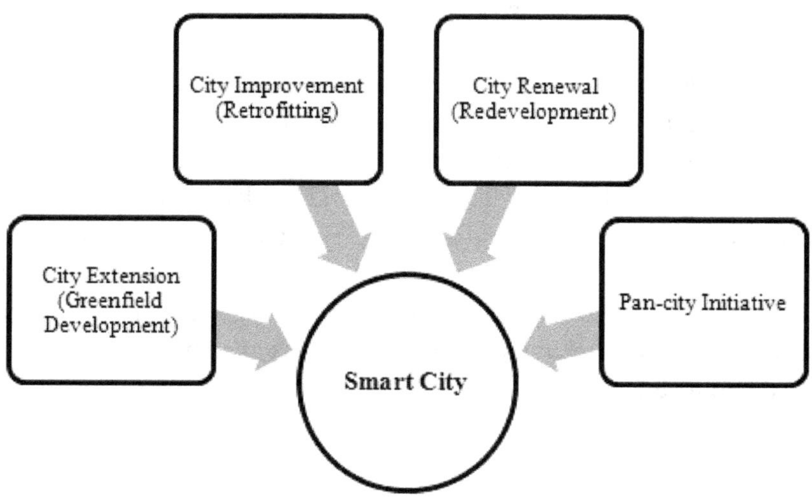

2. **Redevelopment:** In contrast to retrofitting where a good deal of existing setup remains intact and untouched, redevelopment intends significant makeover of an area within the city that exceeds 50 acres, selected by Urban Local Bodies through citizen engagement. In this strategy, the area selected would undergo major transformation as a new landscape with an all-embracing infrastructure, effective usage of land for diverse activities, advanced Floor Space Index and wider ground coverage would be created.
3. **Greenfield Development:** Towards the fulfillment of this strategy, an already unoccupied area exceeding 250 acres would be selected for realization of Smart Solutions by means of ingenious arrangements, premeditated financing and plan execution tools (land reformation through land pooling or land reconstitution) with availability of reasonably priced housing, particularly for the poor. Greenfield developments are a requisite in the vicinity of cities to cater to the needs of the growing populace.
4. **Pan-City Development:** It envisions employment of programmed Smart Solutions to the on hand urban infrastructure. This shall entail the utilization of ICT and data to improve infrastructure and services. For example, smart transportation system that allows diminution in the time required for traveling and commuting along different paths across the city can contribute to the quality of life of citizens.

The smart city proposition for each selected city is projected to incorporate either of the first three aforementioned strategies or a combination of these in addition to a Pan-city element with Smart Solutions. It is critical to note that pan-city initiative is a supplementary aspect to be built-in. Smart city being a miniature efficient area development scheme, it is crucial that inhabitants in the city feel a part of this development wherein there is something in reserve for them too. For that reason, the auxiliary requisite of a minimum 'one across-city smart solution' has been installed to make the scheme all-encompassing and thorough.

SMART TRANSPORTATION IN THE SMART CITY CONTEXT: ANCHORED IN THE INTERNET OF THINGS USING BIG DATA ANALYTICS

One of the essential imperatives of a smart city is 'smart transportation system inclusive of smart traffic management.' This comprises smart vehicles, smart infrastructure, ecologically friendly fuels, intelligent transport system, etc. A smart city is a distinctive arrangement to swap the traditional mode of services to smart and sophisticated ones (Babar & Arif, 2017). In the context of 'smart city,' the idea of smart mobility has been gaining importance in recent years. Smart mobility assumes visibly smart connectivity through digital systems in a manner that people, goods and services are effectively connected by means of internet; it also calls for a considerable increase in efficacy and safety of transport and communication systems in cities (Kourtit, Nijkamp & Steenbruggen, 2017). As an exemplar, sensors and cameras installed on the roads bunch up traffic information relating to the number of vehicles plying on roads at a given time and based on real time processed data, the existing state of affairs is shared with the traffic control unit. Concurrently, the traveler is notified regarding the selection of route to avoid traffic congestion. Citizens can establish how long it may take to get to a destination on the basis of existing traffic intensity and the standard pace of the vehicles (Rathore, Ahmad, Paul & Rho, 2016). The trends acquired from the copious sums of traffic records can help perk up transportation systems with respect to lessening traffic congestion by presenting substitute routes and reducing the frequency of accidents by examining the chronicle of mishaps, as well as aspects such as their cause and the driver speed (Hashem et al., 2016). Thus, the fusion of data sources and big data analytics serves as an expedient solution to ease real-time management of the smart city.

Big data refers to a vast pool of raw data that can be seized, amassed, dissected and managed (Chauhan, Agarwal & Kar, 2016). This voluminous data is not helpful until it is scrutinized and assessed by the big data analytics which reveals the indefinite associations, concealed patterns and other constructive information (Tachizawa, Alvarez-Gil & Montes-Sancho, 2015). The IoT imparts a platform for sensors and connectors to link impeccably within the smart city environment and allows progressively more convenient information exchange across platforms (Gubbi, Buyya, Marusic & Palaniswami, 2013). As illustrated by Babar and Arif (2017), IoT offers the basis for developing smart cities supported by diverse sensors, intelligent meters, closed-circuit television / video surveillance, display monitors, recording devices, and actuators, which contribute massive amounts of data that can be dissected by the techniques of big data analytics. Big data analytics is applied to sizeable datasets to expose unique patterns and relationships for evolving efficient policy framework.

Although numerous constructs go into making a transportation system smart, there are few which may be deemed as the mainstay for the fulfillment of this objective. Table 3 lists the popular constructs of transportation in a typical smart city.

A model smart city possesses all/most of the features described in Table 3. The new-fangled information and communications technologies explored here assist movement of automobiles along roads as securely, effectively, and humanely as possible.

Table 3. Makeup of smart transportation in a smart city

Construct	Description
Automatic Vehicle Location System (AVLS)	• Facilitates commuters with real-time information on arrival and departure times of buses via mobiles and online platforms; enables real-time booking of cabs through mobile applications; examples include London, Seattle and Sydney (Debnath, Chin, Haque and Yuen, 2014). • Enhances dependability and consistency of public transportation (Prabhakar, Gupta and Mehrotra, 2015).
Automatic Fare Collection System (AFCS)	• Collection of fares through contactless smartcards/RFID cards; this sort of a system can be introduced in various forms of public transportation, particularly state buses, to ensure no traveler/passenger undertakes journey without a ticket (Datta, Kaur and Garg, 2015; Chen, Fung, Desai and Sossou, 2012; Pelletier, Trépanier and Morency, 2011; Trépanier, Tranchant & Chapleau, 2007). • Intermodal fare collection mechanism whereby one smartcard can be utilized for all modes of transport; examples include New York, London and Seattle (Debnath, Chin, Haque and Yuen, 2014). • Several cities across the world allow both – cash payments and smartcards – leaving the choice to commuters; however, to encourage increased adoption of smartcards, a higher tariff is collected if the mode of payment is cash (Debnath, Chin, Haque and Yuen, 2014). • Examples: Hong Kong's Octopus Card; London's Oyster Card (Report of the Sub-Committee on Automatic Fare Collection System, 2013).
Real-time Traffic Information	• Enables travelers to have up-to-date traffic status to strategize their trip before traveling, and choose the path accordingly for reduction in travel time and perking up travel safety (Lee, Tseng and Shieh, 2010).
Smart Parking System	• Car owners can locate and attain an unoccupied parking space deemed suitable and expedient to them (Idris, Leng, Tamil, Noor and Razak, 2009). • Such an arrangement helps reduce fuel consumption of vehicles (Rathore, Ahmad, Paul and Rho, 2016). • Instead of moving around the entire area searching for parking space, drivers would receive intimations and alerts over Wi-Fi regarding empty spaces available in vicinity. Furthermore, drivers would be informed about the shortest possible routes to take for reaching the intended destination, from the viewpoint of keeping vehicular emissions under check. Such an arrangement could as well forewarn motorists concerning school zones where there could be hordes of children crossing roads and a substitute path may be suggested (Sherly and Somasundareswari, 2015). • Furthermore, this also helps creation of profit for vendors (merchants/ traders/ retailers) in the city; citizens prefer to go shopping where there is lesser amount of congestion and where parking is readily available, resulting in larger profit for vendors in such areas (Rathore, Ahmad, Paul & Rho, 2016). • Examples: New York and Seattle (Debnath, Chin, Haque and Yuen, 2014).
Bus Rapid Transit Systems (BRTS)	• Dedicated tracks for buses and accelerating their pace through precedence signaling on account of their enhanced capacity to transport a greater bulk of people for the same street space; introduce a system that encourages commuters to use alternative forms of transport than personal vehicles; create bicycle lanes and broad walkways (Rangarajan, 2010).
Smart Bike-sharing System	• Renting bikes for travelling short distances; these bikes are tactically placed throughout the city and can serve as an alternative form of transportation; their real-time availability can be tracked online and by means of mobile applications; this system has been in effect in countries such as Spain, Sweden, France, etc. (Midgley, 2009).
Smart Car-sharing System	• Also known as carpooling that matches supply and demand through connecting people via smartphone applications (Prabhakar, Gupta & Mehrotra, 2015). • Travelers can rent or lease a car for travelling between desired destinations; this resolves issues arising from lacunae in availability of state buses; helps lessen traffic congestion by cutting down vehicular movement on the road, reduces air pollution, and state expenditure; allows increased availability of space for walking and cycling (Shaheen, 2004).
Vehicular Rules Violation Monitoring	• An arrangement for monitoring and registering occurrences of traffic infringement at traffic sites. The framework encompasses a digital camera system installed at a traffic site. This digital camera is distantly tied to a data processing centre. The data processing centre includes a picture processor for assembling photos of motor vehicles and on-sight events generated via digital camera, a corroboration procedure for confirming the authenticity of the motor vehicle snapshots, an image processing arrangement for ascertaining information of the motorist from the captured images of the motor vehicle, and a notification/warning procedure for diffusing information on breaching/contravention to concerned law enforcement bureaus (Ciolli, Whyte, Ercan, and Mack, 2003; Khan, Khan, Zaheer and Khan, 2012). • Automatic Number Plate Recognition System (ANPR) to recognize the number of the vehicle violating traffic rules (Debnath, Chin, Haque and Yuen, 2014; Patel, Shah and Patel, 2013). • The far-reaching CCTV network in Glasgow supervises traffic and street lighting in addition to law-breaking and criminal acts (Marr, 2015).
Coordinated Adaptive Traffic System by Means of Smart Traffic Lights	• Regulating traffic lights corresponding to the flow of traffic and traffic occurrences; example, Sydney Coordinated Adaptive Traffic System (SCATS) which is in force at Sydney and Melbourne, Australia (Debnath, Chin, Haque and Yuen, 2014). • Apply technology to discern traffic scenario and regulate traffic lights to facilitate synchronized flow of traffic (Al Nuaimi, Al Neyadi, Mohamed and Al-Jaroodi, 2015; Prabhakar, Gupta and Mehrotra, 2015; Kanungo, Sharma and Singla, 2014).
Road use payments	Direct fees such as congestion levies and tariffs intended to dissuade use of specific categories of vehicles, fuels or highly polluting motor vehicles on roads. Such charges inhibit peak hour travel and allied traffic jamming or other community/environmental negative externalities (Prabhakar, Gupta and Mehrotra, 2015).
Electric Drive Vehicles	Promote vehicles running on electricity and renewable energy (Prabhakar, Gupta and Mehrotra, 2015).

Source: Various Sources

PUNE SMART CITY PROFILE

Pune City Profile

Pune, fondly referred as the 'Detroit of India' due to the burgeoning automobile industry, and reverentially addressed as the cultural capital of Maharashtra, is a city that has witnessed an august urban tradition: at the outset as a historical seat of pre-colonial urbanism, followed by a major military base during the course of British rule, post-independence as a fast rising advanced industrial hub, and in the present day as an evolving conurbation. Pune is also known as 'Oxford of the East' for the vast number of educational institutions that the city boasts of (*Smart Pune's Step Towards Sustainable Transportation*, 2016); with nine prominent universities and approximately 811 operational higher education centres to meet the demand

Table 4. Pune city profile

Indicator	City (Municipal Corporation)	State (Urban)	India (Urban)
Demographic Indicators			
Total Population*	3124458	50818259	377,106,125
Total Population of Urban Agglomeration (if)*	5057709		
Share of Urban Local Body population in District Urban population (%)*	54.33		
Population Growth Rate (AEGR) 2001-11*	2.08	2.12	2.76
Area (sq. km)**	276.4		
Density of population (person per sq. km)**	11304		
Literacy Rate (%)*	89.56	88.69	84.11
Schedule Caste (%)*	13.42	11.38	12.60
Schedule Tribes (%)*	1.09	2.96	2.77
Youth, 15 - 24 years (%)*	18.51	19.63	19.68
Slum Population (%)*	22.10	2.02	17.36
Working Age Group, 15-59 years (%)*	67.24	66.68	65.27
Economic Indicators			
Per Capita Income (Rs.) at 2004-05 constant price ***	88341	60431	35, 947
Urban Poverty Ratio (% of urban population)#	2.73	9.1	13.7
Unemployment Rate, 2011-12##	3.29	2.3	3.4
Work Participation Rate, 2011-12##	35.07	36.5	35.5
Work Status, 2011-12 (%) ##			
Self-employed: Regular/wage salaried employees: Casual labour:	28.00 66.38 5.62	36.1 54.6 9.4	42.0 43.4 14.6

continued on following page

Table 4. Continued

Indicator	City (Municipal Corporation)	State (Urban)	India (Urban)
Sectoral Distribution of Workers, 2011-12 (%) ##			
Primary Secondary Tertiary	0.34 32.16 67.50	4.8 31.3 64.0	7.5 34.2 58.3
Primary Commodities Manufactured**	Automobile engineering Software Chemical engineering		
Major Industries###	Metallic – Automobile Spares Machine Tools Non metallic – IT Industries & Software Technology Parts Agro Based		
Infrastructure Status			
% of households with access to tap water (from treated source) within premises•	99.20	85.69	84.14
% of households with access to electricity•	98.31	96.16	92.68
% of households having toilet facilities within premises•	76.47	67.28	72.57
% of household waste water outlet connected to drainage•	98.04	91.15	81.77
Type of sewerage system**	Underground sewerage		
Type of solid waste system**	Door to Door		
% households with access to PC /laptop with internet• % households without access to PC /laptop internet•	22.01 16.60	11.74 11.86	8.27 10.40
% of households with access to mobile phones•	64.36	64.16	64.33
Ownership Pattern of Housing (%)			
Owned Rented	66.66 30.63	69.78 27.19	69.16 27.55
% of households living in congested houses•	40.07	44.84	32.94
No. of Hospitals per 1,00,000 people**	0.03		
No of Schools per 1,00,000 people•	90		
Financial Status			
% of households with access to banking facilities•	86.36	76.02	67.77

*Census of India, 2011
**District Census Handbook, Census of India, 2011
***Directorate of Economics and Statistics of respective State Governments and for all India-Central Statistics Office
#Unit Level Data of National Sample Survey Organization, Household Consumer Expenditure in India, 68th Round, 2011-12
##Unit Level Data of National Sample Survey Organization, Employment and Unemployment Situation in India, 68th Round, 2011-12
###District Industrial Profile, Micro, Small and Medium Enterprises, Government of India
•Tables of Houses, Household Amenities and Assets, Census of India, 2011
Source: Various sources, adapted from "cities profile of 20 smart cities – Pune," http://smartcities.gov.in, 2017

of an ever-rising student population (Joshi, 2017), the city is making headway as an academic center with progressing years. It is also widely recognized for its landscape, amiable climate, historical value, vivacious ethos and imminent IT-BT centres. Furthermore, the city has been performing well compared to its counterparts in the country in terms of Mercer Quality of Living Index and Resilient City Index (Chandiramani & Airy, 2017). Table 4 details the profile of Pune city on the basis of vital indicators.

Pune has effective e-governance and digitalization; the city is further enveloped by Atal Mission for Rejuvenation and Urban Transformation (AMRUT), Urban Infrastructure and Governance (UIG) component of Jawaharlal Nehru National Urban Renewal Mission (JNNURM) and National Urban Information System (NUIS) ("Cities Profile of 20 Smart Cities," 2017).

TRANSPORTATION ISSUES AND TRAFFIC CONCERNS IN PUNE

A Prelude

India's mounting traffic congestion has the capacity to let cars creeping at less than 5 kms an hour – thus incentivizing travel by foot than drive; traffic jamming is expected to balloon further with swift growth in urban populace and vehicle ownership (Sanjai, 2017). According to a report by KPIT Technologies Corporation, India incurs losses worth Rs. 60,000 crores annually in traffic jams (*Realizing Smart Cities with Technology - A KPIT Perspective*, 2016). While the state of affairs is true for the nation at large, the city of Pune is no exception when it comes to transportation issues and traffic congestion woes. Numerous reasons are cited for this rapid gush in traffic in the city. One of the prime reasons is rise in population which contributes to a rise in the number of vehicles on roads. Additionally, congestion may be also attributed to sparse capacity of roads, signal delays, deficient information concerning traffic, incompetent transport management, uninhibited demand, etc (Shinde & Jagtap, 2016); thus, it is indispensable to optimize traffic control and ensure it is more active from the standpoint of accommodating the erratic traffic density. Transportation and mobility have evidently come out as the chief pan-city problems to deal with.

Public Transportation in Pune

Pune depends exclusively on buses for public transportation. The city fails to keep pace with other big cities in India with respect to public transportation alternatives. According to the Smart City Proposal for Pune Smart City Challenge – Stage 2,

Pune lacks a mass transit system; rather, it is the only megacity among top eight in the nation without a mass rapid transit system (MRTS). The number of buses per population is lower than the required standards: 37 buses per lakh as against a yardstick of 55 per lakh. Furthermore, buses in Pune are marked by sizeable issues with respect to 'accessibility' – 25 percent fleet off-road more often than not and, 'consistency' – 84 percent routes have a waiting time of more than 20 minutes. For that reason, the public transport trip share has been a meager 18 percent as against the 50 percent target. Additionally, there is a growing perception that natives drawing an income of more than Rs. 20,000 per month scarcely use buses. Complementing to public anguish, congestion has amplified radically in Pune due to supplementary reasons beyond public transportation. A good number of roads are not evenly wide and contribute to restricted access, signal timings are not synchronized, and on-street parking is prominent owing to constrained parking alternatives. There is lesser than standard share of Non-Motorized Transportation (NMT): 33 percent as against 40 to 45 percent, attributable to the paucity of secure NMT infrastructure and pedestrian walkways: 60 percent of walkways have less than 2 m width. A dearth of ring roads adds to the existing set of problems. As an outcome, the average speed is 18 km per hour which is likely to come down considerably to 12 km per hour by 2030 if no measures are initiated; this has been an important projection in the Smart City Proposal for Pune Smart City Challenge – Stage 2. It further assumes that on the basis of positive upshots of the latest BRTS corridors – where there has been a 12 percent modal shift to public transport – there is vivid substantiation that citizens shall utilize more public transport if the quality, consistency and availability of buses are enhanced.

Figure 3. Vehicular density in Pune (1999-2000 to 2013-2014)
Source: Regional Transport Office (RTO), Pune

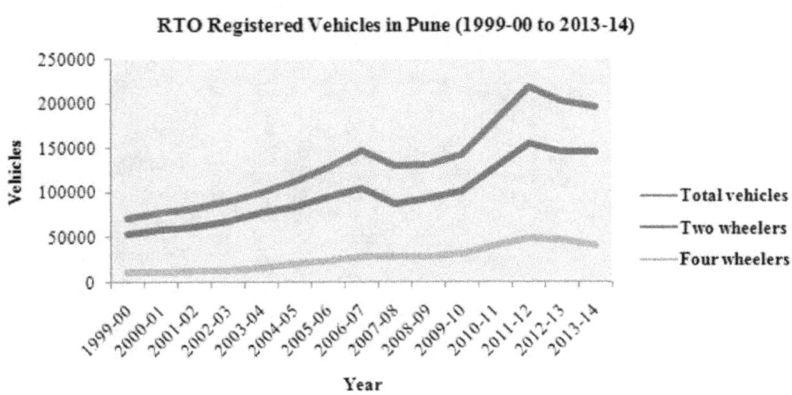

For a more accurate representation see the electronic version.

Vehicular Growth in Pune

One of the foremost causes for traffic congestion in Pune is the soaring number of vehicles. There has been a phenomenal growth in numbers of two-wheelers, four-wheelers and total vehicles in the city. Figure 3 depicts the vehicular density in Pune in recent years, based on the number of vehicles registered with Regional Transport Office, Pune. As vehicles in the city mount, parking becomes a hassle, consuming all accessible spaces meant for other road users in addition to creating a demand for all available open spaces to be converted into parking lots.

It was observed that the average registration of vehicles with RTO, Pune, went up to 22,000 per month, during the financial year 2016-2017 (Dastane, 2017). The Environment Status Report (ESR) for the same year stated that the number of vehicles in the city is equivalent to the number of residents in the city (Khairnar, 2017). Consequently, the city has been experiencing startling escalation in the levels of air and noise pollution (Joshi, 2017). During festive season, the vehicle registration with RTO receives further impetus. In September 2017, during the auspicious occasion of Dussehra – a festival that is reckoned as one among three and a half propitious events for Hindus – the number of vehicles registered with RTO saw an increase by 25%; this was due to special offers run by manufactures and retail outlets to attract more customers (Biswas, 2017a). A similar trend was observed along the same time during the previous year. Figure 4 indicates the numbers for two years in succession.

Cab operators in Pune, who are affiliated to the app-based cab provision services – Ola and Uber – have appreciably contributed to the fleet of cabs in the city. According to Pune Regional Transport Office (RTO) statistics, the number of such

Figure 4. RTO registered vehicles in Pune during Dussehra (2016-2017)
Source: (Biswas, 2017a)

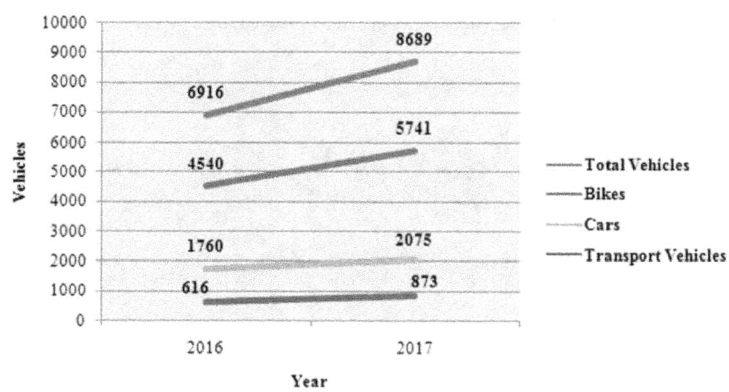

**For a more accurate representation see the electronic version.*

Figure 5. Number of app-based cabs registered in Pune (2014-2017)
Source: (Biswas, 2017b)

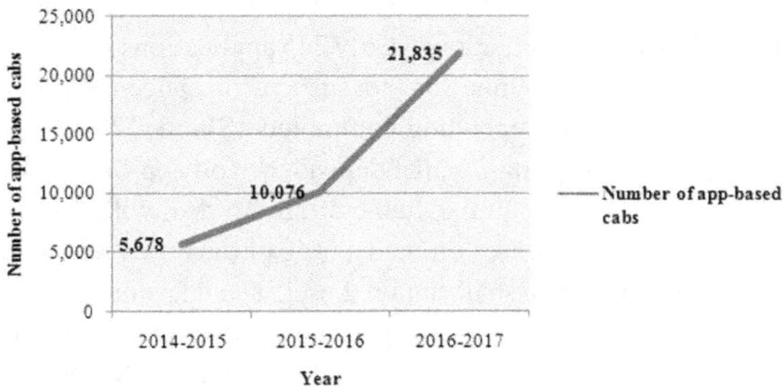

cabs registered during the financial year 2016-2017 nearly doubled – crossed the mark of 21,000 – as compared to the previous financial year when the number was slightly over 10,000 (Biswas, 2017b). Figure 5 reflects these trends.

The massive rise in the number of vehicles has contributed significantly to traffic jams in Pune. It is observed that in some parts of the city, bicycles are ridden on walkways. From 9 am to 11 am in the morning and 5 pm to 7 pm in the evening, the roads are swarmed by vehicles, thus adding to the time of travel required between destinations. For instance, a brief distance of 3.6 kms from MKSSS's Cummins College of Engineering, Karvenagar, to SNDT Arts and Commerce College on Karve Road, Erandwane, takes double the time to travel during peak hours than what it takes during non-peak hours. It is important to examine the raison d'être for private ownership of vehicles in the city. Why do residents of Pune prefer to possess their own vehicles rather than rely on public transportation? Citizens cite a number of reasons for this.

There are locations in the city which are not connected by buses. The number and frequency of Pune Mahanagar Parivahan Mahamandal Limited (PMPML) buses plying between prominent locations in the city are way too less than the demand for them. It is further reported that the allocation of buses along a given route is also subject to a vicious cycle – because there are not enough commuters, buses are cancelled, and because buses are unavailable, commuters have to look up for alternative forms of transport. During peak hours, buses are teeming and filled to capacity. It is a sight to see commuters hanging out from overcrowded buses without much concern for their lives. Alternate forms of transport are either rigid or pricey or both. The nuisance of auto rickshaws plying on the roads but turning down rides to the needy citizens is a concern reported by many. The auto rickshaw pullers are unyielding, uncompromising and express readiness to take a trip only between such destinations where there is probability of getting passengers for the next travel. At times, they charge higher tariffs for trips to remote locations or longer distances. This may not suit everyone's pockets and is particularly unaffordable for the non-working/dependent population. Nearly 245 applications were sent by traffic regulators to RTO from January-June 2017 for suspension of license of auto rickshaw pullers for declining rides to impending commuters (Shinde, 2017). With such a pathetic state of public transportation, the dependence on app-based cab providers and personal vehicles becomes an inevitable affair. To deal with this conundrum, PMPML intends to float new buses on roads; 50% of the buses will be running on CNG, while the remaining 50% shall run on diesel. The decision to buy 50% buses running on CNG is driven by requests and petitions from citizens, civil activists and organizations, from the standpoint of controlling air pollution. Nevertheless, it may take a year or two for the buses to come on roads. Till that time, a further spike in ownership of personal vehicles is expected.

Impending Flyovers

It is a customarily held credence that traffic woes in a growing city can be cracked by instigating actions such as broadening of existing roads, constructing new roads, flyovers, etc. Nevertheless, such solutions are ephemeral and barely sufficient on account of mounting number of vehicles, discussed earlier. As a result, the roads get crammed and jammed again within a short span of time. Furthermore, pedestrians and other travelers have to bear the brunt of vehicle-centric initiatives. Long-drawn-out flyover projects contribute significantly to traffic congestion. As regards Pune city, there are numerous flyovers that have been built recently and few those are presently under construction. The construction of flyovers on Satara Road and Swargate is through. However, construction of flyovers adjacent to College of Engineering, Pune, and the one that stretches along Cummins College Chowk at Karvenagar is

Figure 6. Impending flyover at Karvenagar, Pune, as on September 6, 2017 (A)
Source: Authors

Figure 7. Impending flyover at Karvenagar, Pune, as on September 6, 2017 (B)
Source: Authors

Figure 8. Impending flyover at Karvenagar, Pune, as on September 6, 2017 (C)
Source: Authors

Figure 9. Impending flyover at Karvenagar, Pune, as on September 6, 2017 (D)
Source: Authors

still underway. The latter is a vital junction and the flyover is anticipated to lessen traffic in Warje and contiguous areas. The erection of flyover commenced in 2012, was expected to be completed by the end of 2015, then by 2016, but still remains unfinished. The amount of work finished so far is evident from Figures 6, 7, 8 and 9. The extent of inconvenience to travelers using that road is of great consequence. The Pune Municipal Corporation intends to construct more flyovers in the coming years. Nonetheless, flyovers are by a hair's breadth remedy to moderate tailbacks as they are inclined to simply alter the point of congestion from one place to the other, in conformity with the Comprehensive Mobility Plan (CMP) of the PMC. Studies relating to transportation across the world validate that constructing flyovers only draws in more vehicles (Puri, 2015). Consequently, experts have expressed reservations about the real efficacy of flyovers beyond offering provisional respite from congestion. The only established sustainable remedy to the problem of traffic congestion is imparting a high-quality public transport structure in addition to secure, sufficient and functional amenities for striders and cyclists. This shall make it feasible for citizens to travel expediently without using personal vehicles, thus limiting the number of vehicles on road. Consequently, the traffic congestion woes as well as pollution problems will gradually disentangle.

Parking Issues

One of the key concerns facing the city corresponding to those discussed earlier is the availability of parking space. A typical automobile is functional for one hour on an average and is parked for 23 hours in a day (*Smart Pune's Step Towards Sustainable Transportation*, 2016). Vehicles are randomly parked at roadsides, walkways and vacant spaces in an unorganized fashion. It needs to be realized that creation of additional parking spaces is going to be an uphill task considering the current rate of growth of vehicles. There shall always be a massive shortfall if the present trend continues. The roads are not outfitted to offer parking space for every vehicle. According to administration's ballpark figure, nearly 50% of the city's traffic comprises personal vehicles. A major factor for this colossal number is free parking on roads. Unlawful parking along roadsides and alleys has contributed to unruliness; striders are unable to use walkways due to occupation of those spaces by vehicle users. As an alternative, focused measures need to be initiated to bring down the number of private vehicles. PMC is evolving a new parking policy that recognizes the importance of access to roads for every citizen within the city; this shall involve greater emphasis upon public transportation, availability of space for striders and cyclists, and charging personal vehicle owners for utilizing parking space. Thus, parking would be controlled and closely monitored to give preference for higher priority uses and promote efficiency.

Air Pollution

The rise in number of moving vehicles on roads in Pune has contributed significantly to high concentration of contaminants and noxious wastes. It is setting off pollutants in ambient air to ascend sharply. Of particular significance is the rise in nitrogen dioxide (contingent upon vehicular emissions) and particulate matter (PM); these are perilous to human health and increase the incidences of respiratory and cardiovascular disorders. The trends in NOx and PM10 levels at three prime locations in Pune are depicted in Figures 10, 11, 12, 13, 14 and 15. Environmentalists associated with National Air Quality Monitoring Programme suggest that although the amount of

Figure 10. Interannual variation of NOx at Bhosari, Pune (2000-2017)
Source: Pollution Control Board, Government of India

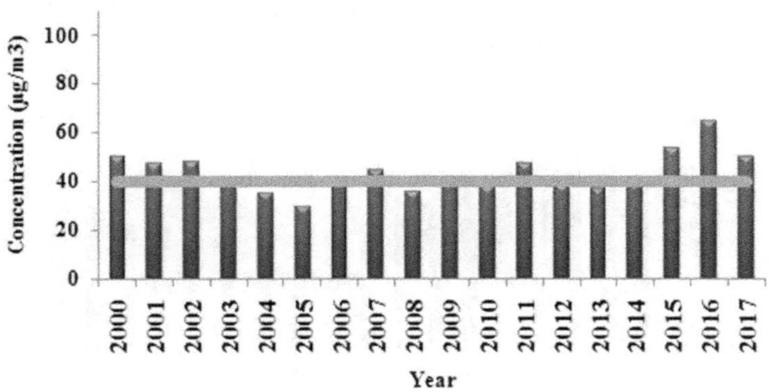

Figure 11. Interannual variation of NOx at Nal Stop, Pune (2001-2017)
Source: Pollution Control Board, Government of India

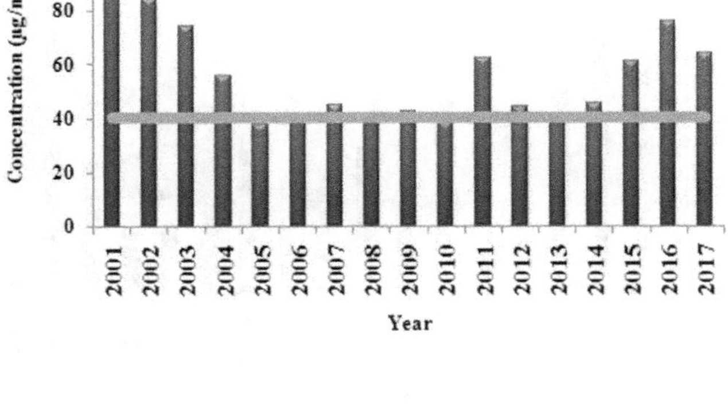

Figure 12. Interannual variation of NOx at Swargate, Pune (2000-2017)
Source: Pollution Control Board, Government of India

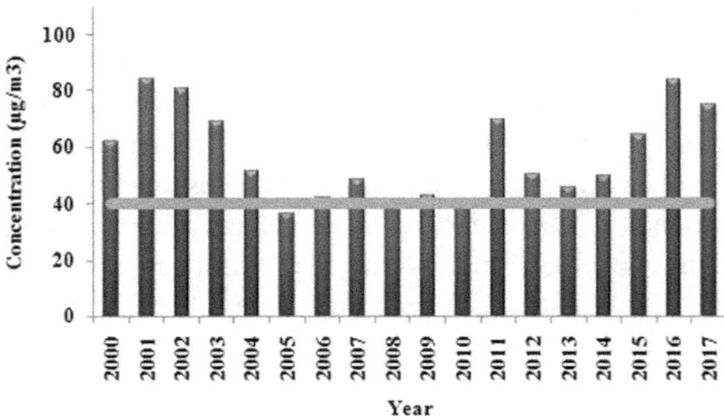

Figure 13. Interannual variation of PM10 at Bhosari, Pune (2000-2017)
Source: Pollution Control Board, Government of India

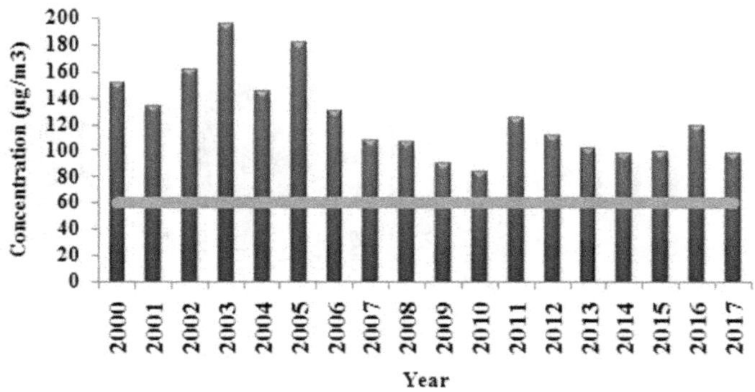

Figure 14. Interannual variation of PM10 at Nal Stop, Pune (2001-2017)
Source: Pollution Control Board, Government of India

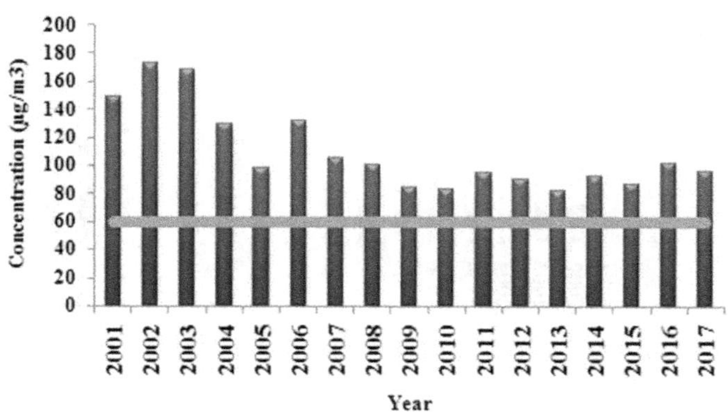

Figure 15. Interannual variation of PM10 at Swargate, Pune (2000-2017)
Source: Pollution Control Board, Government of India

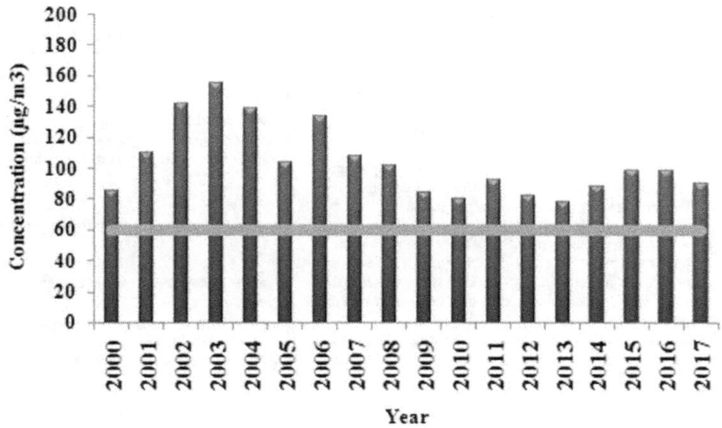

PM10 is showing a declining trend in recent years, it is still above permissible limits and poses considerable concerns to human health. Recent studies point a direct relationship between traffic congestion and air pollution. The higher the traffic jams, the greater is the extent of vehicular emissions and lesser is the ambient air quality (Zhang & Batterman, 2013).

The time is ripe to necessitate actions for controlling air pollution in Pune – a city which has for long enjoyed the reputation of being green city.

REMEDIAL MEASURES: ACTIONS TAKEN TO TACKLE TRANSPORTATION ISSUES IN PUNE SMART CITY – IOT PERSPECTIVE

The ambitious and elaborate smart city project commenced for Pune with a series of planned initiatives in target areas for infusion of technology from the standpoint of improving livability for the Puneties. One of the prime target areas has been effectual, economical & 'smart' multimodal transportation arrangement that is accessible to all residents. It is significant to ensure effective mobility by optimum utilization of prevailing stock and well-planned schemes and investments for highest bang for the buck (Pune Smart City – Vision Document (Version 1.0) Volume – II, n.d.). Table 5 lists the initiatives taken so far to improve transportation and combat traffic congestion as well as scope for further improvement through employment of big data and IoT.

Table 5. Smart city initiatives for transportation in Pune – present scenario and future prospects

Dimension	Initiatives Taken	Further Scope
Supply-side Policies	• Mass Rapid Transit System (MRTS) – Pune Metro Rail project for improved connectivity across the city has begun after significant delays and numerous obstacles spanning across land possessions and environmental consents. • Bus Rapid Transit System (BRTS) called 'Rainbow' stretching over 30 kms is proposed. This is an important step in the milieu of the fact that 16 kms of the stretch out of intended network of 68.80 kms is completed.	• Such additional travel options with real-time tracking and passenger information can significantly deal with supply-side bottlenecks in transportation system.
Automatic Vehicle Location System (AVLS) by means of Commuter Information Plan	• Vehicle Tracking System (VTS) that enables real-time tracking of buses by means of GPS setup; 1,750 buses are presently equipped with GPS tracking system (Bari, 2017) • Information concerning arrival time of buses is available on PMPML website • PMPML website provides details on standard service and nighttime service; planned and actual trips; planned and actual buses running on roads; availability of passes and mobility cards; tariffs for differential routes on the basis of distance, types of commuters, category of service availed; ridership; grievances – received, addressed and closed, etc. • PMPML also ensures availability of information concerning bus routes, schedules and timings through mobile apps.	• Mixed response to mobile app, mostly inclined towards dissatisfactory • Response to mobile app is summarized as follows: → Welcomed by passengers → Contains outdated information and fails to reflect up-to-date information → Not-user friendly • There is further requirement to improve real-time tracking of buses through installation of GPS by extending it to a larger number.
Automatic Fare Collection System (AFCS) through Common Mobility Card	• Mobile Integration (MI) card, which is a contactless smart card, has been introduced in January 2017 and rolled out in phases • Facilitates cashless payments • This is known to be a single payment card for all modes of transport – PMPML buses, metros, railways, cabs, parking, etc. • Forestalls ticketless travel and ensures greater transparency in collection of fares • This arrangement shall run parallel alongside cash payments; nevertheless, to encourage the usage of this card, more discounts and concessions will be given to users	• Execution is taking longer than expected; still in implementation stages • The setting up of requisite infrastructure for electronic ticket issuing has been a challenge. • The usage of this card was enforced upon employees of PMPML and all civic body-run schools; this too involved more time than projected. • Although PMPML has called forth disabled persons to migrate to this form, the response has been disheartening from beneficiaries. • If implemented successfully through extended outreach, this will be a major IoT initiative for Pune smart city.
City Command Control Centre	• Video Management System for close inspection and surveillance • Integrated Traffic Management for traffic detection; traffic re-routing apps founded on real-time traffic information • Transport Management System for easing transportation for citizens such as electronic ticketing system • Event Management System for crisis management and emergency warnings • Flood Monitoring System • Parking Management System	• A huge headway is made in this regard and the role of IoT towards recognizing and responding to the needs of citizens has been impactful. • However, there is vast scope for further improvement as the outreach has been limited and few initiatives have witnessed considerable delays. • Measures for adaptive traffic management for smart traffic project are still underway.
Traffic Mobility Plan	• This is used to collect information on people's whereabouts, activities and travels through Internet of Things (IoT) sensors; data so obtained can be utilized for creating an enhanced traffic mobility plan	• It is yet to be fully implemented and may involve sizeable time considering the delay in execution of other smart city projects.
Smart Traffic System	Gadkari (2017) explains execution of smart traffic system in Pune city: • Implementation to be in phases instead of all at once • Syncing of traffic signals such that the time span of red light and green light can be attuned and in sync with the traffic flow • Updates on signals shall be shared on mobile application • People having access to mobile app shall get alerts about traffic jams, mishaps or mayhem in specific areas • The app shall also suggest options/ other paths to reach the destination	• This is intended to be rolled out in phases to ensure economy and efficiency through trial and error

continued on following page

Table 5. Continued

Dimension	Initiatives Taken	Further Scope
Vehicular Rules Violation Monitoring	Banerjee (2017) discusses effectiveness of linking e-challan system with the city's CCTV camera to check unbridled traffic outlawry. • A traffic lawbreaker receives an SMS notifying his contravention, along with photo confirmation of the same, as well as a link directing him to the nearby police station where the fine can be paid. • Fines may be paid through cards as swipe machines are available with city police to facilitate cashless payments. • Currently, about 1,250-odd CCTV cameras are installed across the city, providing live feed to the main police control room; a screen-grab of infringement is captured from this footage as photographic proof, which is saved online. • Text messages are composed and sent to the mobile number of the lawbreaker, along with the e-challan	• This arrangement has worked wonders with respect to controlling infringement in traffic rules. • The system, however, faces following challenges: → Non-payment of fines owing to technical glitches with swiping machines or non-availability of debit/credit cards with lawbreakers to make timely payments (Bengrut, 2017) → Inability to catch hold of violators of rules by government vehicles as they are not registered in the names of specific individuals Banerjee (2017)
Walkways and Bicycle Tracks in Selected Areas	• Measures for allotment of ample space to striders and delineated tracks for cyclists have commenced in Aundh, Pune.	• There is further scope to have marked out zones to bring about mixed land use for diverse activities and ascertain a sustainable mobile ecosystem underscored by tracking of human movements through IoT sensors.
Parking System and Road Use Payments	• Revised parking policy has come into effect that lays emphasis on supply aspect in terms of limited capacity of roads to provide parking space for all motor vehicles, rather than the demand for the same. • Policy envisages payment for using parking spaces on a graded scale to limit the use of personal vehicles • Parking fees shall be contingent upon area rentals, motor vehicle size and traffic congestion in a given area • There will be vehicle-free zones in busy areas of the city. • Parking shall be prohibited on walkways and cycling tracks.	• Smart parking solutions are yet to be extended across the city. • The city is yet to make great strides with respect to real-time tracking of vacant parking spaces that can save time in searching for them.

Source: Various Sources

RECOMMENDATIONS

The rapid progression of smart city and Internet of Things applications creates numerous logical and designing difficulties that call for astute research endeavors from both scholarly world and industry, particularly for the advancement of proficient, adaptable, and dependable smart city in view of IoT. Pune needs prompt arrangements and urban body's regard for matters concerning transportation through citizen engagement. Citizens' apathy and indifference can do no good for the city. It is important to realize the needs of citizens and tailor the policies suiting their requirements. It is equally important to make them realize their responsibilities in this process. In addition to providing an effective public transport system and pedestrian-friendly roads, the administration needs to take cues from countries around the world, who have struck the chord for betterment of lives for their masses – Beijing for its public bus service; London, Seattle and Sydney for their Automatic Vehicle Location System; Hong Kong and London for Automatic Fare Collection System; New York and Seattle for Smart Parking System; Spain, Sweden, France for Smart Bike-sharing System; Paris for Electric Car-sharing System; and Sydney for its Coordinated Adaptive Traffic System. Another prerequisite for the IoT system to be prepared sooner is the correct sort of skill development and capacity enhancement for data innovation experts. Furthermore, the administration ought to consider the security concerns of employing big data. The gathering of substantial information and the utilization of analytics obviously ensnare privacy concerns. The targets of guaranteeing information security and ensuring protection end up obviously harder as data is duplicated and shared at all times broadly around the globe. Data with respect to people's wellbeing, location, power utilization, and online action is out in the open for scrutiny, raising worries about profiling, separation, exclusion, and loss of control. Policymakers should likewise address the concerns pertaining to individual consent in privacy and protection of confidentiality.

DIRECTIONS FOR FUTURE RESEARCH

Each city has its own established setup and concerns that need to be addressed for citizen welfare. While the smart city initiative is launched in India with the objective of creating 100 smart cities, it is important to identify how the efforts for advancing these cities have fared in terms of administration's capacity to introduce progressive measures and people's responsiveness and acceptance of the same. Future research can be centred on ways and means adopted by the administration of each city towards making the city smart and livable, the time dimension across which these means were adopted, the period taken for attaining the desired outcomes and citizen engagement to make the mission a grand success. The findings of such studies can provide meaningful insights to other cities intending to embark upon similar initiatives. Future researchers can also aim to establish people's willingness to share their personal information in the context of privacy concerns and evolve ways to ensure anonymity and confidentiality. They should come up with a model that strikes a balance between the advantages of data for businesspersons and researchers and potential gains/losses arising out of privacy concerns of individuals.

CONCLUSION

Pune, the city which has witnessed extensive development in diverse domains and arenas in the last three decades, began its smart city journey with targeted initiatives for improving the quality of life of citizens. The rationale has been to improve access to basic services through greater infusion of ICT initiatives. The present chapter, which primarily focused upon transportation sector in Pune and issues surrounding therein, wraps up with the finding that the city has a long way to go with respect to perking up mobility. Transport is one of the few areas where the city lags behind compared to its urban counterparts in the country. Although big data and IoT can make a considerable impact in realizing the smart city objectives for efficient transportation in Pune, they can serve as complementary measures to supply-side policies for expanding travel alternatives for the residents. Public transport has been ineffectual, auto rickshaws have been rigid and pricey, thus making it indispensable to possess personal vehicles or resort to app-based cab services. Traffic congestion and air pollution have been the obvious outcomes as a consequence. Although flyovers are constructed for strategic routing of vehicles and parking policies are redesigned to accommodate a rising number, these are short term solutions to a longstanding problem. The administration realizes its responsibilities and is making efforts to holistically deal with the issue. As mentioned earlier, only when public transportation picks up and a large number of travel options are made available to citizens, ICT can contribute to easing of traffic congestion in the city.

ACKNOWLEDGMENT

The authors express gratitude to Mr. Shashikant Nehul, research scholar at Department of Environmental Sciences, Savitribai Phule Pune University, for providing datasets relating to vehicular density in Pune (RTO data) and air quality in Pune (interannual variation of NOx and PM10 in diverse parts of the city). He is presently associated with National Air Monitoring Program funded by Central Pollution Control Board, Government of India, and may be reached at ssnehul@yahoo.com.

REFERENCES

Al Nuaimi, E., Al Neyadi, H., Mohamed, N., & Al-Jaroodi, J. (2015). Applications of big data to smart cities. *Journal of Internet Services and Applications*, *6*(1), 25. doi:10.118613174-015-0041-5

Babar, M., & Arif, F. (2017). Smart urban planning using Big Data analytics to contend with the interoperability in Internet of Things. *Future Generation Computer Systems*, *77*, 65–76. doi:10.1016/j.future.2017.07.029

Banerjee, S. (2017, April 16). CCTV and e-challans help reign in Pune's unruly traffic. *The Hindu*.

Bari, P. (2017, July 4). One year on, many projects of PM Modi's Smart City initiative still in pipeline. *Hindustan Times*.

Bengrut, D. (2017, June 17). 82% Traffic Violators Haven't Bothered To Pay Up E-Challan Fines. *Pune Mirror.*

Biswas, P. (2017a). *Pune RTO sees 25 per cent rise in vehicle registration on Dussehra. The Indian Express.*

Biswas, P. (2017b). *Pune, cab aggregators leading vehicle growth. The Indian Express.*

Census 2011 - Census of India. (2011). Retrieved 30 December 2017, from http://www.censusindia.gov.in/2011census/PCA/PCA_Highlights/pca_highlights_file/India/Chapter-1.pdf

Chandiramani, J., & Airy, A. (2017). Urbanization and Socio-Economic Growth in South Asia Region. In U. Benna & I. Benna (Eds.), *Urbanization and Its Impact on Socio-Economic Growth in Developing Regions*. Hershey, PA: IGI Global.

Chauhan, S., Agarwal, N., & Kar, A. (2016). Addressing big data challenges in smart cities: A systematic literature review. *Info*, *18*(4), 73–90. doi:10.1108/info-03-2016-0012

Chen, M., Mao, S., & Liu, Y. (2014). Big data: A survey. *Mobile Networks and Applications*, *19*(2), 171–209. doi:10.100711036-013-0489-0

Chen, R., Fung, B., Desai, B. C., & Sossou, N. M. (2012, August). Differentially private transit data publication: a case study on the montreal transportation system. In *Proceedings of the 18th ACM SIGKDD International Conference on Knowledge Discovery and Data Mining* (pp. 213-221). ACM. 10.1145/2339530.2339564

Ciolli, R., Whyte, P., Ercan, G., & Mack, A. (2003). *U.S. Patent No. 6,546,119.* Washington, DC: U.S. Patent and Trademark Office.

Cities Profile of 20 Smart Cities (2017). New Delhi: Ministry of Housing and Urban Affairs, Government of India. Retrieved from http://smartcities.gov.in/upload/uploadfiles/files/Maharashtra_Pune.pdf

Coverage & Duration: Smart Cities Mission, Government of India. (2017). *Smartcities.gov.in*. Retrieved 30 December 2017, from http://smartcities.gov.in/content/innerpage/coverageduration.php

Dargay, J., Gately, D., & Sommer, M. (2007). Vehicle ownership and income growth, worldwide: 1960-2030. *Energy Journal*, 143–170.

Dash, D. (2017, November 29). Deaths in road accidents see steepest decline ever. *The Times of India.*

Dastane, S. (2017, March 23). 4 lakh vehicles add to on-road clutter this fiscal. *The Times of India.*

Datta, P., Kaur, N., & Garg, N. (2015). Automatic Bus Fare Collection System in India. In J. Mandal, S. Satapathy, M. Kumar Sanyal, P. Sarkar, & A. Mukhopadhyay (Eds.), *Information Systems Design and Intelligent Applications* (pp. 681–689). New Delhi: Springer. doi:10.1007/978-81-322-2247-7_69

Debnath, A. K., Chin, H. C., Haque, M. M., & Yuen, B. (2014). A methodological framework for benchmarking smart transport cities. *Cities (London, England), 37*, 47–56. doi:10.1016/j.cities.2013.11.004

Eremia, M., Toma, L., & Sanduleac, M. (2017). The Smart City Concept in the 21st Century. *Procedia Engineering, 181*, 12–19. doi:10.1016/j.proeng.2017.02.357

Gadkari, S. (2017, May 4). Syncing signals for traffic ease. *Pune Mirror.*

Government of India. (2011). Census of India. *Censusindia.gov.in.* Retrieved 30 December 2017 from http://www.censusindia.gov.in/2011census/PCA/PCA_Highlights/pca_highlights_file/India/Chapter-1.pdf

Gubbi, J., Buyya, R., Marusic, S., & Palaniswami, M. (2013). Internet of Things (IoT): A vision, architectural elements, and future directions. *Future Generation Computer Systems, 29*(7), 1645–1660. doi:10.1016/j.future.2013.01.010

Hashem, I., Chang, V., Anuar, N., Adewole, K., Yaqoob, I., Gani, A., ... Chiroma, H. (2016). The role of big data in smart city. *International Journal of Information Management, 36*(5), 748–758. doi:10.1016/j.ijinfomgt.2016.05.002

Idris, M. Y. I., Leng, Y. Y., Tamil, E. M., Noor, N. M., & Razak, Z. (2009). Car park system: A review of smart parking system and its technology. *Information Technology Journal, 8*(2), 101–113. doi:10.3923/itj.2009.101.113

Interstate migration in India doubled between 2001-2011: WEF. (2017, October 25). *The Economic Times*

Joshi, B. (2016, June 29). Smart Cities Mission: Flaws in a flagship programme. *The Hindu.*

Joshi, Y. (2017, July 7). Pune is in a jam: Why flyovers haven't helped traffic congestion. *Hindustan Times.*

Kanungo, A., Sharma, A., & Singla, C. (2014, March). Smart traffic lights switching and traffic density calculation using video processing. In 2014 Recent Advances in Engineering and computational sciences (RAECS) (pp. 1-6). IEEE. doi:10.1109/RAECS.2014.6799542

Khairnar, A. (2017, July 29). Pune has equal number of vehicles, people: PMC's environment status report. *Hindustan Times*.

Khan, R., Khan, S. U., Zaheer, R., & Khan, S. (2012, December). Future internet: the internet of things architecture, possible applications and key challenges. In *2012 10th International Conference on Frontiers of Information Technology (FIT),* (pp. 257-260). IEEE. 10.1109/FIT.2012.53

Khanna, P. (2015). Urbanization, Technology, and the Growth of Smart Cities. *Asian Management Insights, 2*(2).

Kourtit, K., Nijkamp, P., & Steenbruggen, J. (2017). The significance of digital data systems for smart city policy. *Socio-Economic Planning Sciences, 58*, 13–21. doi:10.1016/j.seps.2016.10.001

Krishnamurthy, R., Mishra, R., & Desouza, K. (2016). City profile: Pune, India. *Cities (London, England), 53*, 98–109. doi:10.1016/j.cities.2016.01.011

Kulkarni, P. (2017, September 3). Smart city projects in deadline crisis. *Pune Mirror*.

Lee, W., Tseng, S., & Shieh, W. (2010). Collaborative real-time traffic information generation and sharing framework for the intelligent transportation system. *Information Sciences, 180*(1), 62–70. doi:10.1016/j.ins.2009.09.004

Marr, B. (2015). How Big Data and the Internet of Things Create Smarter Cities. *Forbes.com*. Retrieved 15 November 2017, from https://www.forbes.com/sites/bernardmarr/2015/05/19/how-big-data-and-the-internet-of-things-create-smarter-cities/#27e75b361767

Mathur, N. (2017, March 15). Indian cities perform poorly in Mercer 2017 Quality of Living index. *Livemint*.

Midgley, P. (2009). The role of smart bike-sharing systems in urban mobility. *Journeys, 2*(1), 23–31.

Patel, C., Shah, D., & Patel, A. (2013). Automatic number plate recognition system (ANPR): A survey. *International Journal of Computers and Applications, 69*(9).

Pelletier, M., Trépanier, M., & Morency, C. (2011). Smart card data use in public transit: A literature review. *Transportation Research Part C, Emerging Technologies, 19*(4), 557–568. doi:10.1016/j.trc.2010.12.003

Poole, S. (2014, December 17). The truth about smart cities: 'In the end, they will destroy democracy'. *The Guardian.*

Prabhakar, V., Gupta, S., & Mehrotra, R. (2015, September 2). Smart transportation for Smart Cities. *The Economic Times*.

Pune Smart City – Vision Document (Version 1.0) (Vol. 2). (n.d.). Pune: Pune Municipal Corporation.

Puri, A. (2014, August 15). What are smart cities? *The Hindu*.

Puri, T. (2015, January 6). Parking chaos needs a quick fix first. *The Times of India*.

Rangarajan, A. (2010, January). BRTS-Bus Rapid Transit System in Pune Modeling, Simulation and Feasibility Analysis. In *Proceedings of the 2010 International Conference on Industrial Engineering and Operations Management*, Dhaka.

Rathore, M., Ahmad, A., Paul, A., & Rho, S. (2016). Urban planning and building smart cities based on the Internet of Things using Big Data analytics. *Computer Networks*, *101*, 63–80. doi:10.1016/j.comnet.2015.12.023

Realizing Smart Cities with Technology - A KPIT Perspective. (2016). Pune: KPIT Technologies Ltd. Retrieved from https://www.kpit.com/resources/downloads/whitepapers/realizing-smart-cities-technology-whitepaper.pdf

Reddy, B. S., & Balachandra, P. (2012). Urban mobility: A comparative analysis of megacities of India. *Transport Policy*, *21*, 152–164. doi:10.1016/j.tranpol.2012.02.002

Reimagining Pune: Mission Smart City. *Punesmartcity.in*. Retrieved 30 December 2017 from https://punesmartcity.in/wp-content/uploads/2017/10/SPC.pdf

Report of the Sub-Committee on Automatic Fare Collection System. (2013). New Delhi: Government of India, Ministry of Urban Development.

Rodríguez Bolívar, M. (2015). Smart Cities: Big Cities, Complex Governance? In M. Rodríguez Bolívar (Ed.), *Transforming City Governments for Successful Smart Cities*. Switzerland: Springer.

Sanjai, P. (2017, September 11). India's traffic is so bad, it's changing the cars people buy. *Livemint*.

Sekhsaria, P. (2004, November 1). The future of Pune's public transport. *Indiatogether.org*. Retrieved from http://www.indiatogether.org/pmtplans-economy

Shaheen, S. (2004). Dynamics in behavioral adaptation to a transportation innovation: A case study of Carlink–a smart car-sharing system.

Sherly, J., & Somasundareswari, D. (2015). Internet of Things Based Smart Transportation Systems. *International Research Journal of Engineering and Technology*, *2*(7), 1207–1210.

Shinde, S. (2017, July 6). Spike in autorickshaws without permits on Pune roads. *Hindustan Times*.

Shinde, S., & Jagtap, S. (2016). Big Data Solution for Improving Traffic Management System Effectively for Vehicular Networks. *International Journal of Advanced Research in Computer and Communication Engineering*, *5*(6).

Smart City Features: Smart Cities Mission, Ministry of Housing and Urban Affairs, Government of India. (2017). Smartcities.gov.in. Retrieved 30 December 2017, from http://smartcities.gov.in/content/innerpage/smart-city-features.php

Smart Pune's Step Towards Sustainable Transportation: Public Parking Policy (Draft). (2016). Pune. Retrieved from https://pmc.gov.in/informpdf/Road/PMC_Draft%20 parking%20policy_English%20version.pdf

Strategy: Smart Cities Mission, Ministry of Housing and Urban Affairs, Government of India. (2017). *Smartcities.gov.in*. Retrieved 30 December 2017, from http://smartcities.gov.in/content/innerpage/strategy.php

Tachizawa, E. M., Alvarez-Gil, M. J., & Montes-Sancho, M. J. (2015). How "smart cities" will change supply chain management. *Supply Chain Management*, *20*(3), 237–248. doi:10.1108/SCM-03-2014-0108

Townsend, A. (2014). *Smart Cities: Big Data, Civic Hackers, and the Quest for a New Utopia*. New York: W.W. Norton & Company.

Trépanier, M., Tranchant, N., & Chapleau, R. (2007). Individual Trip Destination Estimation in a Transit Smart Card Automated Fare Collection System. *Journal of Intelligent Transport Systems*, *11*(1), 1–14. doi:10.1080/15472450601122256

Umbrajkar, M. (2012, January 20). In last decade, population of Pune urban agglomeration grew 34%. *The Times of India*.

What is Smart City. (2017). New Delhi: Ministry of Housing and Urban Affairs, Government of India. Retrieved from http://smartcities.gov.in/content/innerpage/what-is-smart-city.php

Yin, S., & Kaynak, O. (2015). Big Data for Modern Industry: Challenges and Trends [Point of View]. *Proceedings of the IEEE*, *103*(2), 143–146. doi:10.1109/JPROC.2015.2388958

Zhang, K., & Batterman, S. (2013). Air pollution and health risks due to vehicle traffic. *The Science of the Total Environment*, *450-451*, 307–316. doi:10.1016/j.scitotenv.2013.01.074 PMID:23500830

ADDITIONAL READING

Ali, A. H., Atia, A., & Mostafa, M. S. M. (2017). Recognizing Driving Behavior and Road Anomaly Using Smartphone Sensors. *International Journal of Ambient Computing and Intelligence*, *8*(3), 22–37. doi:10.4018/IJACI.2017070102

Bhatt, C., Dey, N., & Ashour, A. S. (Eds.). (2017). *Internet of things and big data technologies for next generation healthcare.* doi:10.1007/978-3-319-49736-5

Dey, N., Ashour, A. S., & Bhatt, C. (2017). Internet of things driven connected healthcare. In *Internet of things and big data technologies for next generation healthcare* (pp. 3–12). Cham: Springer. doi:10.1007/978-3-319-49736-5_1

Dey, N., Hassanien, A. E., Bhatt, C., Ashour, A., & Satapathy, S. C. (Eds.). (2018). *Internet of Things and Big Data Analytics Toward Next-Generation Intelligence.* Springer. doi:10.1007/978-3-319-60435-0

Elhayatmy, G., Dey, N., & Ashour, A. S. (2018). Internet of Things Based Wireless Body Area Network in Healthcare. In *Internet of Things and Big Data Analytics Toward Next-Generation Intelligence* (pp. 3–20). Cham: Springer. doi:10.1007/978-3-319-60435-0_1

Matallah, H., Belalem, G., & Bouamrane, K. (2017). Towards a New Model of Storage and Access to Data in Big Data and Cloud Computing. *International Journal of Ambient Computing and Intelligence*, *8*(4), 31–44. doi:10.4018/IJACI.2017100103

Sarkar, M., Banerjee, S., Badr, Y., & Sangaiah, A. K. (2017). Configuring a trusted cloud service model for smart city exploration using hybrid intelligence. *International Journal of Ambient Computing and Intelligence*, *8*(3), 1–21. doi:10.4018/IJACI.2017070101

Solanki, V. K., Katiyar, S., Bhashkar Semwal, V., Dewan, P., Venkatasen, M., & Dey, N. (2016). Advanced automated module for smart and secure city. *Procedia Computer Science*, *78*, 367–374. doi:10.1016/j.procs.2016.02.076

Yang, W., Wang, X., Song, X., Yang, Y., & Patnaik, S. (2018). Design of Intelligent Transportation System Supported by New Generation Wireless Communication Technology. *International Journal of Ambient Computing and Intelligence*, *9*(1), 78–94. doi:10.4018/IJACI.2018010105

Zhang, W., Qi, Q., & Deng, J. (2017). Building Intelligent Transportation Cloud Data Center Based on SOA. *International Journal of Ambient Computing and Intelligence*, *8*(2), 1–11. doi:10.4018/IJACI.2017040101

Chapter 12
Songdo Smart City:
An Aerotropolis and a Ubiquitous City

Somayya Madakam
FORE School of Management, India

Rajesh M. Holmukhe
Bharati Vidyapeeth University (Deemed), India

ABSTRACT

South Korea is one of the most advanced developed countries. South Korea's industrialization, software and hardware products and services, technological advancements, and management styles are world-renowned concepts. It has also achieved many milestones in urbanization to provide a better life for its citizens. In this current global urban scenario, it is taking a grand leap forward to an even higher level in its development of Smart City structures. In this light, this book chapter discusses and explores South Korea's Songdo Smart City. The book chapter will give a 360-degree view of Songdo Smart City project—specifically with regards to planning, design, deployment of IoT technologies, focused on urban analytics. The chapter's major areas of focus will be looking into Songdo from the perspectives of its being (1) an Aerotropolis and (2) a Ubiquitous city. Songdo IBD is a new ubiquitous city built from scratch on 600 hectares of domesticated land along Incheon's seaside. With all the modern amenities and LEED-certified buildings, the Songdo Smart City will also survive in the future.

DOI: 10.4018/978-1-5225-6207-8.ch012

INTRODUCTION

Nowadays, there are lots of discussions happening in conferences; workshops; symposiums; doctoral colloquiums on urban planning, budgeting, designing, and development; and collaborations which lead towards new urban solutions. Real estate builders, civil architects, environmental specialists, software and hardware developers from multinational companies, policymakers across countries, and – last but not the least – urban living aspirants are coming together to build the new cities and also to convert the existing brownfield cities into 24/7 technologically - automated cities. They are planning cities for sustainability i.e. even for future generations as well. The reasons are that many parts of the world, the urban people are desperately looking for quality of life in their day-to-day life. They have lost patience with transportation congestion in the existing cities, and the high increase in air pollution. They are unable to endure the power shortages and power thefts or illegal usages. Some parts of each city lack a 24/ 7 potable water supply. They have lost their trust in the municipality because of the inadequacies of the civic public utility services, and due to the scarcity of urban infrastructure in terms of housing, toilets, schools, hospitals, and research & development (R&D) organizations. Unexpected urban migration has made the government handicapped in terms of providing essential services. Proper urban policies are lacking in the life of the city. In this urban scenario, the entire globe is striving towards building new automated smart cities for more sustainable quality of life. These cities are fully deploying Internet of Things (IoT) technologies. Thus, this chapter is basically describing the Songdo Smart City project. At first it is explaining an answer to the question "What does one mean by the term 'smart cities'?" Then, it addresses the smart cities scenarios in South Korea. Followed by, it is illuminating to see an in-detail view of a representative aerotropolis and ubiquitous city of Songdo City.

SMART CITIES

According to the great urban planner Adam Greenfield, we are currently observing a rare crusade in the olden times of urbanism, then onwards large-scale multinational companies like IT/ITes had never before been so extremely involved with the building up of new urbanism principles (Greenfield, 2013; Townsend, 2013). The reason behind the high growth of urbanization is that urban places, if they have rich enough - consist of all sorts of amenities. These amenities include housing, medical, educational institutions, shopping malls, tourist destinations, parks, religious places, power sector, and other city infrastructures. In many cities, the lakes, mountains, museums, man-made bridges, and woodlands are attractive places for tourists. Of course, the in-built environments really make life pleasant for citizens. However, the sudden migration to cities from the countryside; the natural birthrate increase; the inadequacy in urban infrastructure; the lack of civic sense in the usage of public utilities in terms of water, power, gas, soil, and trees; bureaucratic issues; and so on are causing city life to be miserable. In order to solve these urban issues, the new urban phenomenon came in the form of "Smart Cities." This phenomenon was initiated in practical form by the multinational company M/s. International Business Machines (IBM) as part of the "Smarter Planet" initiative in 2008. Of course, the inception of the smart cities concept emerged in the 1990s in the form of 'digital cities.' The data was in the form of '0' and '1's. Couclelis (2004) defined "Digital City is an all-inclusive of web-based demonstration or replica of several aspects a specific existing city across the world. Much more than being not only a technical but also practical issue, this digital city has many magnitudes that are related to social, cultural, political, ideological, financial and of course also theoretic".

The authors Tiwari and Jain (2014) already reported around the world that there exist many new smart cities which have been systematically designed from the scratch and built on green belt sites-cities such as Putrajaya in Malaysia; Living PlanIT Valley in Portugal; Lavasa, and Nano City in India; Songdo and Incheon in Korea; Gujarat International Finance Tec (GIFT); Dubai Waterfront and Dubai Central in United Arab Emirates; King Abdullah Economic City in Saudi Arabia; Wuxi Huishan, Dongtan and Meixi in China; Neapolis in Cyprus; and Skolkovo in Russia; etc. The smart' nature of these cities is also supported in the literature by authors Madakam & Ramaswamy (2016). On the other hand, several initiatives – mostly in developing countries - have been started in developing completely innovative smart cities, such as Cyberport Hong Kong in China; Skolkovo Innovation Center in Russia; Songdo in S. Korea; Smart City Iskandar in Malaysia; and Masdar in UAE (Angelidou, 2014). In the United States of America (USA), European, and Australian countries, smart city projects are less in number as compared to Asian countries. The reason behind this is that these new urban architectures in Asian countries and is the fact that cities are experiencing population overcrowding, inadequate infrastructure, western metropolitan adapting learnings, increases in both education and economic development due to...economic reforms. Another reason is that in European countries and in the United States, urbanization had already surpassed it. Smart Cities are urban environments newly built up for the purpose of resolving already existing urban issues. These smart cities also exist in many forms, based on their reflecting city developments. The different forms of smart cities include are Digital Cities, Intelligent Cities, Information Cities, Knowledge Cities, Ubiquitous Cities, Resilient Cities, Eco-Cities, Sustainable Cities, Low-Carbon Cities, Liveable Cities, and also mixtures of such as 'Low Carbon Eco-Cities and U-Eco-Cities (De Jong et al., 2015). Even though these cities are in different forms, the ultimate aim of all these city constitutions of cities is providing quality of life for city dwellers, economic development for the region, and sustainable cities for future generations. The cities are smart enough to manage urban components. One set of people say that Smart Cities are very 'Specific,' 'Measurable,' 'Accountable', 'Relevant,' and 'Timely' ('SMART'). Another set of people say that 'Smart Cities' can be understood as an acronym for 'Sustainable Management Action Resource Tools for Cities' ('SMART Cities') (Madakam, & Ramachandran, 2015). However, the above author defines smart cities as follows: "Smart Cities provide quality of life to the city people, besides look for environment balance and economic development of next generations using with completely IoTs. These IoT tools in general tie the knot with city subsystems of mobility, safety, and supremacy, along with other community services to bring the operational efficiency. The IoT technologies include smart cards, radio frequency identification, quick response codes, electronic product codes, Internet Protocol Version 6, sensors, actuators, wireless fidelity, Bluetooth, ZigBee, near field communication, geographical information systems, geographic position system, social media, business intelligence, ambience intelligence, machine learning, cloud computing, telemedicine, Web 3.0, analytics, etc.

Smart Cities in South Korea

Around in the late 1980s, in South Korea, as the housing scarcities situation turn into decidedly eviler and the surviving existing land for large-scale urban growth was became almost exhausted, so the people began to spill over away from the green belt. Challenged with boundaries in the supply of land available for city development, the Korean government began to build a number of new towns in the Seoul Metropolitan Region-towns including places of Jungdong (Bucheon), Bundang (Sungnam), Sanbon (Gunpo), Pyeongchon (Anyang), and Ilsan (Goyang). The government started the establishment of new cities in the empty areas of the country and economic benefits by inviting foreign investors. So, the Korean governors are started following the concept of not exactly "Smart City" but in another name for "Ubiquitous-City." In South Korea, the novel idea of a "U-City" (Ubiquitous-City) pulls a lot of attention then and now. Even the Journalists Association of Korea (JAK), in early April 2017, reported witnessing that now is a good opportunity for seeing Korea's economy and urban development progress. It is tried and trying to deploy advanced modern physical infra and omnipresent info services into pockets of new city infrastructure. In addition, it is thought to bringing innovative urban solutions. At last, it is moving forward towards becoming a Ubiquitous Society by an intelligence-based civilization where humans, computers, and things are combined (Jang & Suh, 2010). Another study also points out that South Korea is currently pursuing a national strategy of applying 'ubiquitous technology' in urban planning, as part of which Sejong, Songdo and Incheon are advancing as 'ubiquitous eco-cities' (Kim et al. 2009; Shwayri 2013). South Korea's Ubiquitous-Eco (U-Eco) city development package contains of (1) An entire team is responsible for tactical marketing, research, planning, designing and testbed construction etc. (2) Main teams are in authority with planting city physical infra and allied urban services. The city estimated whole budget for all U-Eco-city projects starting from 2007 through 2013 was approximately US$130 million (Lee et al., 2011). For instance, the S. Korean smart cities use pervasive computing to measure various environmental indices (Lee et al., 2009). These are used by the municipalities to carry out respected people involvements for the enhancement of urban services (Anthopoulos et al., 2012).

Another study points out that S. Korea's new administrative center, styles itself as being a 'high-tech eco-friendly city' (Republic of Korea 2010). The Dutch Ministry of Economic Affairs, Agriculture, and Innovation signed a MoU (Memorandum of Understanding) with the Korean Prime Minister's Office in April 2010, concerning the sustainable growth of the Saemangeum area. The objective of this MoU is the cooperation and exchange of Information and Communication Technologies (ICT) between both countries in the areas of sustainable agriculture, water, dredging, design and architecture, renewable energy, ecology, and recreation. Joss et al.'s (2013) manuscript saying that Sejong is a brand-new city currently under construction in central South Korea. Another study by Yigitcanlar & Lee (2014) reporting that in S. Korea, there are 5 eco-city projects already developed of (1) Incheon Eco-City (2) Gwanggyo Ekonhill (3) Sejong (4) Magok Seoul and (5) Songdo IBD. These smart cities have been established under the South Korean urban policy program. New Songdo (Hyang-Sook et al., 2007) and Osaka are the first cases of 'u-cities,' in which ICT were deployed during metropolitan expansion relatively just being applied in the city-suggesting that future urban planning will possibly set Information and Communication Technologies as an axis or a dimension of precedence (Anthopoulos, & Fitsilis, 2010). Thus, like other countries which give importance to building smart cities designed to provide quality of life to its citizens. Thus, South Korea is planting a good amount of smart cities in the name of growing Ubiquitous Cities. These cities definitely platforms for civic services.

SONGDO SMART CITY: CASE STUDY

Geographical Developments

The "Songdo International Business District," is also known as "Songdo IBD". Songdo IBD is an entirely upcoming Smart City from scratch. Songdo is too calling as "The Experimental Prototype Community of Tomorrow." The word "Songdo" in

Mandarin is "Sōngdǎo"; and in Japanese, it is "Matsushima." The Songdo Smart city is a ubiquitous city, built from the scratch. The total city area spread crosses, 500 acres that is almost 1 600 hectares. The best part of this ubiquitous city constitution is that it is located along with Incheon's waterfront, in the domesticated land, 40 miles i.e. 65 kilometers in the southwest of Seoul, the principal and administrative city of South Korea. That means the Songdo IBD is a new city which is in final phase. It is being built on flat land of recovered wetland. The city is connected to the Korean mainland by electronically-monitored bridges. It is tactically located close to IIA (Incheon International Airport). The best part is that the IIA is strategically located just 7.6 miles nothing but 12.3 KMs away. The city is connected by Incheon Bridge, a reinforced concrete highway. It is strategically located by taking one hour's drive from southwest of Seoul. Songdo Smart City is a part of the IFEZ (Incheon Free Economic Zone), along with Yeongjong and Cheongna. However, the plan set out back in July 2001 for Songdo Smart City by locating on a man -made island. Former President Mr. Lee Myungbak is the man behind Songdo Smart City. It is an extension of Seoul's commitment to urban innovation. Its development main motto is to push towards regulating Korean economy from an export-driven manufacture to rest of the world which is caused to a low-carbon, and knowledge-driven economy. The below are unique factors of the Songdo Smart City project development (Kim & Ahn, 2006):

- Songdo is a product of complicated interactions between city governments and national;
- Songdo is targeted towards the development of national and regional competitiveness by inviting Foreign Direct Investment (FDI);
- An international private investor, M/s. Gale, a Boston-based real estate company (USA) and Wentworth Company, another US firm - were also involved as key actors - a little that had before been unnoticed in the history of Korea's planning.

As a part of the Incheon Free Economic Zone (IFEZ), Songdo's development has benefited from opportunistic circumstances that have led to its pursuit of becoming a "Ubiquitous Eco-City". This Songdo smart city project acmes the importance of the interaction of local circumstances and other participants, helping to avoid risks and to enhance the future value of the project (Lee & Oh, 2008). Songdo is one of a trio of cities that make up the Incheon Free Economic Zone, a development initiated by the Korean government to attract residents and FDIs.

Sustainable Goals

The Smart City Songdo IBD is a sustainable city that foreshadows the ability of ecological urban growth by blending high-tech IoT with urban principles for effectiveness and environmentally-conscious urban life. The Songdo Smart City offers some of the most forward-thinking solutions for urban infrastructure and electric power issues in the world. The city is connected through underground tubes. Modern transportation modes tracking, water usage monitoring of city, power consumption payments, gas connectivity leakage reporting, fire alarms systems, citizen security information, etc., are controlled by smartphone devices; and city emissions are offset by a forest-sized park. The Songdo Smart City setup new standards for large-scale urban sustainable development. It was master-planned according to the principles of LEED-NC (Leadership in Energy and Environmental Design – Neighborhood Development). In Songdo city, the Cisco fully settings their "Smart + Connected Communities" program, which is nothing but Smart Living. The city produces less carbon dioxide (CO_2) emissions. The Hartford-based UTC's supporting of the Ubiquitous-Life in Northeast Asia Headquarters, which will be the first Leadership in Energy and Environmental Design Platinum-rated building in S. Korea. The six core design goals for Songdo IBD are given below, echo the designers' commitment to sustainability (Grover, 2013):

- Constructing Songdo International Business District in Incheon as a master-planned green city of an unprecedented scale. Open-space connection for all citizens is built into the city's plan, design, and layout. Native/adapted species are utilized all over the city's development;
- Targeting certification of the smart city under the LEED-Certification System (LEED-CS) or Leadership in Energy and Environmental Design–Neighborhood Development (LEED-NC) or the Korean Green Building Certification System (KGBCS) rating systems;
- Creating energy-efficient building projects for comfort life, i.e. built above the standards of the ASHRAE20 i.e. American Society of Heating, Refrigerating and Air-Conditioning facilities;
- Optimizing the use of reducing storm water runoff and of potable water and justifying the urban heat island effect. Encouraging biodiversity and species habitat preservation in the city;
- Recycling the materials i.e. locally-manufactured materials or utilizing locally-produced;
- Typical reutilizing of construction materials waste circa 75% of what ought to be recycled. The less usage of Portland cement by deployment with fly-ash-content concrete.

Collaborations

The Songdo Smart City has been stimulated as a nodal point in the multicenter metropolitan region. It is a gateway to North-East Asia. Living space for U-life is the prime motto for all citizens, ever since the ground-breaking thought process in Korea in 2004 onwards (Kim, 2010). The SIBD, organized through a MoU (Memorandum of Understanding) among the M/s. Gale International, M/s. United Technologies Corporation (UTC), and M/s. Hanjin Group (Chun Go-eun, 2007). The New Songdo International City (NSIC) Development LLC operates as a real estate development company. The company - which was founded way back in 2002 and is based in Incheon. S. Korea-made a MoU to revise the land usage plan, purchasing of land, obtain approval from government and implement the property growth.

The LLC NSIC Expansion comprises M/s. Gale International as a managing partner as mentioned above. The M/s. Pohang Iron and Steel Company (POSCO), which is a Korean based domestic company too. The Songdo Smart City project is highly international in nature, meaning that it includes a foreign company architect and lead developer. That means it is much financed by foreign capital. And it is aimed at international world-class residents or occupants. Besides, structures imaginative conceptual planning, designing both local and international developer partnerships, sophisticated investment and financing techniques. Songdo's master plan, developed in collaboration with the US-based company Gale International, had been projected to take into account the environmental impact of every aspect of the city and upon scheduled completion of the plan since 2015 (O'Donnell, 2010). M/s. Cisco, the networking giant and the San-Jose-based networking company systems have stepped into to wire Songdo City from top to bottom with an amount of US$47 million as an investment. Of course, we know that this effort was made as a part of Cisco's "Smart+Connected Communities" initiative (Strickland, 2011; Singer, 2012). South Korea's largest steel producer, M/s. Posco-is in partnership with the Korean Government with financing operations with the help of the World Bank, ABN AMRO, Morgan Stanley, and KB Kookmin Bank (Korean: KB 국민은행). M/s. Kohn Pederson Fox bagged the master architect business for the New Songdo IBD. However, there are other architects also involved by out sourcing of architecture to take charge of individual projects in Songdo including the HOK New York, KlingStubbins, and Dong-II Architects (Alusi et al., 2011). M/s. Arup, with expertise in sustainability, was engaged to support engineering and design work in the downtown area of the Songdo Smart City. The multinational vendor employed the state-of-the-art technologies in all constructions-deploying a fiber network that connects all the components of the city components including houses, government buildings, research and development, hospitals, colleges and in even public spaces including parks and libraries. Technology companies 3M and Microsoft agreed to create a world-class International City Hospital.

Quality of Life: Amenities

Strickland, (2011) reveals that the primary developer, behind the business district, M/s. Gale International, constructed towers that meet strictly the green building standards; laid out neighborhoods intelligent design; and created an urban oasis model like in Central Park at New York City. The highlights of the Songdo master plan comprise many landmarks such as Korea's tallest building including Northeast Asia Trade Tower, 68-story building with its' most advanced corporate center; Songdo Convensia, Incheon's primary convention center and Korea's largest column-free interior space. And also, the city comprises the Incheon Arts Center - houses a concert hall, an opera house, Asian contemporary art (museum), a music conservatory, a design school, artist-in-residence housing, and a library (Grover, 2013). However, the state-of-the-art Convensia is Asia's first LEED-certified convention center. A unique way to experience culturally and historically is by the Gyeongwonru Banquet Hall, which built in the renowned jump style of the Goryo Dynasty. Cisco's vision of the new urban home is the 63rd-floor penthouse. An electronic screen on the building wall displays a menu regarding music, temperature, lights, and window blinds can be simply controlled (Strickland, 2011). The good news for the high bureaucrats who are likely to send their kids abroad for education is that a foreign school in Songdo, which is open to local children (Jun-Woo & Ahn, 2011). The newly set up Songdo's both public or private schools are the region's best talented school and world class academic education. Besides, the New Songdo International City (NSIC) Development constructed a park, a golf course, and a hotel, focusing on overall quality of life - to attract businesses. The Songdo Smart City will contain luxurious facilities in terms of smart homes, smart buildings, shopping malls, hospitals, and cultural amenities; global universities, with replicas of landmarks such as New York City's Central Park as said above and Venice's waterways to be incorporated as well. Striking views of Central Park, the Yellow Sea, and the Jack Nicklaus Golf Club Korea complement the highest standards of comfort and convenience. The International City Hospital will offer the modernized medical diagnosis and treatment (Grover, 2013). Songdo is one type of residential city provided with a basic social service as its campus especially university education system. Four American universities going to built-up their own offshore campus in Songdo Smart City, therein offering its residents a top-level North American education whilst residing in South Korea itself. Not only Korean students but also International students take advantage of established national as well as global educational opportunities. This education-based system initiative bolsters government efforts to forge a knowledge-based economy. Yonsei University is building full-sized, degree-granting campuses in Songdo, just outside of Seoul (Looser, 2012). Besides, some of the educational institutions listed here are the Chadwick International (PK-12), the Mason Korea - George Mason University, the Ghent University Global Campus, the State University of New York, and the University of Utah Asia Campus to name a few. The State University of New York - Korea is bringing a fashion design and fashion business management curriculum to South Korea. The Songdo considered as being Smart City is making Incheon a special economic zone - a crucial component of Seoul's economic landscape. The Incheon Free Economic Zone actively courts FDIs (Foreign Direct Investment) and significant tax incentives, efficient logistics, are supporting to become one of the world's "smartest" urban regions. Songdo' s novel feature and main objective is its green building program designed for 75,000 residents (Joss, 2011); whereas one study by Moser expects that the Songdo smart city will house up to 252,000 residents and circa 300,000-400,000 commuters in an area of 86 square-kilometers area. This city is basically intended to feature higher education, the free economic zone, and banking etc. (Moser et al., 2015). Songdo has been featured as the filming location of choice for some of KPOP's hottest cine stars. Songdo offers casual and gourmet dining choices for all enjoyments. One can eat, drink, play and shop at a leisurely pace - all while walking along the green perimeter of a public parks and promenades. The International Facility Management in Songdo set up the U-Life Management Center. It is an integrated city command control center designed to offer greater convenience in a 24/7 capacity to the businessmen, residents, and tourists at Songdo within the Incheon Free Economic Zone (IFEZ). The command control room is mounted by screens displaying hundreds of concurrent scenes transmitted by CCTV cameras. The newly-deployed corporate infrastructure will be actively utilized in proving basic public utility services like facility management, coordinated controlling, occupant management, and other value-added intelligent services which are resulting in increase in residents' quality of life.

South Korea's new built-in City Songdo delivers a lot civil services in 24/7, which are being developed as (1) an Aerotropolis as (2) a Ubiquitous City (Kshetri et al., 2014).

Songdo as an Aerotropolis City

The developing smart city in South Korea is an innovative city project with an outstanding scale. Songdo is the most expensive development projects in the history of the world, with a cost of approximately more than $40 billion. We know that this

Figure 1. Quality of Life - Amenities in Songdo

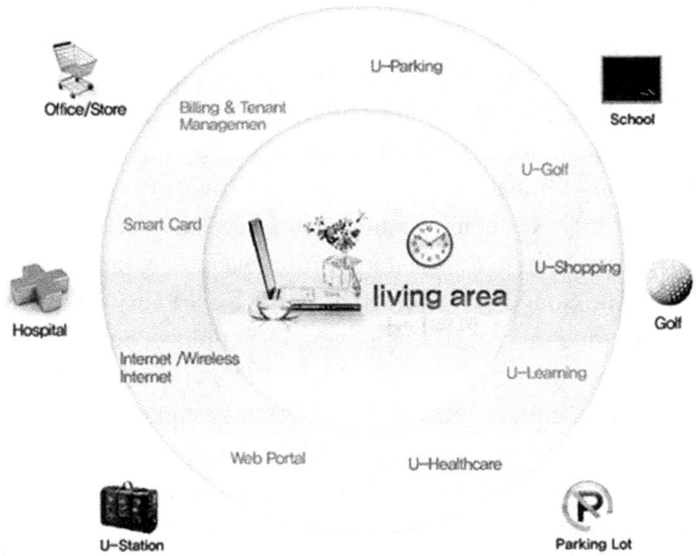

Figure 2. Songdo City

Aerotropolis city is a pan-Asian business district defined by travel distance from Songdo's travel hub. The Air City basic features includes a set of modern commercial multiplexes being established with all the amenities including contemporary metropolitan center, office buildings, firm logistics, retail business areas, high-tech assembly facilities, Information and Communication Technologies functions, recreation events, conference-cum-exhibition center, and etc. Besides, the elaborate national expressways, bridges, and underground tunnels join the international airport to Seoul and also to nearby islands; the latter forming an expensive commercial and residential complex (Kasarda, 2008). Songdo Smart City is playing dual role one is as an Aerotropolis and other as a Ubiquitous. These two characteristics are really bringing inspiration to the minds of many Korean urbanists with the city's innovative connectivity and synchronization of civic services setting a new standard for future city developments. The Incheon International Airport has been being usage from 2001 onwards to resolve the scarcities of Kimpo Airport. Gradually the Airport got increased to many non-Korean flights and handling the complex logistics required of existing airports (Yun, 2015). Chun (2007) says that the city was currently in its 3rd phase of its development; and experts predict that the its capacity will be expanded up to 7, 40,000 flights, will carry for 100 million passengers, and around 10 million tons/year of airborne goods. Perchance Songdo is the most aspiring to develop as an airport city. With this an Aerotropolis city is taking place around S. Korea's Incheon International Airport. The applicability of the Aerotropolis model to Songdo City is immense, due to the geographical dividend of its location, i.e. 1/3rd world population in China. The China is just a 3 and 1/2-hour flight journey from the Songdo City. Hence, the City can power to improve its 24/7 connectivity as the Aerotropolis model and boost the economic development in manifold. Songdo City can benefit by connecting the numerous mega population cities like Shanghai, Beijing, Hong Kong and Tokyo in highly effective way (Chohan, 2014).

Songdo as a Ubiquitous City

As rightly said by Shwayri (2013) in his seminal paper that the green building infrastructure of the new Songdo Smart City is to be boosted by the provision of extra civic services that combine ICT as well as digital networks to create harmonious relations among the natural environment, society (people), and technology. The Smart Cities use technologies to increase the efficiency of public utility services, optimum natural resource consumption and improve urban residents' quality of life. The Songdo city has a unique feature and that is its omnipresent via physical computing infrastructure. To fulfill of this goal, M/s. Gale International enlisted the CISCO, San-Jose-based networking company to develop many of the smart technologies, tools and services then associated infrastructure (Lindsay, 2010; Cortese, 2007).

A combined venture between Gale, POSCO E&C, and tech firm Life is Good (LG) CNS, and Ubiquitous-Life NAH will serve as the computer nerve center of Songdo International Business District. Smart technologies (Internet of Things) deployed in Songdo are shaping the workplace, governance, the home automation, healthcare, education system, citizen security, manufacturing process, transportation facilities, car parking, and the ways in which people use energy. Songdo U-Life –

Figure 3. Songdo as Center: Aerotropolis City

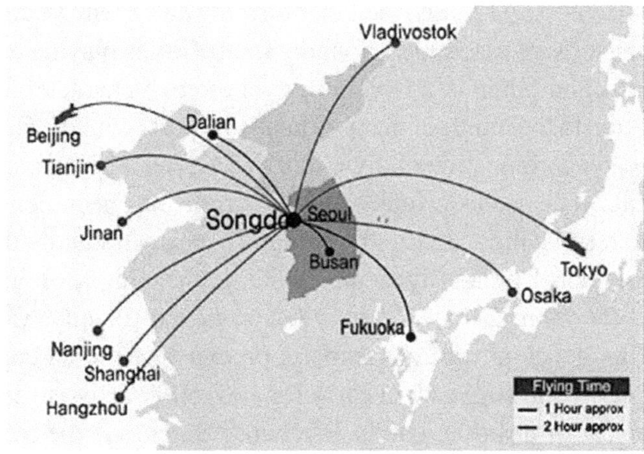

as a partnership, the technological infrastructure will be built and managed by the New Songdo City Development and the South Korean network integrator LG CNS. Currently, the LG CNS is recruiting foreign Information Technology/IT-enabled Services (IT/ITeS) firms as developing partners (O'Connell, 2005). The Ubiquitous City (U-City) system based on IT/ITeS services that allows people to take part in routine works efficiently and quickly wherever and whenever they want. Ever since the pioneering ceremony in 2004, Songdo has been promoted as a node in the multicenter metropolitan region, a living space for U-life and a gateway to Northeast Asia (Kim, 2010). U-Life services provided for residents include services such as home automation and civic security monitoring. Residents from Songdo City can control remotely some of their homes appliances. Every citizen is able to interact from anywhere through "video chat" through Cisco's telepresence system (Cisco, 2010a, 2010b). Huge amount of sensor technology and connected networked infrastructure are integrated throughout Songdo City premises. The beauty of U-Business provides work-related services such as information management, document sharing and video conferencing etc. Besides, Ubiquitous Government also provides services that enable citizens by making them aware of city traffic and citizen health threats. Enables citizens to participate in the governance process i.e. public policy at any time 24/7, and from anywhere (Gil-Castineira et al., 2011).

Songdo will locate the U-city command control center (C4) and provide U-services to the citizens. These includes TOPAZ digital signage for on video wall, interactive board, display series or on-call taxis and digital emergency services for patients. Other ubiquitous features will include the MeION mobile game, the Cyworld on-line social media network, the Moneta mobile credit-card service, and satellite multimedia broadcasting (Lee et al., 2008). Incheon Free Economic Zone is a holding company that is testing and filtering a fleet of commodity cities that are imagined as being mobile, and plastic territories. They each simultaneously involve in development of software, hardware, networking, screen, algorithms, and data (Crang & Graham, 2007) - serving as interfaces with, and channels into, networks linked to other territories. The computers that are connected through wide-area network will link SIBD's structures and businesses via various digital services including individual apartments and each room lighting, temperature, and access to media. The state-of-the-art is the C4 (city command control center) will help accomplish all aspects of urban life - from traffic control, to water and energy use, to recycling (Naphade et al., 2011). The city is proposed as a physical materialization of an immense cloud computing with big data analytics, on how much data the city can produce; and on how capable its high-bandwidth conduits are in being able to circulate these data (Halpern et al., 2013). Ideally, the citizens of Songdo city would never get stuck in traffic; would regularly know where their personal effects are, and they have instant access to medical staff or education facilities. Smart Homes in the city reflect careful attention to detail and reflect the efficiency afforded by the modern sustainable technologies. The automatically-coordinated environments will free to citizens from many obstacles to citizen daily life and, thus improve Songdo residents' quality of life (Benedikt, 2016).

CONCLUSION

Like other big, long-term urban projects, Songdo Smart City in Korea is also facing problems regarding harmonious collaboration, technical issues, financial budgeting consensus, project functional smoothness and in time completion, New Songdo Ubiquitous-City is a Y-generation city which had been initially scheduled for completion in 2014 only. The actual creation of the Songdo Smart City (Korea) has been developing very little by little ever compared to the original plans. The reasons behind are as we said above due to many problems including financial issues, insufficient government support, bureaucratic procedural delays, stakeholder involvement and resistance, and failure to attract foreign capital investment (Lee & Oh, 2008; Williamson, 2013; Shwayri, 2013). The 607 - hectare Songdo IBD, which forms the new urban area at the heart of the city, is re-scheduled for completion in 2018; and developers/builders claim that it will be the greenest, fully connected by cat 6 cables/ wirelessly. So, Songdo City is nowadays more of a rich exurbia of Incheon city and mostly inhabited by the local people (Shwayri, 2013). However, the residents from Seoul city and other crowded S. Korean cities are now desperately investing and applying for flats; and developers are counting on attracting with beautiful businesses amenities. However, by the time 2018 has passed, the authors hope that that entire Songdo Smart City project will be completed successfully. Thus, all the Korean people, other global industry people, and worldwide travelers will reap the fruits of the quality of life in this U-City forever and ever. The Songdo Smart city is an excellent example of what a global smart city ought to look like. These will add the quality of life to citizens. The "aerotropolis" and "ubiquitous" fundamentals are novel and marvelous principles in the urbanisation theory.

REFERENCES

Alusi, A., Eccles, R., Edmondson, A., & Zuzul, T. (2011). *Sustainable cities: oxymoron or the shape of the future?* Cambridge, MA: Harvard Business School.

Angelidou, M. (2014). Smart city policies: A spatial approach. *Cities (London, England)*, *41*, S3–S11. doi:10.1016/j.cities.2014.06.007

Anthopoulos, L., & Fitsilis, P. (2010, July). From digital to ubiquitous cities: Defining a common architecture for urban development. In *Proceedings of the 2010 Sixth International Conference on Intelligent Environments (IE)* (pp. 301-306). IEEE.

Anthopoulos, L. G., & Vakali, A. (2012, May). Urban planning and smart cities: Interrelations and reciprocities. In The Future Internet Assembly (pp. 178-189). Berlin: Springer.

Benedikt, O. (2016). The valuable citizens of smart cities: The case of Songdo City. *Graduate Journal of Social Science*, *12*(1), 17–36.

Chohan, U. W. (2014, May 13). The ubiquitous city – Songdo (송도). *McGill*. Retrieved Sept. 17, 2017, from https://www.mcgill.ca/channels/channels/news/ubiquitous-city-songdo-songdo-236653

Chun, Y. (2007, June 16). Incheon gong hang sale hwaljuro nanyeon open [Incheon airport's new landing strip opens next year], *Munhwa Ilbo*.

Chun, G. (2007). Sustainability effort in Songdo International Business District. *The Korea IT Times*. Retrieved Oct. 19, 2007 from http://www.koreaittimes.com/story/3475/sustainability-effort-songdo-international-business-district

Cisco. (2010a). Smart + connected communities; changing a city, a country, the world. San Jose, CA: Cisco.

Cisco (2010b). Smart + Connected Life [World Expo 2010 pavilion guide]. Shanghai, China: Cisco. Retrieved Oct. 5, 2017 from http://www.cisco.com/web/CN/expo/pdf/Smart_Connected_Life_Guide_EN.pdf

Cisco. (2012). Cities of the Future: Songdo, South Korea, episode 3: working in Songdo. [promotional video] Retrieved from https://www.youtube.com/watch?v=JtWXZGev5M

Cortese, A. (2007, Dec. 30). An Asian hub in the making. *The New York Times*. Retrieved from www.nytimes.com/2007/12/30/realestate/commercial/30sqft.html

Couclelis, H. (2004). The construction of the digital city. *Environment and Planning B Urban Analytics and City Science, 31*(1), 5–19.

Crang, M., & Graham, S. (2007). Sentient Cities: Ambient intelligence and the politics of urban space. *Information Communication and Society, 10*(6), 789–817. doi:10.1080/13691180701750991

De Jong, M., Joss, S., Schraven, D., Zhan, C., & Weijnen, M. (2015). Sustainable – smart – resilient – low carbon – eco-knowledge cities; making sense of a multitude of concepts promoting sustainable urbanization. *Journal of Cleaner Production, 109*, 25–38. doi:10.1016/j.jclepro.2015.02.004

Gil-Castineira, F., Costa-Montenegro, E., Gonzalez-Castano, F., López-Bravo, C., Ojala, T., & Bose, R. (2011). Experiences inside the ubiquitous Oulu smart city. *Computer, 44*(6), 48–55. doi:10.1109/MC.2011.132

Greenfield, A., & Kim, N. (2013). *Against the smart city* [Part 1 of This city is here for you to use] (1.3 ed.). Seattle, WA: Amazon Digital Services.

Grover, A. (2013). Airport business districts — An indispensable reality. *Creative Space, 1*(1), 79–99. doi:10.15415/cs.2013.11005

Halpern, O., LeCavalier, J., Calvillo, N., & Pietsch, W. (2013). Test-bed urbanism. *Public Culture, 25*(2 70), 272-306.

Hyang-Sook, C., Byung-Sun, C., & Woong-Hee, P. (2007). Ubiquitous-city business strategies: The case of South Korea. In *Management of Engineering and Technology* (pp. 1147-1153). IEEE. doi:10.1109/PICMET.2007.4349438

Jang, M., & Suh, S. T. (2010). U-city: New trends of urban planning in Korea based on pervasive and ubiquitous geo-technology and geoinformation. In Computational Science and Its Applications, ICCSA-2010 (pp. 262–270).

Joss, S. (2011). Eco-cities: The mainstreaming of urban sustainability–key characteristics and driving factors. *International Journal of Sustainable Development and Planning, 6*(3), 268–285. doi:10.2495/SDP-V6-N3-268-285

Joss, S., Cowley, R., & Tomozeiu, D. (2013). Towards the 'ubiquitous eco-city': An analysis of the internationalization of eco-city policy and practice. *Urban Research & Practice*, *6*(1), 54–74. doi:10.1080/17535069.2012.762216

Jun-Woo, K., & Ahn, Y. J. (2011). *Songdo Free Economic Zone in South Korea.* 대한지리학회지, *46*(5), 662–672.

Kasarda, J. D. (2008). The evolution of airport cities and the aerotropolis. In *Airport cities: The evolution* (pp. 1–39). London, England: Insight Media. Retrieved from http://www.aerotropolis.com/files/evolutionChapter1.pdf

Kim, C. (2010). Place promotion and symbolic characterization of new Songdo City, South Korea. *Cities (London, England)*, *27*(1), 13–19. doi:10.1016/j.cities.2009.11.013

Kim, J. W., & Ahn, Y.-J. (2006). Background and dynamics of free economic zones in Korea. *Journal of the Korean Urban Geographical Society*, *9*, 79–87.

Kim, Y. M., Kim, H. S., Moon, S. Y., & Bae, S.-Y. (2009). Ubiquitous Eco-City Planning in Korea. A Project for the Realization of Ecological City Planning and Ubiquitous Network Society. In M. Schrenk, V. V. Popovich, D. Engelke, and P. Elisei (Eds.), REAL CORP 2009 Proceedings (pp. 925–930). Schwechat, Austria: Competence Center for Urban and Regional Planning.

Kshetri, N., Alcantara, L. L., & Park, Y. (2014). Development of a smart city and its adoption and acceptance: The case of New Songdo. *Communications & Strategies*, *1*(96), 113–128. Retrieved from https://ideas.repec.org/a/idt/journl/cs9606.html

Lee, J., Baik, S., & Lee, C. C. (2011). Building an integrated service management platform for ubiquitous cities. *Computer*, *44*(6), 56–63. doi:10.1109/MC.2011.131

Lee, J., & Oh, J. (2008). *New Songdo City and the value of flexibility: a case study of implementation and analysis of a mega-scale project* [Doctoral dissertation]. Massachusetts Institute of Technology.

Lee, S., Oh, K., & Jung, S. (2009). The Carrying Capacity Assessment Framework for ubiquitous ecological cities in Korea [conference paper]. In *2nd International Urban Design Conference*, Gold Coast.

Lee, S. H., Han, J. H., Leem, Y. T., & Tan Yigitcanlar, T. (2008). Towards ubiquitous city: Concept, planning, and experiences in the Republic of Korea. In Knowledge-based urban development: Planning and applications in the information era (pp. 148-170). Hershey, PA: IGI Global. doi:10.4018/978-1-59904-720-1.ch009

Lindsay, G. (2010, Feb.). The new urbanism: New Songdo & creating cities from scratch. *Fast Company*. Retrieved from http://www.greglindsay.org/articles/the_new_new_urbanism_creating_cities_from_scratch/

Looser, T. (2012). The global university, area studies, and the world citizen: Neoliberal geography's redistribution of the "world." *Cultural Anthropology*, *27*(1), 97–117. Retrieved from http://onlinelibrary.wiley.com/doi/10.1111/j.1548-1360.2012.01128.x/abstract doi:10.1111/j.1548-1360.2012.01128.x

Madakam, S., & Ramachandran, R. (2015). Barcelona Smart City: The Heaven on Earth (Internet of Things: Technological God). *ZTE Communications*, *4*, 3.

Madakam, S., & Ramaswamy, R. (2016). Sustainable smart city: Masdar (UAE) (A city: Ecologically balanced). *Indian Journal of Science and Technology*, *9*(6). doi:10.17485/ijst/2016/v9i6/87669

Moser, S., Swain, M., & Alkhabbaz, M. H. (2015). King Abdullah Economic City: Engineering Saudi Arabia's post-oil future. *Cities (London, England)*, *45*, 71–80. doi:10.1016/j.cities.2015.03.001

Naphade, M., Banavar, G., Harrison, C., Paraszczak, J., & Morris, R. (2011). Smarter cities and their innovation challenges. *Computer*, *44*(6), 32–39. doi:10.1109/MC.2011.187

O'Connell, P. L. (2005, Oct. 5). Korea's high-tech utopia, where everything is observed. *The New York Times*.

O'Donnell, J. K. (2010, Mar.). The US-ROK alliance: A catalyst for green growth? Centre for US-Korea Policy [a project of the Asia Foundation]. Retrieved from https://www.scribd.com/document/35588274/ODonnellGreenGrowth

Republic of Korea. (2010). Sejong to Become Model Eco-Friendly city. Retrieved from http://www.korea.net/Resources/Useful-Info/Learn-Korean

Shwayri, S. T. (2013). A model Korean ubiquitous eco-city? The politics of making Songdo. *Journal of Urban Technology*, *20*(1), 39–55. doi:10.1080/10630732.2012.735409

Singer, N. (2012, March 3). Mission control, built for cities: I.B.M. takes 'smarter cities' concept to Rio de Janeiro. *The New York Times*.

Strickland, E. (2011). Cisco bets on the South Korean smart city. *IEEE Spectrum*, *48*(8), 11–12. doi:10.1109/MSPEC.2011.5960147

Tiwari, A., & Jain, K. (2014). GIS Steering smart future for smart Indian cities. International. *Journal of Scientific and Research Publications*, *4*(8), 442–446.

Townsend, A. M. (2013). *Smart cities. Big data, civic hackers, and the quest for a new utopia*. New York: W.W. Norton & Company. Retrieved from https://ssir.org/articles/entry/smart_cities_big_data_civic_hackers_and_the_quest_for_a_new_utopia

Williamson, L. (2013). Tomorrow's cities: Just how smart is Songdo? *BBC News*. Retrieved Dec. 16, 2013, from http://www.bbc.com/news/technology-23757738

Yigitcanlar, T., & Lee, S. H. (2014). Korean ubiquitous-eco-city: A smart-sustainable urban form or a branding hoax? *Technological Forecasting and Social Change*, *89*, 100–114. doi:10.1016/j.techfore.2013.08.034

Yun, J. (2015). A new city prototype? Songdo International City as an airport city. *Journal of Asian Architecture and Building Engineering*, *14*(3), 549–556. doi:10.3130/jaabe.14.549

ADDITIONAL READING

Abdoullaev, A. (2011). Keynote: a smart world: a development model for intelligent cities. In *The 11th IEEE International Conference on Computer and Information Technology (CIT)*.

Alkandari, A., Alnasheet, M., & Alshaikhli, I. F. T. (2012). Smart cities: A survey. *Journal of Advanced Computer Science and Technology Research*, *2*(2), 79–90.

Anthopoulos, L., & Fitsilis, P. (2013). Using classification and roadmapping techniques for smart city viability's realization. *Electronic Journal of E-Government*, *11*(2).

Anthopoulos, L. G. (2015). Understanding the smart city domain: A literature review. In *Transforming city governments for successful smart cities* (pp. 9–21). Cham: Springer. doi:10.1007/978-3-319-03167-5_2

Anthopoulos, L. G., & Reddick, C. G. (2016). Understanding electronic government research and smart city: A framework and empirical evidence. *Information Polity*, *21*(1), 99–117. doi:10.3233/IP-150371

Aoun, C. (2013). *The smart city cornerstone: Urban efficiency*. Schneider electric.

Bélissent, J. (2010). Getting clever about smart cities: New opportunities require new business models.

Boulos, M. N. K., & Al-Shorbaji, N. M. (2014). On the Internet of Things, smart cities and the WHO Healthy Cities.

Cardone, G., Foschini, L., Bellavista, P., Corradi, A., Borcea, C., Talasila, M., & Curtmola, R. (2013). Fostering participaction in smart cities: A geo-social crowdsensing platform. *IEEE Communications Magazine*, *51*(6), 112–119. doi:10.1109/MCOM.2013.6525603

Carvalho, L. (2014). Smart cities from scratch? A socio-technical perspective. *Cambridge Journal of Regions, Economy and Society*, *8*(1), 43–60. doi:10.1093/cjres/rsu010

Cheng, B., Longo, S., Cirillo, F., Bauer, M., & Kovacs, E. (2015, June). Building a big data platform for smart cities: Experience and lessons from Santander. In *2015 IEEE International Congress on Big Data (BigData Congress)* (pp. 592-599). *IEEE.*

Doran, D., Gokhale, S., & Dagnino, A. (2013, August). Human sensing for smart cities. In *Proceedings of the 2013 IEEE/ACM International Conference on Advances in Social Networks Analysis and Mining* (pp. 1323-1330). *ACM.*

Glasmeier, A., & Christopherson, S. (2015). Thinking about smart cities.

Henry, C. (2011, Mar.) Songdo International Business District / KPF. *Arch Daily*. Retrieved May 13, 2012 from https://www.archdaily.com/118790/songdo-international-business-district-kpf/

Hodgkinson, S. (2011). Is your city smart enough? Digitally enabled cities and societies will enhance economic, social, and environmental sustainability in the urban century.

Kim, J. I. (2014). Making cities global: The new city development of Songdo, Yujiapu and Lingang. *Planning Perspectives*, *29*(3), 329–356. doi:10.1080/02665433.2013.824370

Kitchin, R. (2015). Making sense of smart cities: Addressing present shortcomings. *Cambridge Journal of Regions, Economy and Society*, *8*(1), 131–136. doi:10.1093/cjres/rsu027

Kostakos, V., Ojala, T., & Juntunen, T. (2013). Traffic in the smart city: Exploring city-wide sensing for traffic control center augmentation. *IEEE Internet Computing*, *17*(6), 22–29. doi:10.1109/MIC.2013.83

Lee, K. H. (2011). Building a New Smart City in Asia: Songdo International City in Incheon, S. Korea.

Marvin, S., Luque-Ayala, A., & McFarlane, C. (Eds.). (2015). *Smart urbanism: Utopian vision or false dawn*. Routledge.

Ojo, A., Curry, E., & Janowski, T. (2014). Designing next generation smart city initiatives-harnessing findings and lessons from a study of ten smart city programs.

Pellicer, S., Santa, G., Bleda, A. L., Maestre, R., Jara, A. J., & Skarmeta, A. G. (2013, July). A global perspective of smart cities: A survey. In *2013 Seventh International Conference on Innovative Mobile and Internet Services in Ubiquitous Computing (IMIS)* (pp. 439-444). IEEE. 10.1109/IMIS.2013.79

Ratti, C., & Townsend, A. (2011). The social nexus. *Scientific American*, *305*(3), 42–49. doi:10.1038cientificamerican0911-42 PMID:21870441

Sadowski, J., & Pasquale, F. A. (2015). The spectrum of control: A social theory of the smart city.

Viitanen, J., & Kingston, R. (2014). Smart cities and green growth: Outsourcing democratic and environmental resilience to the global technology sector. *Environment & Planning A*, *46*(4), 803–819. doi:10.1068/a46242

Walravens, N. (2015). Mobile city applications for Brussels citizens: Smart City trends, challenges and a reality check. *Telematics and Informatics*, *32*(2), 282–299. doi:10.1016/j.tele.2014.09.004

Washburn, D., Sindhu, U., Balaouras, S., Dines, R. A., Hayes, N., & Nelson, L. E. (2009). Helping CIOs understand "smart city" initiatives. *Growth*, *17*(2), 1–17.

Compilation of References

Adrian, A., & Dani, A. W. (2014). *Design of small Home system based on Arduino*. IEEE Electrical Power, Electronics, Communications, Control and Informatics.

Affetti, L., Bresciani, G., & Guinea, S. (2015). ADock: A Cloud Infrastructure Experimentation Environment Based on Open Stack and Docker. *Proceedings - 2015 IEEE 8th International Conference on Cloud Computing, CLOUD 2015*. 10.1109/CLOUD.2015.36

Aguilera, G. (2013). *An Accelerated-Time Simulation for Traffic Flow in a Smart City*. FEMTEC.

Ahmed, K., Liu, J., & Wu, X. (2017). An Energy Efficient Demand-Response Model for High Performance Computing Systems, *Proceedings of the 25th IEEE International Symposium on the Modeling, Analysis, and Simulation of Computer and Telecommunication Systems (MASCOTS 2017)*. 10.1109/MASCOTS.2017.25

Al Nuaimi, Al Neyadi, Mohamed, & Al-Jaroodi. (2013). Understanding the value of (big) data. In *Big Data, 2013 IEEE International Conference on*. IEEE.

Al Nuaimi, E., Al Neyadi, H., Mohamed, N., & Al-Jaroodi, J. (2015). Applications of big data to smart cities. *Journal of Internet Services and Applications*, *6*(1), 25. doi:10.118613174-015-0041-5

Alaa. (2017). A review of smart home applications based on Internet of Things. *Journal of Network and Computer Applications, 97*, 48-65.

Alamri, A., Ansari, W. S., Hassan, M. M., Hossain, M. S., Alelaiwi, A., & Hossain, M. A. (2013). A survey on sensor-Cloud: Architecture, applications and approaches. *International Journal of Distributed Sensor Networks*, *2013*, 1–18.

Ali, S.S.M., George, B., Vanajakshi L. (2011). A simple multiple loop sensor configuration for vehicle detection in an undisciplined traffic. In *2011 Fifth International Conference on Sensing Technology (ICST)*. IEEE. Retrieved from http://ieeexplore.ieee.org/xpl/articleDetails.jsp?tp=&arnumber=6137062&url=http%3A%2F%2Fieeexplore.ieee.org%2Fxpls%2Fabs_all.jsp%3Farnumber%3D6137062

Ali, A. H., Atia, A., & Mostafa, M. S. M. (2017). Recognizing Driving Behavior and Road Anomaly using Smartphone Sensors. *International Journal of Ambient Computing and Intelligence*, *8*(3), 22–37. doi:10.4018/IJACI.2017070102

Altawy, R., & Youssef, A. M. (2016). Security Tradeoffs in Cyber Physical Systems: A Case Study Survey on Implantable Medical Devices. *Access IEEE, 4*, 959-979.

Altman, E., Basar, T., Jimenez, T., & Shimkin, N. (2002). Competitive routing in networks with polynomial costs. *IEEE Transactions on Automatic Control, 47*(1), 92–96.

Alusi, A., Eccles, R., Edmondson, A., & Zuzul, T. (2011). *Sustainable cities: oxymoron or the shape of the future?* Cambridge, MA: Harvard Business School.

Aly, H., Elmogy, M., & Barakat, S. (2015). Big Data on Internet of Things: Applications, Architecture, Technologies, Techniques, and Future Directions. *International Journal of Computer Science Engineering, 4*(6).

Amaral, M., Polo, J., Carrera, D., Mohomed, I., Unuvar, M., & Steinder, M. (2015). Performance evaluation of microservices architectures using containers. *Proceedings - 2015 IEEE 14th International Symposium on Network Computing and Applications, NCA 2015*. 10.1109/NCA.2015.49

Amaral, L. A., Hessel, F. P., Bezerra, E. A., Corr^ea, J. C., Longhi, O. B., & Dias, T. F. (2011). ecloudrfid–a mobile software framework architecturefor pervasive rfid-based applications. *Journal of Network and Computer Applications, 34*(3), 972–979. doi:10.1016/j.jnca.2010.04.005

Anand & Padwalkar. (2014). Internet of Things –A Future of Internet: A Survey. *International Journal of Advance Research in Computer Science and Management Studies, 2*(2).

Anderson, J., Hu, H., Agarwal, U., Lowery, C., Li, H., & Apon, A. (2016). Performance considerations of network functions virtualization using containers. *2016 International Conference on Computing, Networking and Communications, ICNC 2016*. 10.1109/ICCNC.2016.7440668

Angelidou, M. (2014). Smart city policies: A spatial approach. *Cities (London, England), 41*, S3–S11. doi:10.1016/j.cities.2014.06.007

Anisha, Kumar Reddy, Reddy, & Reddy. (2012). *Third Party Data Protection Applied To Cloud and Xacml Implementation in the Hadoop Environment With Sprawl*. Academic Press.

Anthopoulos, L. G., & Vakali, A. (2012, May). Urban planning and smart cities: Interrelations and reciprocities. In The Future Internet Assembly (pp. 178-189). Berlin: Springer.

Anthopoulos, L., & Fitsilis, P. (2010, July). From digital to ubiquitous cities: Defining a common architecture for urban development. In *Proceedings of the 2010 Sixth International Conference on Intelligent Environments (IE)* (pp. 301-306). IEEE.

Apache Foundation. (2017). *Welcome to The Apache Software Foundation!* Retrieved January 2, 2018, from https://www.apache.org/

Apache Hadoop Project. (2015). Retrieved from http://hadoop.apache.org/

Armburst, M. (2009). Above the Clouds: A Berkeley View of Cloud Computing. Berkeley, CA: Univ. of California.

Arnott, R., De Palma, A., & Lindsey, R. (1991). Does providing information to drivers reduce traffic congestion? *Transportation Research Part A, General*, *25*(5), 309–318.

Ashrafi. (2007). *A Data Mining Architecture for Distributed Environments*. London, UK: Springer.

Ashton, K. (2009). That 'internet of things' thing. *RFID journal, 22*(7), 97-114.

Asthmapolis. (2013, September). Asthmapolis Relaunches as Propeller Health to Advance Broader Respiratory Mission [Press Release]. Retrieved from https://www.propellerhealth.com/2013/09/10/asthmapolis-relaunches-as-propeller-health-to-advance-broader-respiratory-mission/

Aulbur, W., & Singh, H. V. (2014). *Next gen manufacturing: Industry 4.0: A look at changing landscapes in manufacturing*. Confederation of Indian Industry. Retrieved from https://www.gita.org.in/Attachments/Reports/CII%20Roland%20Berger%20Industry%20 40.pdf

Ayagi, S., & Agarwal, A. (2016). A conceptual framework for IoT based healthcare system using cloud computing. *Cloud System and Big Data Engineering (Confluence), 2016 6th International Conferenc*.e

Ayyub, B. M. (2003). *Risk Analysis in Engineering and Economics*. Boca Raton: Chapman and Hall/CRC Press. doi:10.1201/9780203497692

Babar, M., & Arif, F. (2017). Smart urban planning using Big Data analytics to contend with the interoperability in Internet of Things. *Future Generation Computer Systems*, *77*, 65–76. doi:10.1016/j.future.2017.07.029

Baines, T., Lightfoot, H., Benedettini, O., & Kay, J. M. (2009). The servitization of manufacturing: A review of literature and reflection on future challenges. *Journal of Manufacturing Technology Management*, *20*(5), 547–567. doi:10.1108/17410380910960984

Bakmaz. (2015). Machine – To – Machine Communications Toward Smart Logistics Systems. *2nd Logistics International Conference*.

Banerjee, S. (2017, April 16). CCTV and e-challans help reign in Pune's unruly traffic. *The Hindu*.

Bari, P. (2017, July 4). One year on, many projects of PM Modi's Smart City initiative still in pipeline. *Hindustan Times*.

Barnaghi, Bermudez-Edo, & Tönjes. (2015). Challenges for Quality of Data in Smart Cities. *J. Data and Information Quality, 6*(2-3), 6:1–6:4.

Baron, P. (2013). *Big Data für IT-Entscheider*. München: Hanser Verlag. doi:10.3139/9783446433922

Barry, P., & Crowley, P. (2012). *Modern embedded computing: designing connected, pervasive, media-rich systems*. Elsevier.

Bartoli, A., Hern, J., Hernández-Serrano, J., & Soriano, M. (2011). Security and Privacy in your Smart City. *Smart City Counc*, 1–6. 10.1109/MCOM.2013.6525606

Bassi, A. (2017). Looking at Smart Cities with an Historical Perspective. In V. Angelakis, E. Tragos, H. C. Pöhls, A. Kapovits, & A. Bassi (Eds.), *Designing, Developing, and Facilitating Smart Cities: Urban Design to IoT Solutions* (pp. 3–15). Cham: Springer International Publishing; doi:10.1007/978-3-319-44924-1_1

Beaty, W. (1998). Traffic Waves —Sometimes one driver can vastly improve traffic. Retrieved from http://www.amasci.com/amateur/traffic/traffic1.html

Bello, L. M., Mitchell, P., & Grace, D. (2014). Application of Q-learning for RACH access to support M2M traffic over a cellular network. *European Wireless Conference*, 1–6.

Benedikt, O. (2016). The valuable citizens of smart cities: The case of Songdo City. *Graduate Journal of Social Science*, *12*(1), 17–36.

Bengrut, D. (2017, June 17). 82% Traffic Violators Haven't Bothered To Pay Up E-Challan Fines. *Pune Mirror*.

Berger, R. (2014). THINK ACT - Industry 4.0. The new industrial revolution How Europe will succeed.

Bertot, J. C., & Choi, H. (2013). Big Data and e-government: issues, policies, and recommendations. In *Proceedings of the 14th Annual International Conference on Digital Government Research*. ACM. 10.1145/2479724.2479730

Bhardwaj, A., Subrahmanyam, G. V. B., Avasthi, V., & Sastry, H. (2016). Security algorithms for cloud computing. *Procedia Computer Science*, *85*, 535–542. doi:10.1016/j.procs.2016.05.215

Bhatt, C., Dey, N., & Ashour, A. S. (Eds.). (2017). Internet of things in Health care. Internet of things and big data technologies for next generation healthcare.

Bhatt, C. M., Dey, N., & Ashour, A. (2017). *Internet of things and big data technologies for next generation healthcare*. Springer. doi:10.1007/978-3-319-49736-5

Big Data Analytics, Machine Learning and Artificial Intelligence in the Smart Grid: Introduction, Benefits, Challenges and Issues. (2017). *IEEE Smart Grid Big Data Analytics, Machine Learning and Artificial Intelligence in the Smart Grid Working Group*. Retrieved from: https://smartgrid.ieee.org/images/files/pdf/big_data_analytics_white_paper.pdf

Big Data. (n.d.). Retrieved from: wikipedia.org/wiki/Big Data

Bill Burke, R. M.-H. (2006). *Enterprise Java Beans 3.0*. O'Reilly.

Bioenabletech. (n.d.). Smart traffic management system. Retrieved from https://www.bioenabletech.com/smart-traffic-management-system

Biswas, P. (2017a). *Pune RTO sees 25 per cent rise in vehicle registration on Dussehra. The Indian Express*.

Biswas, P. (2017b). *Pune, cab aggregators leading vehicle growth. The Indian Express*.

Bitam, S., Mellouk, A., & Zeadally, S. (2015). Vanet-cloud: A generic cloud computing model for vehicular ad hoc networks. *IEEE Wireless Communications*, *22*(1), 96–102. doi:10.1109/MWC.2015.7054724

Boccuti, C., & Casillas, G. (2017, March). Aiming for fewer hospital U-turns: The medicare hospital readmission reduction program. KFF. Retrieved from https://www.kff.org/medicare/issue-brief/aiming-for-fewer-hospital-u-turns-the-medicare-hospital-readmission-reduction-program/

Bohu, L., Lin, Z., & Shilong, W. (2010). Cloud Manufacturing: A New Service-oriented Networked Manufacturing Model. *Jisuanji Jicheng Zhizao Xitong*, *16*(1), 1–7, 16.

Botta, A., Donato, W. d., Persico, V., & Pescapé, A. (2014). On the Integration of Cloud Computing and Internet of Things. *2014 International Conference on Future Internet of Things and Cloud (FiCloud)*, 23–30. 10.1109/FiCloud.2014.14

Brenke, K. (2012). Engineers in Germany: No Shortage in Sight. *DIW Economic Bulletin, 5*. Retrieved from https://econpapers.repec.org/article/diwdiwdeb/2012-5-1.htm

Brettel, M., Friederichsen, N., Keller, M., & Rosenberg, M. (2014). How virtualization, decentralization and network building change the manufacturing landscape: An industry 4.0 perspective. *World Academy of Science, Engineering and Technology*, *8*(1), 37–44.

Brewer, E. (2012). CAP twelve years later: How the "rules" have changed. *Computer*, *45*(2), 23–29. doi:10.1109/MC.2012.37

Bui, T. (2015). Analysis of Docker Security. *Computing Research Repository*. Retrieved from http://arxiv.org/abs/1501.02967

Buyya, R., Chee Shin, Y., & Venugopal, S. (2008). Market-Oriented Cloud Computing: Vision, Hype, and Reality for Delivering IT Services as Computing Utilities. *10th IEEE Conference on High Performance Computing and Communications*. 10.1109/HPCC.2008.172

Calvert, K., & Simandan. (2010). Energy, space, and society: A reassessment of the changing landscape of energy production, distribution and use. *Journal of Economic and Business Research*, *1*, 13–37.

Cammarano, A., Petrioli, C., & Spenza, D. Pro-energy: A novel energy prediction model for solar and wind energy-harvesting wireless sensor networks. *Proc. IEEE 9th Int. Conf. Mobile Ad-Hoc Sensor Syst. (MASS)*, 75–83. 10.1109/MASS.2012.6502504

Cand'es, E., Li, X., Ma, Y., & Wright, J. (2011). Robust principal component analysis? *Journal of the Association for Computing Machinery*, *58*(3), 1–37. doi:10.1145/1970392.1970395

Cao, X., Wu, C., Lan, J., Yan, P., & Li, X. (2011). Vehicle detection and motion analysis in low-altitude airborne video under urban environment. *IEEE Transactions on Circuits and Systems for Video Technology*, *21*(10), 1522–1533.

Capkun, S., Buttyan, L., & Hubaux, J.-P. (2003, January/March). Self-organized publickey management for mobile ad hoc networks. *IEEE Transactions on Mobile Computing*, *2*(1), 52–64. doi:10.1109/TMC.2003.1195151

Cardenas, Amin, & Sastry. (2008). Secure control: Towards survivable cyber-physical systems. *Proc. 28th Int. Conf. Distrib. Comput. Syst. Workshops*, 495–500.

Center for Disease Control and Prevention (CDC). (2014). Morbidity and Mortality Weekly Report. Retrieved from https://www.cdc.gov/mmwr/preview/mmwrhtml/su6304a2.htm

Center for Disease Control and Prevention (CDC). (2017). Sepsis: Data & Reports. Retrieved from https://www.cdc.gov/sepsis/datareports/index.html

Centers for Disease Control and Prevention (CDC). (2017). Epi Info for Web & Cloud. Retrieved from https://www.cdc.gov/epiinfo/cloud.html

Chandiramani, J., & Airy, A. (2017). Urbanization and Socio-Economic Growth in South Asia Region. In U. Benna & I. Benna (Eds.), *Urbanization and Its Impact on Socio-Economic Growth in Developing Regions*. Hershey, PA: IGI Global.

Chaturvedi, K. K., & Sing, V. B. (2013). Tools in Mining Software Repositories. *Computational Science and Its Applications (ICCSA) 2013 13th International Conference on*, 89-98.

Chauhan, S., Agarwal, N., & Kar, A. (2016). Addressing big data challenges in smart cities: A systematic literature review. *Info*, *18*(4), 73–90. doi:10.1108/info-03-2016-0012

Chen, M., Mao, S., & Liu, Y. (2014). Big data: A survey. In Mobile Networks and Applications (Vol. 19, pp. 171–209). Academic Press. doi:10.100711036-013-0489-0

Cheng. (2015). Building a big data platform for smart cities: experience and lessons from santander. *Proceedings of the 4th IEEE International Congress on Big Data.*

Chen, R., Fung, B., Desai, B. C., & Sossou, N. M. (2012, August). Differentially private transit data publication: a case study on the montreal transportation system. In *Proceedings of the 18th ACM SIGKDD International Conference on Knowledge Discovery and Data Mining* (pp. 213-221). ACM. 10.1145/2339530.2339564

Chen, S. (2014). A Vision Of Iot: Applications, Challenges, An Opportunities With China Perspective. *IEEE Internet of Tings Journal, 1*(4), 349–359. doi:10.1109/JIOT.2014.2337336

Chen, S., Li, M., Ren, K., Fu, X., & Qiao, C. (2015). Rise of the Indoor Crowd: Reconstruction of Building Interior View via Mobile Crowdsourcing. *Proc. of ACM SenSys*, 5971. 10.1145/2809695.2809702

Chiang, C. C., Ho, M. C., Liao, H. S., Pratama, A., & Syu, W. C. (2011). Detecting and recognizing traffic lights by genetic approximate ellipse detection and spatial texture layouts. International Journal of Innovative Computing, Information and Control, 7(12), 6919-6934.

Chinyere, O. U., Francisca, O. O., & Amano, O. E. (2011). Design and simulation of an intelligent traffic control system. International journal of advances in engineering & technology, 1(5), 47-57.

Chohan, U. W. (2014, May 13). The ubiquitous city – Songdo (송도). *McGill.* Retrieved Sept. 17, 2017, from https://www.mcgill.ca/channels/channels/news/ubiquitous-city-songdo-songdo-236653

Chun, G. (2007). Sustainability effort in Songdo International Business District. *The Korea IT Times.* Retrieved Oct. 19, 2007 from http://www.koreaittimes.com/story/3475/sustainability-effort-songdo-international-business-district

Chun, Y. (2007, June 16). Incheon gong hang sale hwaljuro nanyeon open [Incheon airport's new landing strip opens next year], *Munhwa Ilbo.*

Ciolli, R., Whyte, P., Ercan, G., & Mack, A. (2003). *U.S. Patent No. 6,546,119.* Washington, DC: U.S. Patent and Trademark Office.

Cisco (2010b). Smart + Connected Life [World Expo 2010 pavilion guide]. Shanghai, China: Cisco. Retrieved Oct. 5, 2017 from http://www.cisco.com/web/CN/expo/pdf/Smart_Connected_Life_Guide_EN.pdf

Cisco. (2010a). Smart + connected communities; changing a city, a country, the world. San Jose, CA: Cisco.

Cisco. (2012). Cities of the Future: Songdo, South Korea, episode 3: working in Songdo. [promotional video] Retrieved from https://www.youtube.com/watch?v=JtWXZGev5M

Cisco. (2014). *Cisco Visual Networking Index: Global Mobile Data Traffic Forecast Update, 2013–2018.* Retrieved July 7, 2014, from http://www.cisco.com/c/en/us/solutions/collateral/service-provider/visual-networking-index-vni/white_paper_c11-520862.html

Cities Profile of 20 Smart Cities (2017). New Delhi: Ministry of Housing and Urban Affairs, Government of India. Retrieved from http://smartcities.gov.in/upload/uploadfiles/files/Maharashtra_Pune.pdf

Claverie-Berge, I. (2012). *Solutions Big Data IBM.* Retrieved August 11, 2014, from http://www-05.ibm.com/fr/events/netezzaDM_2012/Solutions_Big_Data.pdf

Coetzee, J., Louw, W., & Bullock, P. (2008). Solar powered traffic signals: myth or marvel? In *The 27th Southern African Transport Conference*, Pretoria.

Combe, T., Martin, A., & Di Pietro, R. (2016). To Docker or Not to Docker: A Security Perspective. *IEEE Cloud Computing, 3*(5), 54–62. doi:10.1109/MCC.2016.100

Coppolino. (2010). An Intrusion Detection System for Critical Information Infrastructures Using Wireless Sensor Network Technologies. *IEEE Critical Infrastructure.*

Core, O. S. (2017). *rkt | rkt Container Engine with CoreOS.* Retrieved October 4, 2017, from https://coreos.com/rkt/

Cortese, A. (2007, Dec. 30). An Asian hub in the making. *The New York Times*. Retrieved from www.nytimes.com/2007/12/30/realestate/commercial/30sqft.html

Couclelis, H. (2004). The construction of the digital city. *Environment and Planning B Urban Analytics and City Science*, *31*(1), 5–19.

Coverage & Duration: Smart Cities Mission, Government of India. (2017). *Smartcities.gov.in*. Retrieved 30 December 2017, from http://smartcities.gov.in/content/innerpage/coverageduration.php

Crang, M., & Graham, S. (2007). Sentient Cities: Ambient intelligence and the politics of urban space. *Information Communication and Society*, *10*(6), 789–817. doi:10.1080/13691180701750991

Daki, H. (2017). *Big Data management in smart grid: Concepts, requirements and implementation*. Big Data, Open Data and Data Development. doi:10.1002/9781119285199.index

Dameri, R. P., & Rosenthal-Sabroux, C. (2014). *Smart City and Value Creation*. Academic Press. 10.1007/978-3-319-06160-3_1

Daniel, J. (2009). Data Management in the Cloud: Limitations and Opportunities. *A Quarterly Bulletin of the Computer Society of the IEEE Technical Committee on Data Engineering*.

Dargay, J., Gately, D., & Sommer, M. (2007). Vehicle ownership and income growth, worldwide: 1960-2030. *Energy Journal*, 143–170.

Dash, D. (2017, November 29). Deaths in road accidents see steepest decline ever. *The Times of India*.

Dastane, S. (2017, March 23). 4 lakh vehicles add to on-road clutter this fiscal. *The Times of India*.

Dastjerdi, A. V., & Buyya, R. (2016). Fog Computing: Helping the Internet of Things Realize its Potential. *Computers*, *49*(8), 112–116. doi:10.1109/MC.2016.245

Datta, P., Kaur, N., & Garg, N. (2015). Automatic Bus Fare Collection System in India. In J. Mandal, S. Satapathy, M. Kumar Sanyal, P. Sarkar, & A. Mukhopadhyay (Eds.), *Information Systems Design and Intelligent Applications* (pp. 681–689). New Delhi: Springer. doi:10.1007/978-81-322-2247-7_69

De Jong, M., Joss, S., Schraven, D., Zhan, C., & Weijnen, M. (2015). Sustainable – smart – resilient – low carbon – eco-knowledge cities; making sense of a multitude of concepts promoting sustainable urbanization. *Journal of Cleaner Production*, *109*, 25–38. doi:10.1016/j.jclepro.2015.02.004

Dealy, G. (2016). The importance of business rules in actionable healthcare data. *Health IT Outcomes*. Retrieved from https://www.healthitoutcomes.com/doc/the-importance-of-business-rules-actionable-healthcare-data-0001

Dean, J., & Ghemawat, S. (2004). Mapreduce: Simplified data processing on large clusters. OSDI 2004, 137–50.

Dean, J., & Ghemawat, S. (2008). MapReduce: Simplified dataprocessing on large clusters. *Communications of the ACM, 51*(1), 107–113. doi:10.1145/1327452.1327492

Debnath, A. K., Chin, H. C., Haque, M. M., & Yuen, B. (2014). A methodological framework for benchmarking smart transport cities. *Cities (London, England), 37*, 47–56. doi:10.1016/j.cities.2013.11.004

Della Corte, V., & Sciarelli, M. (2012). Can coopetition be source of competitive advantage for strategic networks? *Cooperate Ownership & Control, 10*(21), 363–379.

Della Corte, V., Zamparelli, G., & Micera, R. (2013). Innovation in tradition-based firms: Dynamic knowledge for international competitiveness. *European Journal of Innovation Management, 16*(4), 405–439. doi:10.1108/EJIM-06-2012-0065

Demchenko, Y., De Laat, C., & Membrey, P. (2014). Defining architecture components of the Big Data Ecosystem. In *2014 International Conference on Collaboration Technologies and Systems, CTS 2014* (pp. 104–112). Academic Press. 10.1109/CTS.2014.6867550

Department of Infrastructure, Energy, and Resources (DIER). (2007). Traffic Impact Assessments (TIA) Guidelines. Retrieved from http://www.transport.tas.gov.au/_data/assets/pdf_file/0005/108491/TIA_Framework_Edition_1_09-2007_FINAL.pdf

Desai, D., & Somani, S. (2014). Instinctive traffic control and vehicle detection techniques. *Int. J. Sci. Eng. Res, 5*, 2192–2195.

Devi, T.J.B., Subramani, A., & Solanki, V. (2017). *Smart City: IOT Based Prototype for Parking Monitoring and Management System Commanded by Mobile App*. Volume 10. Poland: ACIS. doi:10.15439/2017R23

Dey, N., Hassanien, A. E., Bhatt, C., Ashour, A., & Satapathy, S. C. (2018). *Internet of things and big data analytics toward next-generation intelligence*. Academic Press.

Dey, A. K. (2001). Understanding and using context. *Personal and Ubiquitous Computing, 5*(1), 4–7. doi:10.1007007790170019

Dey, N., Ashour, A. S., & Bhatt, C. (2017). Internet of things driven connected healthcare. In *Internet of things and big data technologies for next generation healthcare* (pp. 3–12). Cham: Springer. doi:10.1007/978-3-319-49736-5_1

Dey, N., Hassanien, A. E., Bhatt, C., Ashour, A., & Satapathy, S. C. (Eds.). (2018). *Internet of Things and Big Data Analytics Toward Next-Generation Intelligence*. Springer. doi:10.1007/978-3-319-60435-0

Dhall & Solanki. (n.d.). An IoT Based Predictive Connected Car Maintenance Approach. *International Journal of Interactive Multimedia and Artificial Intelligence, 4*(3). doi: 10.9781/ijimai.2017.433

Diaz, M., Cerri, P., Pirlo, G., Ferrer, M. A., & Impedovo, D. (2015, September). A survey on traffic light detection. In *International Conference on Image Analysis and Processing* (pp. 201-208). Cham: Springer. 10.1007/978-3-319-23222-5_25

Diaz-Cabrera, M., Cerri, P., & Medici, P. (2015). Robust real-time traffic light detection and distance estimation using a single camera. *Expert Systems with Applications*, *42*(8), 3911–3923. doi:10.1016/j.eswa.2014.12.037

Dickson, B. (2015). Why IoT security is so critical. *Techcrunch*. Retrieved from https://techcrunch.com/2015/10/24/why-iot-security-is-so-critical/

DIN ISO 27000. (2011). Retrieved February 1, 2016, from https://www.beuth.de/de/norm-entwurf/din-iso-iec-27000/243433889

Ding, G., Wang, L., & Wu, Q. (2013). *Big Data Analytics in Future Internet of Things*. arXiv:1311.4112v1 [cs.DC]

Dinh, T., Kim, Y., & Lee, H. (2016). A location-based interactive model for Internet of Things and cloud (IoT-cloud). *Ubiquitous and Future Networks (ICUFN) 2016 Eighth International Conference on*, 444-447.

Dipak, K.D. (2012, May 31). India loses Rs 60,000 crore due to traffic congestion: Study. *Times Of India*. Retrieved from http://articles.timesofindia.indiatimes.com/2012-05-31/india/31920307_1_toll-plazas-road-space-stoppage

Distefano, S., Merlino, G., & Puliafito, A. (2012). Enabling the Cloud of Things. In *Innovative Mobile and Internet Services in Ubiquitous Computing (IMIS), 2012 Sixth International Conference on*. IEEE.

Docker. (2017a). *Docker - Build, Ship, and Run Any App, Anywhere*. Retrieved October 4, 2017, from https://www.docker.com/

Docker. (2017b). *Swarm mode overview | Docker Documentation*. Retrieved October 4, 2017, from https://docs.docker.com/engine/swarm/

Domingos, P., & Hulten, G. (2000). Mining high-speed data streams. In *Proceedings of the sixth ACM SIGKDD international conference on Knowledge discovery and data mining*. ACM. 10.1145/347090.347107

Doukas, C., & Maglogiannis, I. (2012). Bringing iot and cloud computing towardspervasive healthcare," in Innovative Mobile and Internet Servicesin Ubiquitous Computing (IMIS). *Sixth International Conferenceon. IEEE, 2012*, 922–926.

Du, L., Chen, S., & Han, L. (2015a). Coordinated online in-vehicle navigation guidance based on routing game theory. *Transportation Research Record: Journal of the Transportation Research Board*, (2497): 106–116.

Du, L., Han, L., & Chen, S. (2015b). Coordinated Online In-Vehicle Routing Balancing User Optimality and System Optimality through Information Perturbation. *Transportation Research Part B: Methodological*, *79*, 121–133. doi:10.1016/j.trb.2015.05.020

Du, L., Han, L., & Li, X. Y. (2014). Distributed coordinated in-vehicle online routing using mixed-strategy congestion game. *Transportation Research Part B: Methodological*, *67*, 1–17.

Du, L., Peeta, S., & Kim, Y. H. (2012). An adaptive information fusion model to predict the short-term link travel time distribution in dynamic traffic networks. *Transportation Research Part B: Methodological*, *46*(1), 235–252.

Du, L., Peeta, S., & Kim, Y. H. (2013). Online stochastic routing incorporating real-time traffic information. *Transportation Research Record: Journal of the Transportation Research Board*, 2334.

Dustdar, S., Nastić, S., & Šćekić, O. (2017). *Smart Cities - The Internet of Things, People and Systems*. Cham: Springer International Publishing; doi:10.1007/978-3-319-60030-7

Effects of global warming. (n.d.). Retrieved from www.livescience.com

Eisenhauer, M., Rosengren, P., & Antolin, P. (2009). A development platform for integrating wireless devices and sensors into ambient intelligence systems. In *Sensor, Mesh and Ah Hoc Communications and Networks Workshops, SECON 2009, IEEE Communication Society Conference*. Rome: IEEE. 10.1109/SAHCNW.2009.5172913

Elechi, P., Odeyemi, F. M., & Yellowe, K. M. (2014). Improved traffic control in Port-Harcourt using solar dependent traffic light system. *IACSIT International Journal of Engineering and Technology, 4(2), 93-100.*

Electrical Engineering and Automation Department, Aalto University. (2011). Lighting Technologies. In *Guidebook* (Ch. 5). Retrieved from www.lightinglab.fi/IEAAnnex45/guidebook/5_lighting%20technologies.pdf

Elhayatmy, G., Dey, N., & Ashour, A. S. (2018). *Internet of Things Based Wireless Body Area Network in Healthcare*. Academic Press. 10.1007/978-3-319-60435-0_1

Elhayatmy, G., Dey, N., & Ashour, A. S. (2018). Internet of Things Based Wireless Body Area Network in Healthcare. In *Internet of Things and Big Data Analytics Toward Next-Generation Intelligence* (pp. 3–20). Cham: Springer.

Elhayatmy, G., Dey, N., & Ashour, A. S. (2018). *Internet of Things Based Wireless Body Area Network in Healthcare. In Internet of Things and Big Data Analytics Toward Next-Generation Intelligence* (pp. 3–20). Springer.

Elliott, T. E., Holmes, J. H., Davidson, A. J., La Chance, P.-A., Nelson, A. F., & Steiner, J. F. (2013). Data warehouse governance programs in healthcare settings: A literature review and a call to action. *eGEMs. The Journal for Electronic Health Data and Methods*, *1*(1), 1010. PMID:25848561

Elmaghraby, A. S., & Losavio, M. M. (2014). Cyber security challenges in smart cities: Safety, security and privacy. *Journal of Advanced Research*, *5*(4), 491–497. doi:10.1016/j.jare.2014.02.006 PMID:25685517

Elmesalawy & Ali. (2015). A Grouped System Architecture for Smart Grids Based AMI Communications Over LTE. *International Journal of Wireless & Mobile Networks, 7*(6).

Engineeringforchange.org. (2015). *Engineering for Change*. Available at: http://www.engineeringforchange.org/home

Eremia, M., Toma, L., & Sanduleac, M. (2017). The Smart City Concept in the 21st Century. *Procedia Engineering*, *181*, 12–19. doi:10.1016/j.proeng.2017.02.357

Erl, T. (2005). *Service-Oriented Architecture: Concepts, Technology, and Design*. City. doi:10.1109/SAINTW.2003.1210138

Ettorchi-Tardy, A., Levif, M., & Michel, P. (2012). Benchmarking: A method for continuous quality improvement in health. *Healthcare Policy*, *7*(4), 101–119. PMID:23634166

European Commission. (2013). Factories of the future - Multi-annual roadmap for the contractual PPP under. *Horizon*, *2020*, •••. Retrieved from https://www.scribd.com/document/271903700/Factories-of-the-Future-2020-Roadmap#

European Commission. (2017). *European Union, Trade in goods with India*. Author.

European Lamp Companies Federation. (2009). Environmental Aspects of Lamps. Retrieved from http://www.lightingeurope.org/uploads/files/0908011_ele_brochure_environmental_aspects_lamps_updated_final.Pdf

Evans, D. (2011). The Internet of Things: how the next evolution of the Internet is changing everything. *Cisco Systems*. Retrieved from https://www.cisco.com/c/dam/en_us/about/ac79/docs/innov/IoT_IBSG_0411FINAL.pdf

Evans, E. (2012). *Internet of everything: How more relevant and valuable connections will change the world*. Retrieved from https://www.slideshare.net/CiscoIBSG/internet-of-everything

Evans, R. S., Lloyd, J. F., & Pierce, L. A. (2012). Clinical use of an enterprise data warehouse. *AMIA ... Annual Symposium Proceedings* (pp. 189–198). PMID:23304288

Fadlullah, Z. M., Fouda, M. M., Kato, N., Takeuchi, A., Iwasaki, N., & Nozaki, Y. (2011). Toward intelligent machine-to-machine communications in smart grid. *IEEE Communications Magazine*, *49*(4), 60–65. doi:10.1109/MCOM.2011.5741147

Fan, W. (2013). Mining big data: current status, and forecast to the future. *ACM SIGKDD Explor Newsl., 14*(2).

Faria, R. (2017). *Smart Mobility: A Survey*. IEEE.

Farooq, M. U., Waseem, M., Khairi, A., & Mazhar, S. (2015). A critical analysis on the security concerns of Internet of Things (IoT). *International Journal of Computers and Applications*, *111*(7), 1–6. doi:10.5120/19547-1280

Fazio, M., Celesti, A., Ranjan, R., Liu, C., Chen, L., & Villari, M. (2016). Open Issues in Scheduling Microservices in the Cloud. *IEEE Cloud Computing*, *3*(5), 81–88. doi:10.1109/MCC.2016.112

Features, S. C. Smart Cities Mission, Ministry of Housing and Urban Affairs, Government of India. (2017). *Smartcities.gov.in*. Retrieved 30 December 2017 from http://smartcities.gov.in/content/innerpage/smart-city-features.php

Federal Highway Administration, US Department of Transportation. (1996). Investigation Of Highway Workzone Crashes - Summary Report. Retrieved from http://ntl.bts.gov/DOCS/96100/index.html

Federal Highway Administration, US Department of Transportation. (2003). 21st Century operations Using 21st Century Technologies. Retrieved fromhttp://www.ops.fhwa.dot.gov/aboutus/opstory.htm

Federal Highway Administration, US Department of Transportation. (2006, October). Traffic Detector Handbook (3rd ed., Vol. 1). Retrieved from http://www.fhwa.dot.gov/publications/research/operation s/its/06108/

Feki, M. A., Kawsar, F., Boussard, M., & Trappeniers, L. (2013). TheInternet of Things: The Next Technological Revolution. *Computers*, *46*(2), 24–25. doi:10.1109/MC.2013.63

Fernandez, Río, & Herrera. (2017). Fuzzy Rule Based Classification Systems for Big Data with MapReduce: Granularity Analysis. *Advances in Data Analysis and Classification, 11*(4). doi: 10.100711634-016-0260-z

Fernandez, A., Carmona, C. J., del Jesus, M. J., & Herrera, F. (2016). A View on Fuzzy Systems for Big Data: Progress and Opportunities. *International Journal of Computational Intelligence Systems*, *9*(1), 69–80. doi:10.1080/18756891.2016.1180820

Fernandez, A., Río, S., López, V., Bawakid, A., del Jesus, M. J., Benítez, J. M., & Herrera, F. (2014). Big Data with Cloud Computing: An Insight on the Computing Environment, MapReduce and Programming Frameworks. *WIREs Data Mining and Knowledge Discovery*, *4*(5), 380–409. doi:10.1002/widm.1134

Fogliatto, F. S., Da Silveira, G. J. C., & Borenstein, D. (2012). The mass customization decade: An updated review of the literature. *International Journal of Production Economics*, *138*(1), 14–25. doi:10.1016/j.ijpe.2012.03.002

Foresight. (2013). *The future of manufacturing: A new era of opportunity and challenge for the UK: Summary Report*. The Government Office for Science.

Furestenau, N., Schmidt, M., Horack, H., Goetze, W., & Schmidt, W. (1997). Extrinsic fabryperot interferometer vibration acoustic senso systems for airport ground traffic monitoring. *Proc. IEE on Optoelectronics*, 144, 134–144. 10.1049/ip-opt:19971268

Gadkari, S. (2017, May 4). Syncing signals for traffic ease. *Pune Mirror.*

Galdon-Clavell, G. (2013). (Not so) smart cities?: The drivers, impact and risks of surveillance-enabled smart environments. *Science & Public Policy, 40*(6), 717–723. doi:10.1093cipolct070

Gale, L., Burdick, J., Wagstaff, P., Flowers, S., Plott, P. C., Furlong, D., & Daniels, M. (2013). The enterprise data warehouse: Moving from retrospective to proactive analytics. *Journal of Healthcare Information Management, 27*(3), 38–44.

Galpert, Río, Herrera, Ancede-Gallardo, Antunes, & Agüero-Chapin. (2015). An Effective *Big Data Supervised Imbalanced Classification Approach for Ortholog Detection in Related Yeast Species*. BioMed Research International. doi: 10.1155/2015/748681

Gamba, M., Gonella, A., & Palazzi, C. E. (2015). Design issues and solutions in a modern home automation system. *Proceedings of International Conference on Computing Networking and Communications (ICNC 2015)*, 1111-1115. 10.1109/ICCNC.2015.7069505

Gandomi, A., & Haider, M. (2015). Beyond the hype: Big data concepts, methods, and analytics. *International Journal of Information Management, 35*(2), 137–144. doi:10.1016/j.ijinfomgt.2014.10.007

Ganti, R., Ye, F., & Lei, H. (2011). Mobile Crowd Sensing: Current State and Future Challenges. *IEEE Communications Magazine, 49*(11), 32–39. doi:10.1109/MCOM.2011.6069707

Garber, N. J., & Hoel, A. L. (2010). *Traffic and Highway Engineering* (4th ed.). Boston, MA: Cengage Learning.

Gil. (2016). Internet of Things: A Review of Surveys Based on Context Aware Intelligent Services. *Sensors — Open Access Journal.*

Gilbert, S., & Lynch, N. (2002). Brewer's conjecture and the feasibility of consistent, available, partition-tolerant web services. *ACM SIGACT News, 33*(2), 51–59. doi:10.1145/564585.564601

Gil-Castineira, F., Costa-Montenegro, E., Gonzalez-Castano, F., López-Bravo, C., Ojala, T., & Bose, R. (2011). Experiences inside the ubiquitous Oulu smart city. *Computer, 44*(6), 48–55. doi:10.1109/MC.2011.132

Giusto, D., Iera, A., Morabito, G., & Atzori, L. (Eds.). (2010). The Internet of Things. Springer.

Global Sensor Networks. (2011). *International Conference on Computational Intelligence and Communication Networks*. Author. doi: 10.1109/CICN.2011.172

Gondchawar & Kawitkar. (2016). IoT based Smart Agriculture. *International Journal of Advanced Research in Computer and Communication Engineering, 5*(6).

Google Fit. (2015). Retrieved from https://developers.google.com/fit/

Government of India. (2011). Census of India. *Censusindia.gov.in*. Retrieved 30 December 2017 from http://www.censusindia.gov.in/2011census/PCA/PCA_Highlights/pca_highlights_file/India/Chapter-1.pdf

Greenfield, A., & Kim, N. (2013). *Against the smart city* [Part 1 of This city is here for you to use] (1.3 ed.). Seattle, WA: Amazon Digital Services.

Grover, A. (2013). Airport business districts — An indispensable reality. *Creative Space*, *1*(1), 79–99. doi:10.15415/cs.2013.11005

Grund, M., Schaffner, J., Krueger, J., Brunnert, J., & Zeier, A. (2010). The Effects of Virtualization on Main Memory Systems. In *Proceedings of the Sixth International Workshop on Data Management on New Hardware* (pp. 41–46). New York, NY: ACM. 10.1145/1869389.1869395

Grzybek, A., Danoy, G., Bouvry, P., & Seredynski, M. (2015). Mitigating flash crowd effect using connected vehicle technology. *Vehicular Communications*, *2*(4), 238–250.

Gubbi, J., Buyya, R., Marusic, S., & Palaniswami, M. (2013). Internet of Things (IoT): A vision, architectural elements, and future directions. *Future Generation Computer Systems*, *29*(7), 1645–1660. doi:10.1016/j.future.2013.01.010

Gupta, Nath, Chakraborty, & Ghosh. (2016). *SDFog: A Software Defined Computing Architecture for QoS Aware Service Orchestration over Edge Devices*. arXiv.org

Gupta, R., Gupta, H., & Mohania, M. (2012). Cloud computing andbig data analytics: what is new from databases perspective? *1st International Conference on Big DataAnalytics (BDA)*, 42–61.

Gurav, U., & Shaikh, R. (2010). Virtualization: a key feature of cloud computing. In *Proceedings of the International Conference and Workshop on Emerging Trends in Technology*. ACM. 10.1145/1741906.1741957

Haller, S. & Kubach, U. (2004, May 27). US7245220 B2. Retrieved from http://www.google.com/patents/US7245220

Halpern, O., LeCavalier, J., Calvillo, N., & Pietsch, W. (2013). Test-bed urbanism. *Public Culture, 25*(2 70), 272-306.

Han, Haihong, Le, & Du. (2011). Survey on NoSQL database. *Pervasive computing and applications (ICPCA), 2011 6th international conference on*, 363–366.

Han, L. D., Wang, H., & Mackey, W. F. (2006). Finding shortest paths under time-bandwidth constraints by using elliptical minimal search area. *Transportation Research Record: Journal of the Transportation Research Board*, (1977), 225-233.

HAProxy. (2017). *HAProxy - The Reliable, High Performance TCP/HTTP Load Balancer*. Retrieved October 9, 2017, from http://www.haproxy.org/

Harrison, G. (2015). *Next generation databases: NoSQL, NewSQL, and Big Data*. Academic Press.

Hart, D. W. (2011). *Power Electronics*. New York: McGraw-Hill.

Hashem, I., Chang, V., Anuar, N., Adewole, K., Yaqoob, I., Gani, A., ... Chiroma, H. (2016). The role of big data in smart city. *International Journal of Information Management*, *36*(5), 748–758. doi:10.1016/j.ijinfomgt.2016.05.002

Hawas, Y. E. (2012). A cooperative distributed system for real-time route guidance. *Journal of Transportation Technologies, 2(3), 230.*

Hay, S. I., Abajobir, A. A., Abate, K. H., Abbafati, C., Abbas, K. M., Abd-Allah, F., ... Aboyans, V. (2017). Global, regional, and national disability-adjusted life-years (DALYs) for 333 diseases and injuries and healthy life expectancy (HALE) for 195 countries and territories, 1990–2016: A systematic analysis for the Global Burden of Disease Study 2016. *Lancet*, *390*(10100), 1260–1344.

Health Expenditures, C. D. C. (2017). *Center for Disease Control and Prevention*. Retrieved from https://www.cdc.gov/nchs/fastats/health-expenditures.htm

Healthcare Performance Benchmarks. (2017). *McKesson*. Retrieved from http://www.mckesson.com/providers/health-systems/enterprise-intelligence/healthcare-performance-benchmarks/

HealthIT.Gov. (2016). Electronic Health Records Infographic. Retrieved from https://www.healthit.gov/patients-families/electronic-health-records-infographic

Hegyi, A., Bellemans, T., & De-Schutter, B. (2009). Freeway traffic management and control. In R. A. Meyers (Ed.), *Encyclopedia of Complexity and Systems Science*. New York: Springer.

Hevner, A., & Chatterjee, S. (2010). *Design Research in Information Systems. Design Research in Information Systems* (Vol. 22). Boston, MA: Springer US; doi:10.1007/978-1-4419-5653-8

He, W., Yan, G., & Xu, L. (2014). Developing vehicular data cloud services in the iot environment. *IEEE Transactions on Industrial Informatics*, *10*(2), 1587–1595. doi:10.1109/TII.2014.2299233

High Blood Pressure Fact SheetC. D. C. (2016). Retrieved from https://www.cdc.gov/dhdsp/data_statistics/fact_sheets/fs_bloodpressure.htm

HIMSS Asia Pacific. (n.d.). Big Data Improves Chronic Disease Management. Retrieved from http://www.himssasiapac.org/content-library/exclusive-articles/big-data-improves-chronic-disease-management

Hiskey, T. (2017). Preparing for Manufacturing's Future with Industry 4.0. *Industry Week*.

Hoang, B., & Caudill, A. (2012). RFID. IEEE Emerging Technology portal, 2006-2012.

Hollands, R. G. (2008). Will the real smart city please stand up? *City*, *12*(3), 303–320. doi:10.1080/13604810802479126

Höller, J., Tsiatsis, V., Mulligan, C., Karnouskos, S., Avesand, S., & Boyle, D. (2014). *From Machine-to-Machine to the Internet of Things: Introduction to a New Age of Intelligence*. Elsevier.

Hong, I., Park, S., Lee, B., Lee, J., Jeong, D., & Park, S. (2014). IoT-Based Smart Garbage System for Efficient Food Waste Management. *The Scientific World Journal*, *2014*, 646953. doi:10.1155/2014/646953 PMID:25258730

Hougen, D. F. (2000). A miniature robotic system for reconnaissance and surveillance. *Proc. 2000 IEEE International Conference on Robotics and Automation*, 501–507. 10.1109/ROBOT.2000.844104

https://www.cisco.com/c/en_in/about/knowledge-network/smart-traffic.html

Huang, B., & Wu, Q. (2006). A spatial indexing approach for high-performance location based services. *Journal of Navigation, 60(1), 83-93.*

Huang, B., Wu, Q., & Zhang, F. (2006). The shortest path algorithm with novel heuristics for dynamic transportation networks. *International Journal of Geographical Information Science, 21(6), 625-644.*

Huang, H. J., Liu, T. L., & Yang, H. (2008). Modeling the evolutions of day-to-day route choice and year-to-year ATIS adoption with stochastic user equilibrium. *Journal of Advanced Transportation, 42(2), 111-127.*

Huang, Y., & Chung, T. (2008). Modeling and analysis of urban traffic lights control systems using timed CP-nets. *Journal of Information Science and Engineering, 24(3).*

Humblet, P. A. (1991). Another adaptive distributed shortest path algorithm. *IEEE Transactions on Communications*, *39*(6), 995–1003.

Hyang-Sook, C., Byung-Sun, C., & Woong-Hee, P. (2007). Ubiquitous-city business strategies: The case of South Korea. In *Management of Engineering and Technology* (pp. 1147-1153). IEEE. doi:10.1109/PICMET.2007.4349438

HYDRA. (2010). Retrieved from http://hydramiddleware.eu

Idris, M. Y. I., Leng, Y. Y., Tamil, E. M., Noor, N. M., & Razak, Z. (2009). Car park system: A review of smart parking system and its technology. *Information Technology Journal*, *8*(2), 101–113. doi:10.3923/itj.2009.101.113

Ijaz, S., Shah, M. A., Khan, A., & Ahmed, M. (2016). Smart Cities: A Survey on Security Concerns. *International Journal of Advanced Computer Science and Applications*, *7*(2), 612–625. doi:10.14569/IJACSA.2016.070277

Imanuel. (2017). Top 50 Big Data Platforms and Big Data Analytics Software. *Predictive Analytics Today*. Retrieved from https://www.predictiveanalyticstoday.com/bigdata-platforms-bigdata-analytics-software/

Interstate migration in India doubled between 2001-2011: WEF. (2017, October 25). *The Economic Times*

IOTUK. (2016). *Li-Fi Internet of Things*. Retrieved from https://iotuk.org.uk/li-fi-internet-things/

Islam, S. M. R. (2015). The internet of things for health care: A comprehensive survey. *IEEE Access: Practical Innovations, Open Solutions*, *3*, 678–708. doi:10.1109/ACCESS.2015.2437951

Iyengar, S. G., Varshney, P. K., & Damarla, T. (2011, May). A parametric copula-based framework for hypothesis testing using heterogeneous data. *IEEE Transactions on Signal Processing*, *59*(5), 2308–2319. doi:10.1109/TSP.2011.2105483

Jahn, Jentsch, Prause, Pramudianto, Al-Akkad, & Reiners. (2010). The Energy Aware Smart Home. *Proceedings of the Fifth International Conference on Future Information Technology (FutureTech),* 1–8.

Jain, A. K., Murty, M. N., & Flynn, P. J. (1999). Data clustering: A review. *ACM Computing Surveys, 31*(3), 264–323. doi:10.1145/331499.331504

Jain, A., & Bhatnagar, V. (2017). Concoction of Ambient Intelligence and Big Data for Better Patient Ministration Services. *International Journal of Ambient Computing and Intelligence, 8*(4), 19–30. doi:10.4018/IJACI.2017100102

Jang, M., & Suh, S. T. (2010). U-city: New trends of urban planning in Korea based on pervasive and ubiquitous geo-technology and geoinformation. In Computational Science and Its Applications, ICCSA-2010 (pp. 262–270).

Jassas, Qasem, & Mahmoud. (2015). A Smart System Connecting e-Health Sensors and the Cloud A Smart System Connecting e-Health Sensors and the Cloud. *Proceeding of the IEEE 28th Canadian Conference on Electrical and Computer Engineering Halifax*, 712-716.

Jayakrishnan, R., Mahmassani, H. S., & Hu, T. Y. (1994). An evaluation tool for advanced traffic information and management systems in urban networks. *Transportation Research Part C, Emerging Technologies, 2*(3), 129–147.

Jensen, P., Jensen, R., & Brunak, S. (2012). Mining electronic health records: Towards better research applications and clinical care. *Nature Reviews. Genetics, 13*(6), 395–405. doi:10.1038/nrg3208 PMID:22549152

Jesdanun, A. (2015). Challenges for doctors using fitness trackers and apps. *Science, 10.* Retrieved from https://phys.org/news/2015-02-doctors-trackers-apps.html

Jha, I. S., Sen, S., & Kumar, R. (2014). Smart grid development in India — A case study. *2014 Eighteenth National Power Systems Conference (NPSC)*, 1-6. 10.1109/NPSC.2014.7103866

Jiang, L., Da Xu, L., Cai, H., Jiang, Z., Bu, F., & Xu, B. (2014). An iot-oriented data storage framework in cloud computing platform. *IEEE Trans. Ind. Inform., 10*(2), 1443–1451.

Jiang, L., Da Xu, L., Cai, H., Jiang, Z., Bu, F., & Xu, B. (2014). "An IoTOriented Data Storage Framework in Cloud ComputingPlatform," Ind. Informatics. *IEEE Trans., 10*(2), 1443–1451.

Jin, J. (2014). *An information framework for creating a smart city through internet of things* (Vol. 1). IEEE Internet of Things Journal.

Jin, J., Gubbi, J., Marusic, S., & Palaniswami, M. (2014). An information framework for creating a smart city through internet of things. *InternetofThings Journal, IEEE, 1*(2), 112–121. doi:10.1109/JIOT.2013.2296516

Joshi, B. (2016, June 29). Smart Cities Mission: Flaws in a flagship programme. *The Hindu.*

Joshi, Y. (2017, July 7). Pune is in a jam: Why flyovers haven't helped traffic congestion. *Hindustan Times*.

Joss, S. (2011). Eco-cities: The mainstreaming of urban sustainability–key characteristics and driving factors. *International Journal of Sustainable Development and Planning*, *6*(3), 268–285. doi:10.2495/SDP-V6-N3-268-285

Joss, S., Cowley, R., & Tomozeiu, D. (2013). Towards the 'ubiquitous eco-city': An analysis of the internationalization of eco-city policy and practice. *Urban Research & Practice*, *6*(1), 54–74. doi:10.1080/17535069.2012.762216

Julian, S., Shuey, M., & Cook, S. (2016). Containers in Research: Initial Experiences with Lightweight Infrastructure. *Proceedings of the XSEDE16 Conference on Diversity, Big Data, and Science at Scale*. 10.1145/2949550.2949562

Jun-Woo, K., & Ahn, Y. J. (2011). *Songdo Free Economic Zone in South Korea*. 대한지리학회지, *46*(5), 662–672.

Jurčić, I., & Gotovac, S. (2016). New approach in mobile telecom analysis – The eight key field analysis. *WICT III, Workshop on Information and Communication Technologies III, SoftCOM 2016*. Retrieved from http://marjan.fesb.hr/SoftCOM/2016/files/apk/final_program_2016.pdf

Kamal, M. S., Dey, N., & Ashour, A. S. (2017). Large Scale Medical Data Mining for Accurate Diagnosis: A Blueprint. In *Handbook of Large-Scale Distributed Computing in Smart Healthcare* (pp. 157–176). Cham: Springer. doi:10.1007/978-3-319-58280-1_7

Kamal, S., Ripon, S. H., Dey, N., Ashour, A. S., & Santhi, V. (2016). A MapReduce approach to diminish imbalance parameters for big deoxyribonucleic acid dataset. *Computer Methods and Programs in Biomedicine*, *131*, 191–206. doi:10.1016/j.cmpb.2016.04.005 PMID:27265059

Kamilaris, A., Iannarilli, N., Trifa, V., & Pitsillides, A. (2011). Bridging the Mobile Web and the Web of Things in Urban Environments. *First Workshop on the Urban Internet of Things, in Proc. of IoT 2010*.

Kantarci, B., & Mouftah, H. T. (2014). Trustworthy sensing for public safety in cloud-centric internet of things. *Internet of Things Journal, IEEE*, *1*(4), 360–368. doi:10.1109/JIOT.2014.2337886

Kanungo, A., Sharma, A., & Singla, C. (2014, March). Smart traffic lights switching and traffic density calculation using video processing. In 2014 Recent Advances in Engineering and computational sciences (RAECS) (pp. 1-6). IEEE. doi:10.1109/RAECS.2014.6799542

Karagiannis, V., Chatzimisios, P., Vazquez-Gallego, F., & Alonso-Zarate, J. (2015). *A Survey on Application Layer Protocols for the Internet of Things. Transaction on IoT and Cloud Computing*.

Kasarda, J. D. (2008). The evolution of airport cities and the aerotropolis. In *Airport cities: The evolution* (pp. 1–39). London, England: Insight Media. Retrieved from http://www.aerotropolis.com/files/evolutionChapter1.pdf

Kaufman, D. E., Nonis, J., & Smith, R. L. (1998). A mixed integer linear programming model for dynamic route guidance. *Transportation Research Part B: Methodological*, *32*(6), 431–440.

Kayyali, B., Knott, D., & Kuiken, S. V. (2013). The big-data revolution in US health care: Accelerating value and innovation. *McKinsey & Company*. Retrieved from https://www.mckinsey.com/industries/healthcare-systems-and-services/our-insights/the-big-data-revolution-in-us-health-care

Kecskemeti, G., Marosi, A. C., & Kertesz, A. (2016). The ENTICE approach to decompose monolithic services into microservices. *2016 International Conference on High Performance Computing and Simulation, HPCS 2016*. 10.1109/HPCSim.2016.7568389

Khairnar, A. (2017, July 29). Pune has equal number of vehicles, people: PMC's environment status report. *Hindustan Times*.

Khan, M. (2014). Seven V's of Big Data understanding Big Data to extract value. In American Society for Engineering Education, 2014 Conference of the. IEEE.

Khan, R., Khan, S. U., Zaheer, R., & Khan, S. (2012, December). Future internet: the internet of things architecture, possible applications and key challenges. In *2012 10th International Conference on Frontiers of Information Technology (FIT)*, (pp. 257-260). IEEE. 10.1109/FIT.2012.53

Khan, B. N. S., & Han, K. (2016). Internet of things based energy aware smart home control system. *IEEE Access: Practical Innovations, Open Solutions*, *4*, 7556–7566. doi:10.1109/ACCESS.2016.2621752

Khanna, P. (2015). Urbanization, Technology, and the Growth of Smart Cities. *Asian Management Insights, 2*(2).

Khan, Z. (2013). Cloud Based Big Data Analytics for Smart Future Cities. In *6th International Conference on Utility and Cloud Computing*. IEEE Computer Society. 10.1109/UCC.2013.77

Kholidy, H. A., & Baiardi, F. (2012). CIDS: A framework for intrusion detection in cloud systems. *Proceedings of 9th IEEE International Conference on Information Technology-New Generations*, 379–85. doi: 10.1109/ITNG.2012.94

Kilbridge, P. (2015). Introducing the acute care patient portal. *Advisory Board Company*. Retrieved from https://www.advisory.com/research/health-care-it-advisor/research-notes/2015/introducing-the-acute-care-patient-portal

Kim & Scott. (2012). Robust kernel density estimation. *Journal of Machine Learning Research*, *13*(1), 2529–2565.

Kim, Y. M., Kim, H. S., Moon, S. Y., & Bae, S.-Y. (2009). Ubiquitous Eco-City Planning in Korea. A Project for the Realization of Ecological City Planning and Ubiquitous Network Society. In M. Schrenk, V. V. Popovich, D. Engelke, and P. Elisei (Eds.), REAL CORP 2009 Proceedings (pp. 925–930). Schwechat, Austria: Competence Center for Urban and Regional Planning.

Kim, C. (2010). Place promotion and symbolic characterization of new Songdo City, South Korea. *Cities (London, England)*, *27*(1), 13–19. doi:10.1016/j.cities.2009.11.013

Kim, G. H. (2014). *Big-data applications in the government sector*. ACM.

Kim, J. W., & Ahn, Y.-J. (2006). Background and dynamics of free economic zones in Korea. *Journal of the Korean Urban Geographical Society*, *9*, 79–87.

Klein, L. A., Kelley, M. R., & Mills, M. K. (1997). Evaluation of overhead and in-ground vehicle detector technologies for traffic flow measurement. *Journal of Testing and Evaluation, 25(2), 205-214.*

Kohler, J., & Specht, T. (2014). Vertical Query-Join Benchmark in a Cloud Database Environment. In *Proceedings of 2nd World Conference on Complex Systems 2014* (pp. 143–150). Agadir, Morocco: IEEE. 10.1109/ICoCS.2014.7060950

Kohler, J., Lorenz, R., Gumbel, M., Specht, T., & Simov, K. (2017). A security by distribution approach to manage big data in a federation of untrustworthy clouds. In Privacy and Security Policies in Big Data. Hershey, PA: IGI-Global.

Kohler, J., Lorenz, R., Gumbel, M., Specht, T., & Simov, K. (2017). A Security by Distribution Approach to Manage Big Data in a Federation of Untrustworthy Clouds. In N. D. S. Tamane & V. K. Solanki (Eds.), *Privacy and Security Policies in Big Data*. IGI Global. doi:10.4018/978-1-5225-2486-1.ch005

Kohler, J., Simov, K., Fiech, A., & Specht, T. (2015). On The Performance Of Query Rewriting In Vertically Distributed Cloud Databases. *Proceedings of The International Conference Advanced Computing for Innovation ACOMIN 2015.*

Kohler, J., Simov, K., & Specht, T. (2015). Analysis of the Join Performance in Vertically Distributed Cloud Databases. *International Journal of Adaptive, Resilient and Autonomic Systems*, *6*(2), 65–87. doi:10.4018/IJARAS.2015070104

Kohler, J., & Specht, T. (2012). SeDiCo - Towards a Framework for a Secure and Distributed Datastore in the Cloud. *Proceedings of Chip-to-Cloud Security Forum 2012.*

Kohler, J., & Specht, T. (2015). A Performance Comparison Between Parallel And Lazy Fetching in Vertically Distributed Cloud Databases. In *International Conference on Cloud Computing Technologies and Applications - CloudTech 2015*. Marrakesh, Morocco: IEEE Computer Society. 10.1109/CloudTech.2015.7337005

Kourtit, K., Nijkamp, P., & Steenbruggen, J. (2017). The significance of digital data systems for smart city policy. *Socio-Economic Planning Sciences*, *58*, 13–21. doi:10.1016/j.seps.2016.10.001

Kozhirbayev, Z., & Sinnott, R. O. (2017). A performance comparison of container-based technologies for the Cloud. *Future Generation Computer Systems*, *68*, 175–182. doi:10.1016/j.future.2016.08.025

Kramers, A., Höjer, M., Lövehagen, N., & Wangel, J. (2014). Smart sustainable cities–Exploring ICT solutions for reduced energy use in cities. *Environmental Modelling & Software*, *56*, 52–62. doi:10.1016/j.envsoft.2013.12.019

Krebs, P., & Duncan, D. T. (2015). Health App Use Among US Mobile Phone Owners: A National Survey. *JMIR mHealth and uHealth*, *3*(4), e101. doi:10.2196/mhealth.4924 PMID:26537656

Krishnamurthy, R., Mishra, R., & Desouza, K. (2016). City profile: Pune, India. *Cities (London, England)*, *53*, 98–109. doi:10.1016/j.cities.2016.01.011

Kshetri, N., Alcantara, L. L., & Park, Y. (2014). Development of a smart city and its adoption and acceptance: The case of New Songdo. *Communications & Strategies*, *1*(96), 113–128. Retrieved from https://ideas.repec.org/a/idt/journl/cs9606.html

Kubernetes. (2017). *Kubernetes - Production-Grade Container Orchestration*. Retrieved October 4, 2017, from https://kubernetes.io/

Kuhne, R. D. (1991). Freeway control using a dynamic traffic flow model and vehicle reidentification techniques. *Transportation Research Record: Journal of the Transportation Research Board*, 1320.

Kulkarni, P. (2017, September 3). Smart city projects in deadline crisis. *Pune Mirror*.

Kumar. (2016). Role of Big Data and Analytics in Smart Cities. *International Journal of Science and Research, 5*(2).

Kumar, N., Swami, G. C., & Nagadarshini, K. N. (2014). Efficient Garbage Disposal Management in Metropolitan Cities Using VANETs. *Journal of Clean Energy Technologies*, *2*(3).

Kune, R. (2016). The anatomy of big data computing. *Software: Practice and Experience*, *46*, 79–105.

Kunzmann, K. R. (2014). *Smart Cities: A New Padigm of Urban Development* (pp. 9–20). Crios; doi:10.7373/77140

Kuprianov & Petukhov. (2016). Distributed data mining based on actors for Internet of Things. *2016 5th Mediterranean Conference on Embedded Computing (MECO),* 480-484. doi: 10.1109/MECO.2016.7525698

Kuzmyak, J. R. (2012). *Land Use and Traffic Congestion (Vol. 618).* Arizona Department of Transportation Research Center.

Lam, C. (2011). *Hadoop in Action* (1st ed.). Greenwich, CT: Manning.

Landi, H. (2016). The business case for increasing patient portal adoption. *Healthcare Informatics*.

Laney, D. (2001). 3D Data Management: Controlling Data Volume, Velocity, and Variety. *Gartner*. Retrieved from http://blogs.gartner.com/doug-laney/files/2012/01/ad949-3D-Data-Management-Controlling-Data-Volume-Velocity-and-Variety.pdf

Lasi, H., Fettke, P., Kemper, H., & Hoffman, M. (2014). Industry 4.0. Business and Information Systems Engineering. *The International Journal of WIRTSCHAFTSINFORMATIK*, *6*(4), 239–242.

Lee, C. H. (2013). Building a generic platform for big sensor data application. *IEEE International Conference on IEEE*. 10.1109/BigData.2013.6691559

Lee, J., & Oh, J. (2008). *New Songdo City and the value of flexibility: a case study of implementation and analysis of a mega-scale project* [Doctoral dissertation]. Massachusetts Institute of Technology.

Lee, J., Bagheri B., & Kao H. (2014). A cyber-physical systems architecture for Industry 4.0-based manufacturing systems. *Manufacturing Letters, 3*(2015), 18-23.

Lee, S. H., Han, J. H., Leem, Y. T., & Tan Yigitcanlar, T. (2008). Towards ubiquitous city: Concept, planning, and experiences in the Republic of Korea. In Knowledge-based urban development: Planning and applications in the information era (pp. 148-170). Hershey, PA: IGI Global. doi:10.4018/978-1-59904-720-1.ch009

Lee, S., Oh, K., & Jung, S. (2009). The Carrying Capacity Assessment Framework for ubiquitous ecological cities in Korea [conference paper]. In *2nd International Urban Design Conference*, Gold Coast.

Lee, C. H., & Yoon, H.-J. (2017). Medical big data: Promise and challenges. *Kidney Research and Clinical Practice*, *36*(1), 3–11. doi:10.23876/j.krcp.2017.36.1.3 PMID:28392994

Lee, J., Baik, S., & Lee, C. C. (2011). Building an integrated service management platform for ubiquitous cities. *Computer*, *44*(6), 56–63. doi:10.1109/MC.2011.131

Lee, J., Kao, H., & Yang, S. (2014). Service innovation and smart analytics for Industry 4.0 and big data environment. *Procedia CIRP*, *16*, 3–8. doi:10.1016/j.procir.2014.02.001

Lee, W., Tseng, S., & Shieh, W. (2010). Collaborative real-time traffic information generation and sharing framework for the intelligent transportation system. *Information Sciences*, *180*(1), 62–70. doi:10.1016/j.ins.2009.09.004

Leitheiser, R. L. (2001). Data quality in health care data warehouse environments. In *Proceedings of the 34th Annual Hawaii International Conference on System Sciences*, Maui, HI (pp. 1-10). 10.1109/HICSS.2001.926576

Leopold, G. (2015). AWS rates highest on cloud reliability. *Enterprise Tech*. Retrieved from https://www.enterprisetech.com/2015/01/06/aws-rates-highest-cloud-reliability/

Li, W., & Kanso, A. (2015). Comparing containers versus virtual machines for achieving high availability. *Proceedings - 2015 IEEE International Conference on Cloud Engineering, IC2E 2015*. 10.1109/IC2E.2015.79

Li, Xu, & Zhao. (2014). *The Internet Of Things: A Survey*. Springer Science+Business Media.

Liao, Y. (2014). SnowFort: An Open Source Wireless Sensor Network for Data Analytics in Infrastructure and Environmental Monitoring. *IEEE Sensors Journal*, *14*(12), 4253–4263. doi:10.1109/JSEN.2014.2358253

Li, H., Lu, H., Huang, B., & Huang, Z. (2005). Two Ellipse based Pruning Methods for Group Nearest Neighbor Queries. In *GIS '05 Proceedings of the 13th annual ACM international workshop on Geographic information systems*.

Li, J. (2013). An efficient night traffic light recognition method. *Journal of Information and Computational Science*, *10*(9), 2773–2781.

Lindsay, G. (2010, Feb.). The new urbanism: New Songdo & creating cities from scratch. *Fast Company*. Retrieved from http://www.greglindsay.org/articles/the_new_new_urbanism_creating_cities_from_scratch/

Liu, C. H., Fan, J., Branch, J. W., & Leung, K. K. (2014). Toward QoI and Energy-Efficiency in Internet-of-Things Sensory Environments. *IEEE Transactions on Emerging Topics in Computing*, *2*(4), 473–487. doi:10.1109/TETC.2014.2364915

Lo, H. K., & Szeto, W. Y. (2004). Modeling advanced traveler information services: Static versus dynamic paradigms. *Transportation Research Part B: Methodological*, *38*(6), 495–515.

Looser, T. (2012). The global university, area studies, and the world citizen: Neoliberal geography's redistribution of the "world." *Cultural Anthropology*, *27*(1), 97–117. Retrieved from http://onlinelibrary.wiley.com/doi/10.1111/j.1548-1360.2012.01128.x/abstract doi:10.1111/j.1548-1360.2012.01128.x

Loveday, G. C. (1995). *Electronic Testing and Fault Diagnosis*. London: Longman.

Luan, Gao, Li, Xiang, Wei, & Sun. (2016). *Fog Computing: Focusing on Mobile Users at the Edge*. arXiv.org

Lumpkins, W. (2013). The Internet of Things Meets Cloud Computing. *IEEE Consum. Electron. Mag.*, *2*(2), 47–51. doi:10.1109/MCE.2013.2240615

Madakam, S., & Ramachandran, R. (2015). Barcelona Smart City: The Heaven on Earth (Internet of Things: Technological God). *ZTE Communications*, *4*, 3.

Madakam, S., & Ramaswamy, R. (2016). Sustainable smart city: Masdar (UAE) (A city: Ecologically balanced). *Indian Journal of Science and Technology*, *9*(6). doi:10.17485/ijst/2016/v9i6/87669

Mahmassani, H. S., & Jayakrishnan, R. (1991). System performance and user response under real-time information in a congested traffic corridor. *Transportation Research Part A, General*, *25*(5), 293–307. doi:10.1016/0191-2607(91)90145-G

Maillo, J., Ramírez-Gallego, S., Triguero, I., & Herrera, F. (in press). kNN-IS: An Iterative Spark-based design of the k-Nearest Neighbors classifier for big data. *Knowledge-Based Systems*. doi:10.1016/j.knosys.2016.06.012

Manyika, J., & Chui, M. (2015). By 2025, Internet of things applications could have $11 trillion impact. *McKinsey & Company*. Retrieved from https://www.mckinsey.com/mgi/overview/in-the-news/by-2025-internet-of-things-applications-could-have-11-trillion-impact

Mardani, M., Mateos, G., & Giannakis, G. B. (2013, February). Dynamic Anomalography: Tracking Network Anomalies Via Sparsity and Low Rank. *IEEE Journal of Selected Topics in Signal Processing*, *7*(1), 50–66. doi:10.1109/JSTSP.2012.2233193

Mari, D., & Kotz, S. (2001). *Correlation and Dependence*. London, UK: Imperial College Press. doi:10.1142/p226

Marotta, Both, Rochol, Granville, & Tarouco. (2014). *Evaluating management architectures for Internet of Things devices*. 2014 IFIP Wireless Days (WD), Rio de Janeiro, Brazil. doi: 10.1109/WD.2014.7020811

Marr, B. (2015). How Big Data and the Internet of Things Create Smarter Cities. *Forbes.com*. Retrieved 15 November 2017, from https://www.forbes.com/sites/bernardmarr/2015/05/19/how-big-data-and-the-internet-of-things-create-smarter-cities/#27e75b361767

Matallah, H., Belalem, G., & Bouamrane, K. (2017). Towards a New Model of Storage and Access to Data in Big Data and Cloud Computing. *International Journal of Ambient Computing and Intelligence*, *8*(4), 31–44. doi:10.4018/IJACI.2017100103

Mathur, N. (2017, March 15). Indian cities perform poorly in Mercer 2017 Quality of Living index. *Livemint*.

Mayer-Schönberger, V., & Cukier, K. (2013). Big Data: A revolution that will transform how we live, work, and think. *American Journal of Epidemiology*, *179*(9), 1143–1144.

McAuley. (2011). The Dataware manifesto. *Communication Systems and Networks (COMSNETS), 2011 Third International Conference*.

McIntire-Strasburg, J. (2015). An "Internet of Things" Approach to cleaner, greener public space.

Medicare hits hospitals with readmission penalties; does the regulation need to be tweaked? (2014). *Real Time Medicare Data*. Retrieved from http://rtmd.org/topics/2014/11/medicare-hits-hospitals-with-readmission-penalties-does-the-regulation-need-to-be-tweaked/#.WfQixxNSyqB

Mehmood, Y., Ahmad, F., Yaqoob, I., Adnane, A., Imran, M., & Guizani, S. (2017). Internet-of-Things-Based Smart Cities: Recent Advances and Challenges. *IEEE Communications Magazine*, *55*(9), 16–24. doi:10.1109/MCOM.2017.1600514

Mell, P., & Grance, T. (2011, September). *The NIST Definition of Cloud Computing*. Retrieved February 1, 2016, from http://csrc.nist.gov/publications/nistpubs/800-145/SP800-145.pdf

Meng, Wu, Muvianto, & Gray. (2017). A Data-Oriented M2M Messaging Mechanism for Industrial IoT Applications. *IEEE Internet of Things Journal, 4*(1), 236-246. doi: .10.1109/JIOT.2016.2646375

Michalik, P., Stofa, J., & Zolotova, I. (2014). Concept definition for Big Data architecture in the education system. *Applied Machine Intelligence and Informatics (SAMI), 2014 IEEE 12th International Symposium*, 331–334. 10.1109/SAMI.2014.6822433

Midgley, P. (2009). The role of smart bike-sharing systems in urban mobility. *Journeys*, *2*(1), 23–31.

Miles, R. N., Robert, R., & Hoy, R. R. (1995). Mechanically coupled ears for directional hearing in parasitoid fly ornia ochracea. *The Journal of the Acoustical Society of America*, *98*(6), 3059–3070. doi:10.1121/1.413830 PMID:8550933

Ministry of Human Resource Development (MHRD). (2016). *All India survey on higher education (2015-16)*. Government of India.

Misra, S. M. D. (2015). New report finds more than 165,000 mobile health apps now available, takes close look at characteristics & use. *iMedicalApps*. Retrieved from https://www.imedicalapps.com/2015/09/ims-health-apps-report/

Mitton, N. (2012). *Combining Cloud and Sensors in a Smart City Environment* (Vol. 1). EURASIP J. Wireless Commun. and Net.

Mobile Crowdsensing. (n.d.). Retrieved from: wikipedia.org/wiki /Crowdsensing

Moghbelli, H., Ellithy, K., Eslami, Z., & Vartanian, R. (2009). Investigation of solar energy applications with design and implementation of photovoltaic traffic light signal system for Qatar. In *International Conference on Renewable Energies and Power Quality (ICREPQ)*, Valencia,

Mohamed, K., & Zabani, I. (2016). Utilizing health analytics in improving the performance of healthcare services: A case study on a tertiary care hospital. *Journal of Infection and Public Health*, *9*(6), 757–765. doi:10.1016/j.jiph.2016.08.016 PMID:27663517

Mohamed, N. (2014). *Real-time big data analytics: Applications and challenges. High Performance Computing & Simulation*. HPCS.

Morabito, R., Kjaellman, J., & Komu, M. (2015). Hypervisors vs. lightweight virtualization: A performance comparison. *Proceedings - 2015 IEEE International Conference on Cloud Engineering, IC2E 2015*. 10.1109/IC2E.2015.74

Moser, S., Swain, M., & Alkhabbaz, M. H. (2015). King Abdullah Economic City: Engineering Saudi Arabia's post-oil future. *Cities (London, England)*, *45*, 71–80. doi:10.1016/j.cities.2015.03.001

Munir, A., Kansakar, P., & Khan, S. U. (2017). *IFCIoT: Integrated Fog Cloud IoT Architectural Paradigm for Future Internet of Things*. arXiv preprint arXiv: 1701.08474

Murdoch, T. B., & Detsky, A. S. (2013). The inevitable application of big data to health care. *Journal of the American Medical Association*, *309*(13), 1351–1352. doi:10.1001/jama.2013.393 PMID:23549579

Naphade, M., Banavar, G., Harrison, C., Paraszczak, J., & Morris, R. (2011). Smarter cities and their innovation challenges. *Computer*, *44*(6), 32–39. doi:10.1109/MC.2011.187

National Center for Chronic Disease Prevention and Health Promotion. (2017). Retrieved from https://www.cdc.gov/chronicdisease/overview/index.htm

Neirotti, P., De Marco, A., Cagliano, A. C., Mangano, G., & Scorrano, F. (2014). Current trends in Smart City initiatives: Some stylised facts. *Cities (London, England)*, *38*, 25–36. doi:10.1016/j.cities.2013.12.010

Nelsen, R. B. (1999). *An Introduction to Copulas*. New York: Springer-Verlag. doi:10.1007/978-1-4757-3076-0

Ngu, Gutierrez, Metsis, Nepal, & Sheng. (2017). IoT Middleware: A Survey on Issues and Enabling Technologies. *IEEE Internet of Things Journal, 4*(1), 1-20. doi: 10.1109/JIOT.2016.2615180

Ni, H., & Xu, J. (2016). COPD-related Mortality by Sex and Race Among Adults Aged 25 and Over: United States, 2000–2014. *NCHS Data Brief*, *256*, 1–8. PMID:27648679

Nimmy, Kamal, Hossain, Dey, Ashour, & Shi. (2017). Neural Skyline Filtering for Imbalance Features Classification. *International Journal of Computational Intelligence and Applications, 16*(3).

Nine Examples of Big Data Analytics in Healthcare That Can Save People. (2017). *Elogic Square*. Retrieved from http://elogicsquare.com/blog/9-examples-of-big-data-analytics-in-healthcare-that-can-save-people/

Niyato, D., Xiao, L., & Wang, P. (2011). Machine-to-Machine Communications for Home Energy Management System in Smart Grid. *IEEE Communications Magazine*, *49*(April), 53–59. doi:10.1109/MCOM.2011.5741146

Nyeem, H., Boles, W., & Boyd, C. (2013). Utilizing least siginificant bit-planes of RONI pixels for medical image watermarking. In *Proceedings of the International Conference on Digital Image Computing:Techniques and Applications* (pp. 1–8). IEEE.

O'Connell, P. L. (2005, Oct. 5). Korea's high-tech utopia, where everything is observed. *The New York Times*.

O'Donnell, J. K. (2010, Mar.). The US-ROK alliance: A catalyst for green growth? Centre for US-Korea Policy [a project of the Asia Foundation]. Retrieved from https://www.scribd.com/document/35588274/ODonnellGreenGrowth

Ongaro, D., & Ousterhout, J. (2014). In Search of an Understandable Consensus Algorithm. Atc '14, 22(2), 305–320. doi:10.1145/1529974.1529978

Onuu, M. U., & Nkanu, P. I. (2006). On the electric signal direction indicator for teh control of road traffic. *Nigerian Journal of Physics*, *18*(1), 67–72.

Open Source PlatformF. I. W. A. R. E. (n.d.). Retrieved from https://www.fiware.org/

Osigbemeh, M. S. (2012). Design and construction of an electronic multiplexer circuit with diagnostic simulator software. In UNILAG Golden Jubilee Research Conference, Lagos.

OWASP. (2017). *Open Web Application Security Project (OWASP) Web Site*. Retrieved September 29, 2017, from https://www.owasp.org/index.php/Main_Page

Pääkkönen, P., & Pakkala, D. (2015). *Reference Architecture and Classification of Technologies, Products and Services for Big Data Systems*. Big Data Research. doi:10.1016/j.bdr.2015.01.001

Pahl, C., & Lee, B. (2015). Containers and clusters for edge cloud architectures-A technology review. *Proceedings - 2015 International Conference on Future Internet of Things and Cloud, FiCloud 2015 and 2015 International Conference on Open and Big Data, OBD 2015*. 10.1109/FiCloud.2015.35

Pan. (2015). An Internet of Things framework for smart energy in buildings: designs, prototype, and experiments. *IEEE Internet of Things Journal, 2*(6), 527–537.

Pang, Y.-C., Chao, S.-L., Lin, G.-Y., & Wei, H.-Y. (2014). Network access for M2M/H2H hybrid systems: A game theoretic approach. *IEEE Communications Letters, 18*(5), 845–848. doi:10.1109/COML.2014.6845397

Patchava, V., Kumar, S. D., Hussain, S. R., & Naidu, K. R. (2015). Raspberry PI controlled SMS-Update-Notification (Sun) system. *Proceeding of IEEE International Conference on Electrical Computer and Communication Technologies (ICECCT 2015)*, 1-4.

Patel, C., Shah, D., & Patel, A. (2013). Automatic number plate recognition system (ANPR): A survey. *International Journal of Computers and Applications, 69*(9).

Patil, S., Polte, M., Ren, K., Tantisiriroj, W., Xiao, L., López, J., … Rinaldi, B. (2011). YCSB++. In *Proceedings of the 2nd ACM Symposium on Cloud Computing - SOCC '11* (pp. 1–14). New York: ACM Press. 10.1145/2038916.2038925

Patil. (2010). Intelligent Intrusion Detection System (In2DS) using Wireless Sensor Networks. *IEEE Networking, Sensing and Control.*

Pearson, S. (2013). Privacy, Security and Trust in Cloud Computing. In *Privacy and Security for Cloud Computing SE - 1* (pp. 3–42). Springer London. doi:10.1007/978-1-4471-4189-1_1

Pelletier, M., Trépanier, M., & Morency, C. (2011). Smart card data use in public transit: A literature review. *Transportation Research Part C, Emerging Technologies, 19*(4), 557–568. doi:10.1016/j.trc.2010.12.003

Peng, M., Yan, S., Zhang, K., & Wang, C. (2016). Fog Computing Based Radio Access Networks: Issues and Challenges. *IEEE Network, 30*(4), 46–53. doi:10.1109/MNET.2016.7513863

Pennic, F. (2015). Novant Health To Integrate Epic MyChart with FitBit & HealthKit. *HIT Consultant*. Retrieved from http://hitconsultant.net/2015/08/21/novant-health-to-integrate-epic-mychart-with-fitbit-healthkit/

Peralta, Río, Ramírez-Gallego, Triguero, Benítez, & Herrera. (2015). Evolutionary Feature Selection for Big Data Classification: A MapReduce Approach. *Mathematical Problems in Engineering*. doi: 10.1155/2015/246139

Perera, C., Liu, C. H., & Jayawardena, S. (2015). The Emerging Internet of Things Marketplace from an Industrial Perspective: A Survey. *IEEE Transactions on Emerging Topics in Computing*, *3*(4), 585–598. doi:10.1109/TETC.2015.2390034

Perera, C., Liu, C. H., Jayawardena, S., & Chen, M. (2014). A Survey on Internet of Things from Industrial Market Perspective. *IEEE Access: Practical Innovations, Open Solutions*, *2*, 1660–1679. doi:10.1109/ACCESS.2015.2389854

Philip Chen, C. L., & Zhang, C.-Y. (2014). Data-intensive applications, challenges, techniques and technologies: A survey on Big Data. *Information Sciences*, *275*, 314–347. doi:10.1016/j.ins.2014.01.015

Plattner, H. (2013). *A Course in In-Memory Data Management: The Inner Mechanics of In-Memory Databases*. Berlin, Germany: Springer. doi:10.1007/978-3-642-36524-9

Plattner, H., & Zeier, A. (2012). *In-Memory Data Management. Springer*. Berlin, Germany: Springer; doi:10.1007/978-3-642-29575-1

Ponomarkenko, N., Keivenko, S., Egiazarian, K., Astola, J., & Lukin, V. (2010). Weighted MSE based metrics for characterization of visual quality of image denosing methods. *Proceeding of the 8th Inte.Workshop on Video Processing and Quality Metrics for consumer Electronics.*

Poole, S. (2014, December 17). The truth about smart cities: 'In the end, they will destroy democracy'. *The Guardian.*

Prabhakar, V., Gupta, S., & Mehrotra, R. (2015, September 2). Smart transportation for Smart Cities. *The Economic Times*.

Preeth, E. N., Mulerickal, J. P., Paul, B., & Sastri, Y. (2016). Evaluation of Docker containers based on hardware utilization. *2015 International Conference on Control, Communication and Computing India, ICCC 2015*. 10.1109/ICCC.2015.7432984

Press, G. (2014). A Very Short History Of The Internet Of Things. *Forbes*. Retrieved from https://www.forbes.com/sites/gilpress/2014/06/18/a-very-short-history-of-the-internet-of-things/3/#52cb61da350a

Prinsloo, J., & Malekian, R. (2016). Accurate Vehicle Location System Using RFID, an Internet of Things Approach. *Sensors (Basel)*, *16*(6), 825. doi:10.339016060825 PMID:27271638

Public Policy Institute. (n.d.). *Livable communities*. Retrieved from: www.aarp.org/ppi/issues/livable-communities/info-2015/

Pune Smart City – Vision Document (Version 1.0) (Vol. 2). (n.d.). Pune: Pune Municipal Corporation.

Puri, A. (2014, August 15). What are smart cities? *The Hindu*.

Puri, T. (2015, January 6). Parking chaos needs a quick fix first. *The Times of India*.

Purkayashta, S. (n.d.). Smart traffic management system for emergency services. *IBM*. Retrieved from https://www.ibm.com/blogs/bluemix/2015/11/pubnub-smart-traffic-management-system-for-emergency-services/

Qin, J., Liu, Y., & Grosvenor, R. (2017). A categorical framework of manufacturing for Industry 4. 0 and beyond. *Procedia CIRP*, *52*, 173–178.

Quoc, H. N. M., & Le Phuoc, D. (2015). An Elastic and Scalable Spatiotemporal Query Processing for Linked Sensor Data. In *Proceedings of the11th International Conference on Semantic Systems, ser. SEMANTICS '15*. New York, NY: ACM. 10.1145/2814864.2814869

Raggett, D. (2015, May). The Web of Things: Challenges and Opportunities. *IEEEComputer*, *48*(5), 26–32.

Raghupathi W, & Raghupathi V. (2014). Big data analytics in healthcare: promise and potential. *Health Information Science and Systems, 2*(3).

Rahman, M., Carbunar, B., & Topkara, U. (2016). Secure management of low power fitness trackers. *IEEE Transactions on Mobile Computing*, *15*(2), 447–459. doi:10.1109/TMC.2015.2418774

Ramírez-Gallego, S., García, S., Mouriño Talín, H., Martínez-Rego, D., Bolón-Canedo, V., Alonso-Betanzos, A., ... Herrera, F. (2016). Data discretization: Taxonomy and big data challenge. *Wiley Interdisciplinary Reviews. Data Mining and Knowledge Discovery*, *6*(1), 5–21. doi:10.1002/widm.1173

Rangarajan, A. (2010, January). BRTS-Bus Rapid Transit System in Pune Modeling, Simulation and Feasibility Analysis. In *Proceedings of the 2010 International Conference on Industrial Engineering and Operations Management*, Dhaka.

Rani, Ahmed, Talwar, Malhotra, & Song. (2017). IoMT: A Reliable Cross Layer Protocol for Internet of Multimedia Things. *IEEE Internet of Things Journal*. doi:.10.1109/JIOT.2017.2671460

Rao, B., Saluia, P., Sharma, N., Mittal, A., & Sharma, S. (2012). Cloud computing for internet of things & sensing based applications. *Sensing Technology (ICST), 2012 Sixth International Conference on. IEEE*, 374–380. 10.1109/ICSensT.2012.6461705

Rao, R. V., & Selvamani, K. (2015). Data security challenges and its solutions in cloud computing. *Procedia Computer Science*, *48*, 204–209. doi:10.1016/j.procs.2015.04.171

Rathore, Ahmad, Paul, & Rho. (n.d.). Urban planning and building smart cities based and internet of things. *Computer Network*. DOI:10.1016/jcomnet.2015.12.023

Rathore, M., Ahmad, A., Paul, A., & Rho, S. (2016). Urban planning and building smart cities based on the Internet of Things using Big Data analytics. *Computer Networks*, *101*, 63–80. doi:10.1016/j.comnet.2015.12.023

Realizing Smart Cities with Technology - A KPIT Perspective. (2016). Pune: KPIT Technologies Ltd. Retrieved from https://www.kpit.com/resources/downloads/whitepapers/realizing-smart-cities-technology-whitepaper.pdf

Red Hat. (2017). *Ceph Ceph storage - Ceph*. Retrieved October 4, 2017, from http://ceph.com/ceph-storage/

Reddy, B. S., & Balachandra, P. (2012). Urban mobility: A comparative analysis of megacities of India. *Transport Policy*, *21*, 152–164. doi:10.1016/j.tranpol.2012.02.002

Reimagining Pune: Mission Smart City. *Punesmartcity.in*. Retrieved 30 December 2017 from https://punesmartcity.in/wp-content/uploads/2017/10/SPC.pdf

Ren, K., Wang, C., & Wang, Q. (2012). Security challenges for the public cloud. *IEEE Internet Computing*, *16*(1), 69–73. doi:10.1109/MIC.2012.14

Report of the Sub-Committee on Automatic Fare Collection System. (2013). New Delhi: Government of India, Ministry of Urban Development.

Republic of Korea. (2010). Sejong to Become Model Eco-Friendly city. Retrieved from http://www.korea.net/Resources/Useful-Info/Learn-Korean

Rios, M. (2014Fueled by Aging Baby Boomers, Nation's Older Population to Nearly Double in the Next 20 Years. *Census Bureau Reports*. Retrieved from https://www.census.gov/newsroom/press-releases/2014/cb14-84.html

Rodrigues, J. (2010). Health Information Systems: Concepts, Methodologies, Tools, and Applications (Vol. 1). Hershey, PA: IGI Global. doi:10.4018/978-1-60566-988-5

Rodrigues, J., de la Torre, I., Fernández, G., & López-Coronado, M. (2013). Analysis of the security and privacy requirements of cloud-based electronic health records systems. *Journal of Medical Internet Research*, *15*(8), e186. doi:10.2196/jmir.2494 PMID:23965254

Rodríguez Bolívar, M. (2015). Smart Cities: Big Cities, Complex Governance? In M. Rodríguez Bolívar (Ed.), *Transforming City Governments for Successful Smart Cities*. Switzerland: Springer.

Rosenthal, E. (2014). The Health Care Waiting Game. Long Waits for Doctors' Appointments Have Become the Norm. *The New York Times*. Retrieved from https://www.nytimes.com/2014/07/06/sunday-review/long-waits-for-doctors-appointments-have-become-the-norm.html

Rumsfeld, J. S., Joynt, K. E., & Maddox, T. M. (2016). Big data analytics to improve cardiovascular care: Promise and challenges. *Nature Reviews. Cardiology*, *13*(6), 350–359. doi:10.1038/nrcardio.2016.42 PMID:27009423

Rupa. (2016). Performance Evaluation of Message Security Using Fractal Sieve with MMD. *International Journal of Digital Crimes and Security, 34*(4).

Rupa, C. (2010). Performance Evaluation of Message Security Using Fractal Sieve with MMD. *International Journal of Security, Privacy and Trust Management*, 2(4), 27–32.

Sadalage, P. J., & Fowler, M. (2013). *NoSQL distilled : a brief guide to the emerging world of polyglot persistence*. Addison-Wesley.

Sagar, B. S., & Praveen, D. J. (2017). A study on impact of Industry 4.0 in India. *International Advanced Research Journal in Science, Engineering and Technology, 4*(7), 24-28.

Salahuddin, M. A., Al-Fuqaha, A., & Guizani, M. (2015). Software-defined networking for rsu clouds in support of the internet of vehicles. *Internet of Things Journal, IEEE, 2*(2), 133–144. doi:10.1109/JIOT.2014.2368356

Sanjai, P. (2017, September 11). India's traffic is so bad, it's changing the cars people buy. *Livemint.*

Sarkar, M., Banerjee, S., Badr, Y., & Sangaiah, A. K. (2017). Configuring a trusted cloud service model for smart city exploration using hybrid intelligence. *International Journal of Ambient Computing and Intelligence, 8*(3), 1–21. doi:10.4018/IJACI.2017070101

Savazzi, S., Rampa, V., & Spagnolini, U. (2014). Wireless CloudNetworks for the Factory of Things: Connectivity Modeling andLayout Design. *IEEE Internet Things J., 1*(2), 180–195. doi:10.1109/JIOT.2014.2313459

Says. (n.d.). *6.4 Billion Connected Things' Will Be in Use in 2016, Up 30 Percent From 2015.* Gart. Inc.

Schaffers, H., Sallstrom, A., Pallot, M., Hernandez-Munoz, J. M., Santoro, R., & Trousse, B. (2011). Smart Cities and the Future Internet: Towards Cooperation Frameworks for Open Innovation. *2011 17th International Conference on Concurrent Enterprising*, 1–11. 10.1007/978-3-642-20898-0

Schneeweiss, S. (2014). Learning from big health care data. *The New England Journal of Medicine, 370*(23), 2161–2163. doi:10.1056/NEJMp1401111 PMID:24897079

Schniederman, R. (2013). Machine-to-Machine Connectivity isbooming. *IEEE Signal Processing Magazine, 30*(4), 10–13. doi:10.1109/MSP.2013.2255940

SCOPE A Smart-city Cloud-based Open Platform and Ecosystem. (n.d.). Retrieved from https://www.bu.edu/hic/research/highlighted-sponsoredprojects/scope/

Sekhsaria, P. (2004, November 1). The future of Pune's public transport. *Indiatogether.org.* Retrieved from http://www.indiatogether.org/pmtplans-economy

Seredynski, M., & Grzybek, A. (2016, June). Coping with non-recurring congestion with distributed hybrid routing strategy. In *2016 IEEE Intelligent Vehicles Symposium (IV)* (pp. 1121-1127). IEEE. 10.1109/IVS.2016.7535530

Settles, C. (2014). Integrating Patient Devices into Health IT Systems. *Medical Design Technology.* Retrieved from https://www.mdtmag.com/article/2014/12/integrating-patient-devices-health-it-systems

Shaheen, S. (2004). Dynamics in behavioral adaptation to a transportation innovation: A case study of Carlink–a smart car-sharing system.

Sharma, P., & Rajput, S. (2017). *Sustainable smart cities in India: Challenges and future perspectives.* Cham: Springer International Publishing. doi:10.1007/978-3-319-47145-7

Shcherbina, A., Mattsson, C. M., Waggott, D., Salisbury, H., Christle, J. W., Hastie, T., ... Ashley, E. A. (2017). Accuracy in wrist-worn, sensor-based measurements of heart rate and energy expenditure in a diverse Cohort. *Journal of Personalized Medicine*, *7*(2), 3. doi:10.3390/jpm7020003 PMID:28538708

Sherly, J., & Somasundareswari, D. (2015). Internet of Things Based Smart Transportation Systems. *International Research Journal of Engineering and Technology*, *2*(7), 1207–1210.

Shinde, S. (2017, July 6). Spike in autorickshaws without permits on Pune roads. *Hindustan Times*.

Shinde, S., & Jagtap, S. (2016). Big Data Solution for Improving Traffic Management System Effectively for Vehicular Networks. *International Journal of Advanced Research in Computer and Communication Engineering*, *5*(6).

Shruti, G. (2016, March). Implementation of Smart Home Automation System through E-mail using Raspberry Pi and Sensors. *International Journal of Innovative Research in Electrical, Electronics, Instrumentation and Control Engineering*, *4*(3).

Shukla. (2016). The big data approach towards smart cities. *The Engineering Journal of Application & Scopes, 1*(2).

Shwayri, S. T. (2013). A model Korean ubiquitous eco-city? The politics of making Songdo. *Journal of Urban Technology*, *20*(1), 39–55. doi:10.1080/10630732.2012.735409

Silva. (2017). *Big Data Analytics Embedded Smart City Architecture for Performance Enhancement through Real-Time Data Processing and Decision-Making*. Hindawi Wireless Communications and Mobile Computing.

Singer, N. (2012, March 3). Mission control, built for cities: I.B.M. takes 'smarter cities' concept to Rio de Janeiro. *The New York Times*.

Sinha, T., Kumar, K. M., & Saisharan, P. (2015). smart Dustbin. *International Journal of Industrial Electronics and Electrical Engineering*, *3*(5), 101–104.

Smart City Mission. (n.d.) Retrieved from: http://smartcities.gov.in/content/

Smart Pune's Step Towards Sustainable Transportation: Public Parking Policy (Draft). (2016). Pune. Retrieved from https://pmc.gov.in/informpdf/Road/PMC_Draft%20parking%20policy_English%20version.pdf

Smits, T., & Ryter, A. (2015). Chronic Conditions and Fall Risk. *Philips*. Retrieved from https://www.lifeline.philips.com/resources/blog/2015/09/report-chronic-conditions-and-the-high-risk-of-falling.html

Sniderman, B., Mahto, M., & Cotteleer, M. J. (2016). *Industry 4.0 and manufacturing ecosystems: Exploring the world of connected enterprises*. Deloitte University Press. Retrieved from http://dupress.com/articles/industry-4-0-manufacturing-ecosystems- exploring-world-connected-enterprises/

Solanki, V. K., Venkatesan, M., & Katiyar, S. (2016). Think Home: A Smart Home as Digital Ecosystem in Circuits and Systems. *Circuits and Systems, 7*(8). doi:10.4236/cs.2016.78172

Solanki, V. K., Venkatesan, M., & Katiyar, S. (n.d.). Conceptual Model for Smart Cities for Irrigation and Highway Lamps using IoT. *International Journal of Interactive Multimedia and Artificial Intelligence, 4*(3). doi:10.9781/ijimai.2017.435

Solanki, V. K., Katiyar, S., BhashkarSemwal, V., Dewan, P., Venkatasen, M., & Dey, N. (2016). Advanced automated module for smart and secure city. *Procedia Computer Science*, *78*, 367–374. doi:10.1016/j.procs.2016.02.076

Solberg, V. (1996). *Design Guidelines for Surface Mount and Fine- pitch Technology* (2nd ed.). New York: McGraw-Hill.

Soliman, M., Abiodun, T., Hamouda, T., Zhou, J., & Lung, C.-H. (2013). Smart home: Integrating internet of things with web services and cloudcomputing. *Cloud Computing Technology and Science (CloudCom), 2013 IEEE 5th International Conference on,* 2, 317–320.

Song, H., Srinivasan, R., Sookoor, T., & Jeschke, S. (2016). *Smart cities: Foundations, principles, and applications*. Academic Press.

Speller, K. E., & Yu, D. (2004). A low-noise MEMS accelerometer for unattended ground sensor applications. *Proc. SPIE 5417, Unattended/Unmanned Ground. Ocean, and Air Sensor Technologies and Applications*, *4*. doi:10.1117/12.540337

Stankovic, J. (2014). Research Directions for the Internet of Things. *Internet of Things Journal, IEEE.*, *1*(1), 3–9. doi:10.1109/JIOT.2014.2312291

Stojanovic, D., Predic, B., & Stojanovic, N. (2016). Mobile crowd sensing for smart urban mobility. In European Handbook of Crowdsourced Geographic Information (pp. 371–382). London: Ubiquity Press. doi:10.5334/bax.aa

Strategy: Smart Cities Mission, Ministry of Housing and Urban Affairs, Government of India. (2017). *Smartcities.gov.in*. Retrieved 30 December 2017, from http://smartcities.gov.in/content/innerpage/strategy.php

Strickland, E. (2011). Cisco bets on the South Korean smart city. *IEEE Spectrum*, *48*(8), 11–12. doi:10.1109/MSPEC.2011.5960147

Su, K. (2011). Smart city and the applications. In *Electronics, Communications and Control (ICECC), 2011 International Conference on*. IEEE. 10.1109/ICECC.2011.6066743

Sun, Z., Liu, C. H., Bisdikian, C., Branch, J. W., & Yang, B. (2012). QoIaware energy management in internet-of-things sensory environments. *2012 9th Annual IEEE Communications Society Conference on Sensor, Mesh and Ad Hoc Communications and Networks*, 19–27.

Sundmaeker, Guillemin, Friess, & Woelfflé. (2010). Vision and challenges for realizing the Internet of Things. *Cluster of European Research Projects on the Internet of Things - CERP IoT*.

Sun, E., Zhang, X., & Li, Z. (2012). The internet of things (iot) and cloud computing (cc) based tailings dam monitoring and pre-alarm system in mines. *Safety Science*, *50*(4), 811–815. doi:10.1016/j.ssci.2011.08.028

Sun, Y., Song, H., Jara, A. J., & Bie, R. (2016). *Internet of Things and Big Data Analytics for Smart and Connected Cities*. IEEE.

Survey of Patient Appointment Wait Times. (2017). *Merritt Hawkins*. Retrieved from https://www.merritthawkins.com/2014-survey/patientwaittime.aspx

Swoyer, S. (2012). *Big Data - Why the 3Vs Just Don't Make Sense*. Retrieved August 11, 2014, from http://tdwi.org/articles/2012/07/24/big-data-4th-v.aspx

Tachizawa, E. M., Alvarez-Gil, M. J., & Montes-Sancho, M. J. (2015). How "smart cities" will change supply chain management. *Supply Chain Management*, *20*(3), 237–248. doi:10.1108/SCM-03-2014-0108

Tamane, Solanki, & Joshi. (2017). The Basics of Big Data & Security Concerns. In S. Tamane, V. K. Solanki, & N. Dey (Eds.), *Privacy and Security Policies in Big Data*. Hershey: IGI Global.

Tan, Sooriyabandara, & Fan. (2011). M2M Communications in the Smart Grid: Applications, Standards, Enabling Technologies, and Research Challenges. *International Journal of Digital Multimedia Broadcasting*. doi:10.1155/2011/289015

Tanenbaum, A. S., & Van Steen, M. (2007). *Distributed Systems: Principles and Paradigms, 2/E. Communication*. Retrieved from http://www.pearsonhighered.com/academic/product/0,0132392275,00+en-USS_01DBC.html

Technology Strategy Board. (2010). *Accelerating business innovation across the UK: The Technology Strategy Board, regional development agencies and devolved administrations working in partnership*. Author.

The impact of medical coding on a hospital's bottom line. (2014). *Real Time Medicare Data*. Retrieved from http://rtmd.org/topics/2014/07/the-impact-of-a-wrong-medical-coding-on-a-hospitals-bottom-line/#.WfQdGxNSyqB

The smart Grid, department of energy. (n.d.). Retrieved from: energy.gov/sites/prod/files/ oeprod/

Tiwari. (2010). A Secure Hash Function MD-192 With Modified Message Expansion. *International Journal of Computer Science and Information Security, 7*(2).

Tiwari, A., & Jain, K. (2014). GIS Steering smart future for smart Indian cities. International. *Journal of Scientific and Research Publications*, *4*(8), 442–446.

Tokheim, R. L. (1999). *Digital Electronics: Principles and Applications* (5th ed.). New York: McGraw-Hill.

Townsend, A. (2014). *Smart Cities: Big Data, Civic Hackers, and the Quest for a New Utopia*. New York: W.W. Norton & Company.

Townsend, A. M. (2013). *Smart cities. Big data, civic hackers, and the quest for a new utopia.* New York: W.W. Norton & Company. Retrieved from https://ssir.org/articles/entry/smart_cities_big_data_civic_hackers_and_the_quest_for_a_new_utopia

TPC. (2016). *TCP Website.* Retrieved February 1, 2016, from http://www.tpc.org/

Trépanier, M., Tranchant, N., & Chapleau, R. (2007). Individual Trip Destination Estimation in a Transit Smart Card Automated Fare Collection System. *Journal of Intelligent Transport Systems, 11*(1), 1–14. doi:10.1080/15472450601122256

Triguero, I., Río, S., López, V., Bacardit, J., Benítez, J. M., & Herrera, F. (2015). ROSEFW-RF: The winner algorithm for the ECBDL'14 Big Data Competition: An extremely imbalanced big data bioinformatics problem. *Knowledge-Based Systems, 87*, 69–79. doi:10.1016/j.knosys.2015.05.027

Uddin, A. (2009, March 23). Traffic congestion in Indian cities: Challenges of a Rising power.

Umachandran, K. (2012). *Exploitation of human rights in Indian industrial sector (Automobile Segment).* Paper presented at the 3rd International Seminar of the Regional Network on Poverty Eradication.

Umachandran, K., & Ferdinand-James, D. (2017). Affordances of Data Science in Agriculture, Manufacturing, and Education. In S. Tamane (Ed.), *Privacy and Security Policies in Big Data.* IGI Global. doi:10.4018/978-1-5225-2486-1.ch002

Umbrajkar, M. (2012, January 20). In last decade, population of Pune urban agglomeration grew 34%. *The Times of India.*

United Nations Industrial Development Organization. (2015). *Industrial Development Report 2016. The Role of Technology and Innovation in Inclusive and Sustainable.* Industrial Development.

United States Department of Labor. (n.d.). *Recommended practices for safety and health programs – Hazard prevention and control.* Retrieved from https://www.osha.gov/shpguidelines/hazard-prevention.html

Urbanization in India. (n.d.). Retrieved from: wikipedia.org/wiki/Urbanisation_in_India

Varghese, B., Subba, L. T., Thai, L., & Barker, A. (2016). Container-based cloud virtual machine benchmarking. In *Proceedings - 2016 IEEE International Conference on Cloud Engineering, IC2E 2016: Co-located with the 1st IEEE International Conference on Internet-of-Things Design and Implementation, IoTDI 2016.* IEEE. 10.1109/IC2E.2016.28

Vermesan & Friess. (2013). Internet of Things: Converging Technologies for Smart Environments and Integrated Ecosystems. IERC.

Viennot, N., Lécuyer, M., Bell, J., Geambasu, R., & Nieh, J. (2015). Synapse: A Microservices Architecture for Heterogeneous-Database Web Applications. *Proceedings of the Tenth European Conference on Computer Systems - EuroSys '15.* 10.1145/2741948.2741975

Viles, E., Ormazabal, M., & Lleo, A. (Eds.). (2018). *Closing the gap between practice and research in industrial engineering*. Gewerestrasse, Switzerland: Springer. doi:10.1007/978-3-319-58409-6

Vos, T., Barber, R. M., Bell, B., Bertozzi-Villa, A., Biryukov, S., Bolliger, I., ... Murray, C. J. L. (2015). Global, regional, and national incidence, prevalence, and years lived with disability for 301 acute and chronic diseases and injuries in 188 countries, 1990–2013: A systematic analysis for the Global Burden of Disease Study 2013. *Lancet*, *386*(9995), 743–800. doi:10.1016/S0140-6736(15)60692-4 PMID:26063472

Vujoic, V., & Makshimovic, M. (2015). Raspberry Pi as a sensor Web node for home automation. *Elsevier Transaction on Computer and Engineering*, *42*, 153–171.

Wallace, W. A., Wojtowicz, J. M., Torrey, D., Renna, N., & Tan, J. (2009). *Guidelines for Traffic Signal Energy Back-up Systems* (No. C-06-08).

Wang, H., Liu, W., & Soyata, T. (2014). Accessing big data in the cloud using mobile devices. In Handbook of Research on Cloud Infrastructures for Big Data Analytics (pp. 444-470). Hershey, PA: IGI Global.

Wang, F. (2015). Adaptive semi-supervised recursive tree partitioning: The ART towards large scale patient indexing in personalized healthcare. *Journal of Biomedical Informatics*, *55*, 41–54. doi:10.1016/j.jbi.2015.01.009 PMID:25656756

Wang, P., Ali, A., & Kelly, W. (2015). Data security and threat modeling for smart city infrastructure. *2015 International Conference on Cyber Security of Smart Cities, Industrial Control System and Communications, SSIC 2015 - Proceedings*. 10.1109/SSIC.2015.7245322

Wang, S., Wan, J., Zhang, D., Li, D., & Zhang, C. (2016). Towards smart factory for industry 4.0. *The International Journal of Computer and Telecommunications Networking*, *101*(C), 158–168.

Want, R., Schilit, B. N., & Jenson, S. (2015). Enabling the Internet of Things. *IEEE Computer*, *48*(1), 28–35. doi:10.1109/MC.2015.12

Weldon, D. (2017). St. Luke's University uses data warehouse to improve performance. *Health Data Management*. Retrieved from https://www.healthdatamanagement.com/news/st-lukes-university-uses-data-warehouse-to-improve-performance

What is IoT. (n.d.). Retrieved from: www.happiestminds.com

What is Smart City. (2017). New Delhi: Ministry of Housing and Urban Affairs, Government of India. Retrieved from http://smartcities.gov.in/content/innerpage/what-is-smart-city.php

What is the smart grid. (n.d.). Retrieved from: www.smartgrid.gov

White, T. (2012). *Hadoop, The Definitive Guide*. Sebastopol, CA: O'Reilly Media, Inc.

Wikipedia. (n.d.). Health Level 7. Retrieved from https://en.wikipedia.org/wiki/Health_Level_7

Williamson, L. (2013). Tomorrow's cities: Just how smart is Songdo? *BBC News*. Retrieved Dec. 16, 2013, from http://www.bbc.com/news/technology-23757738

Wolfgang, W. (2016). *Industrie 4.0: Cyber-Physical Production System for Mass Customization.* Retrieved from 16/Industrie_4_0_Cyber-Physical_Production_Systems_for_Mass_Customizations.pdf

World Economic Forum. (2013). *Manufacturing for growth strategies for driving growth and employment - Volume 1: Globally competitive policy - A World Economic Forum Report in collaboration with Deloitte Touche Tohmatsu Ltd.* Author.

World Economic Forum. (2016). *Global Challenge Insight Report - The Future of Jobs, Employment: Employment, Skills and Workforce Strategy for the Fourth Industrial Revolution.* Author.

World Health Organization (WHO). (2016). Chronic obstructive pulmonary disease (COPD) Retrieved from http://www.who.int/mediacentre/factsheets/fs315/en/

Yang, W., Wang, X., Song, X., Yang, Y., & Patnaik, S. (2018). Design of Intelligent Transportation System Supported by New Generation Wireless Communication Technology. *International Journal of Ambient Computing and Intelligence*, *9*(1), 78–94. doi:10.4018/IJACI.2018010105

Yang, Y. (2013). A Self-Learning Sensor Fault Detection Framework for Industry Monitoring IoT. *Mathematical Problems in Engineering*, *15*, 2013.

Yang, Z. (2011). *Study and application on the architecture and key technologies for IOT*. IEEE Multimedia Technology.

Yavari, A. (2017). ConTaaS: An Approach to Internet-Scale Contextualisation for Developing Efficient Internet of Things Applications. *Proceedings of the 50th Hawaii International Conference on System Sciences*. 10.24251/HICSS.2017.715

Yavuz, Hasan, Serkan, & Duygu. (2007). Safe and Secure PIC Based Remote Control Application for Intelligent Home. *International Journal of Computer Science and Network Security, 7*(5).

Yigitcanlar, T., & Lee, S. H. (2014). Korean ubiquitous-eco-city: A smart-sustainable urban form or a branding hoax? *Technological Forecasting and Social Change*, *89*, 100–114. doi:10.1016/j.techfore.2013.08.034

Yin, S., & Kaynak, O. (2015). Big Data for Modern Industry: Challenges and Trends [Point of View]. *Proceedings of the IEEE, 103*(2), 143–146. doi:10.1109/JPROC.2015.2388958

Yin, Y., & Yang, H. (2003). Simultaneous determination of the equilibrium market penetration and compliance rate of advanced traveler information systems. *Transportation Research Part A, Policy and Practice*, *37*(2), 165–181.

Yu, H., & Huang, W. (2015). Building a Virtual HPC Cluster with Auto Scaling by the Docker. *Computing Research Repository*. Retrieved from http://arxiv.org/abs/1509.08231

Yun, J. (2015). A new city prototype? Songdo International City as an airport city. *Journal of Asian Architecture and Building Engineering*, *14*(3), 549–556. doi:10.3130/jaabe.14.549

Yu, S., Wang, C., Ren, K., & Lou, W. (2010). Achieving Secure, Scalable, and Fine-grained Data Access Control in Cloud Computing. In *Proceedings of IEEE International Conference on Computer Communications*, San Diego, CA (pp. 1-9). 10.1109/INFCOM.2010.5462174

Yuvaraj, S., & Sangeetha, M. (2016). Smart supply chain management using internet of things(IoT) and low power wireless communication systems. *Wireless Communications Signal Processing and Networking (WiSPNET) International Conference on*, 555-558. 10.1109/WiSPNET.2016.7566196

Zanella, N., Bui, A., Castellani, L., Vangelista, & Zorzi. (2014). Internet of things for smart cities. *IEEE Internet of Things Journal, 1*(1), 22–32.

Zhang, K., & Batterman, S. (2013). Air pollution and health risks due to vehicle traffic. *The Science of the Total Environment, 450-451*, 307–316. doi:10.1016/j.scitotenv.2013.01.074 PMID:23500830

Zhang, K., Ni, J., Yang, K., Liang, X., Ren, J., & Shen, X. S. (2017). Security and Privacy in Smart City Applications: Challenges and Solutions. *IEEE Communications Magazine, 55*(1), 122–129. doi:10.1109/MCOM.2017.1600267CM

Zhang, W., Qi, Q., & Deng, J. (2017). Building Intelligent Transportation Cloud Data Center Based on SOA. *International Journal of Ambient Computing and Intelligence, 8*(2), 1–11. doi:10.4018/IJACI.2017040101

Zheng, Y., Capra, L., Wolfson, O., & Yang, H. (2014). Urban Computing: Concepts, Methodologies, and Applications. *ACM Trans. Intell. Syst. Technol., 5*(3), 38:1-38:55. 10.1145/2629592

About the Contributors

Nilanjan Dey, PhD., is an Asst. Professor in the Department of Computer Science and Engineering in Techno India College of Technology, Rajarhat, Kolkata, India. He holds an honorary position of Visiting Scientist at Global Biomedical Technologies Inc., CA, USA and Research Scientist of Laboratory of Applied Mathematical Modeling in Human Physiology, Territorial Organization Of- Scientific And Engineering Unions, BULGARIA, Associate Researcher of Laboratoire RIADI, University of Manouba, TUNISIA. He is the Editor-in-Chief of the International Journal of Ambient Computing and Intelligence; the International Journal of Rough Sets and Data Analysis (IGI Global, USA); Series Editor of Advances in Geospatial Technologies (AGT) Book Series (IGI Global, USA); the Executive Editor of the International Journal of Image Mining (IJIM) (Inderscience); Regional Editor-Asia of the International Journal of Intelligent Engineering Informatics (IJIEI) (Inderscience); and the Associated Editor of the International Journal of Service Science, Management, Engineering, and Technology (IGI Global, USA). He has 15 books and 250 international conferences and journal papers. https://www.researchgate.net/profile/Nilanjan_Dey3

Sharvari Tamane, Ph.D, is working as the Head of Department of Information Technology in MGMs Jawaharlal Nehru Engineering College, Aurangabad, Maharashtra, India. She has 21 years of teaching experience in Jawaharlal Nehru Engineering College, Aurangabad. She is the active member of "Board of Studies" for "Computer Science & Engineering' for Dr. Babasaheb Ambedkar Marathwada University, Aurangabad. Her research interests include: Big Data Analytics, Watermarking, Neural Network, Fuzzy Logic, Wavelet analysis, etc. She is working as an Editorial Board Member of the International Journal of Machine Learning and Networked Collaborative Engineering (IJMLNCE) ISSN 2581-3242. Dr. Tamane edited a book on "Privacy and Security Policies in Big Data" with IGI Global. She is author of a book, "Data Structures Using C" and published more than 40 research papers in international conferences and journals. She has also worked as a Reviewer for many SCIE, Scopus indexed journals. She has been invited as a panelist for the

International Conference at Mauritius in the panel discussion Towards "Sustainable Development Goals in India". She has also been invited to many institutes to deliver expert talks on various research areas.

***

R. Babu is a research scholar at Anna University. He has completed his Master's in Software Engineering with a merit of 2nd rank in the University and a Bachelor's in Computer Science and Engineering from Rajalakshmi Engineering College affiliated to Anna University. His areas of interest include web services, service-oriented architecture, cloud computing, big data analytics and Internet of Things. He is a life member of CSI and served as a Management Committee member for two years. He has received Active Participation Award – Youth from CSI. He also received Faculty Excellence Award from Infosys in the Faculty Enablement Program for two successive years.

Jyoti Chandiramani, Prof., has more than 30 years of experience in teaching and research. She is presently the Director of Symbiosis School of Economics and the Dean of the Faculty of Humanities and Social Sciences at the Symbiosis International (Deemed University), Pune – India. She teaches Urban Economic Development, Trade, Aid & Development, and International Organization and Regional Cooperation at the master's level. In 2007, she was awarded a short-term scholarship at the University of DePauw, Indianapolis, USA, to study a Liberal Arts Education. In 2013, she was a part of the Indian delegation to Pakistan (Lahore) for the Tenth South Asian Economic Students Meet (SAESM). Recently, she was invited to the 5th ASEM Rectors' Conference and as a Mentor for the Students' Forum (ARC5), organized by Charles University in Prague and the Asia-Europe Foundation (ASEF) at Prague in 2016.Besides writing a couple of textbooks, she has jointly edited a book with Ramanath Jha titled *Perspectives in Urban Development: Issues in Infrastructure, Planning, and Governance*. Her areas of research interest include urban infrastructure, urban transport, urban poverty and is presently working on Pune-centric cases. Additionally, she keenly studies and follows trends in International Development Cooperation, from the MDGs to the SDGs.

Lokesh Chouhan, Dr., has completed his Bachelor of Engineering (B.E.) in Computer Engineering from SGSITS, Indore, India. He did his M.Tech. (with distinction) in Information Technology from Bhopal. At present, he is working as an Assistant Professor of the Computer Science & Engineering (CSE) Department at the National Institute of Technology, Hamirpur (H.P.), India. Before joining to this institute, he also worked as an Assistant Professor in the Computer Science & Engineering (CSE) Department at the Maulana Azad National Institute of Technology (MANIT), Bhopal (M.P.), India. He has completed his Ph.D. in the area of cognitive radio in the Department of Information and Communication Technology at the Indian Institute of Information Technology and Management, Gwalior, India. His areas of interest are computer networks, cloud computing, cyber security and forensics, Internet of Things, smart city, and applications of Android. He has also teaching and industrial experience. He has published various papers in national and international journals and conferences and also filed Indian Patents in the field of IoT. He has completed his Ph.D. in the area of cognitive radio in the Department of Information and Communication Technology at the Indian Institute of Information Technology and Management, Gwalior, India. His areas of interest are computer networks, cloud computing, cyber security and forensics, Internet of Things, smart city, and applications of Android. He has also teaching and industrial experience. He has published various papers in national and international journals and conferences and also filled few Indian Patents in the field of IoT.

Valentina Della Corte focuses her research on strategic management and in particular on resource-based theory; corporate finance and its relations with firm's strategies; business networks and competition; dynamic capabilities; value creation and appropriation both intra-firm and inter-firm. She is a promoter of several research projects with many foreign universities.

Debra Ferdinand-James completed her doctoral studies in workforce education and development while working in distance education for six years in the areas of online course design and development, faculty training and support for WebCT, and helpdesk enquiries for WebCT. Her post-graduate studies were funded by scholarships and awards include two Organisation of American States Fellowships. Currently serving as the SOE's blended learning champion and an educational technologist with over fifteen years of adult education experience locally and internationally (F2F and Online), she facilitates courses in technology and workforce development for under- and post-graduate students. Her blended learning (BL) champion service includes hosting workshops and one-to-one sessions for faculty as needed; in preparing their course content; and using tools for online delivery via MyeLearning (customized Moodle platform); and Blackboard Collaborate Ultra. Her research supervision of post-graduate students focuses on topics related to teaching with technology and workforce development. She publishes and presents widely in these areas at local, regional, and international conferences and in peer-reviewed journals.

K. Jayashree, Dr. is an engineer by qualification, having done her doctorate in the area of Web Services Fault Management from Anna University, Chennai and Master's in Embedded System Technologies from Anna University and Bachelor's in Computer Science and Engineering from Madras University. She is presently an Associate Professor in the Department of Computer Science and Engineering at Rajalakshmi Engineering College, affiliated to Anna University Chennai. Her areas of interest include Web services, Cloud Computing, Data Mining and distributed computing. She is a member of ACM, CSI.

Igor Jurčić is currently an Assistant Professor in the Department of Mechanical Engineering and Computing at Mostar University with over 20 years of experience in telecom operations to include management and marketing and sales positions; research interests include digital marketing, e-commerce, and Internet of Things with over 15 related scientific and professional articles and white papers.

Vrushali Kadam pursued a Bachelor of engineering in Information Technology from Jawaharlal Nehru engineering college, Aurangabad. Currently working as a Lab Programmer at Jawaharlal Nehru Engineering College, Aurangabad.

Neelu Khare, Dr., is presently working as an Associate Professor in School of Information Technology and Engineering at VIT University, Vellore, India. She completed her PhD degree from MANIT Bhopal, India. She has published 32 papers in international journals and conferences. Her areas of interest are data mining: association, classification, soft computing techniques, IoT and bio-informatics.

Jens Kohler is a researcher at the Faculty of Informatics at the Institute for Enterprise Computing in Mannheim. His research focus is Cloud Computing, Distributed Systems and Distributed Databases.

Ganesh Kshirsagar is currently working as a senior lecturer at NIT Polytechnic, Nashik, India. He has more than 7 years of experience in teaching field, completed Master of Engineering in E&TC with specialization of VLSI & Embedded Systems.

Pankaj Kumar obtained his B.Tech degree in computer science engineering from IEET, Baddi (H.P.), India. He did his M.Tech in Mobile Computing at the Department of Computer Science and Engineering, NIT Hamirpur (H.P.), India in 2015. He is a research scholar at the NIT Hamirpur (H.P.), India and is currently working on the integration of wireless sensor networks (WSNs) with the internet of things (IoT).

Navin Kumar received his PhD in Information Systems from University of Maryland, Baltimore County in 2006. His main research interests are in big data analytics, knowledge discovery and business intelligence. He holds more than 15 years of experience and is currently leading the work at the Advisory Board Company to build rich data solutions for the health care industry. He is the author of multiple journal and conference articles, a prominent guest speaker, and an active member of editorial review boards of multiple Journals.

Somayya Madakam (somu4smart@gmail.com) received his B.Tech (CSE) and MBA (IT) from Andhra University, India. He later worked for TISS, Mumbai for six years as a systems analyst cum programmer. Due to keen interest in teaching and research, he joined NITIE, Mumbai in 2012 to pursue the Fellow Program. His research topic is on "Internet of Things Technologies in Smart Cities: An Exploratory Study in India." During these five years, he presented and published circa 20 articles and posters in both national and International conferences. The article "Smart Cities: Six Dimensions" was presented at the ACIT 2014 conference. The poster "The Lavasa Smart City is in Indian Clouds" got one of the best posters at the ICCC 2015 conference. Besides, the article "A ReView on Internet of Things" was presented at the ICNGCCT 2014 conference. Further works include Smart Homes (IEEE CPS), 100 New Smart Cities (NSITNSW 2015), and "GIFT Smart City: A Business Model" at (ICWR 2015).

Sushma Nayak is a visiting faculty member at Symbiosis School of Economics (SSE), a department of Symbiosis International (Deemed University). In addition to M.Phil. (Economics), she also holds a Post Graduate Diploma in Supply Chain Management. She has 8 years of teaching experience and has taught 14 different courses relating to Economics, Banking, Finance and Political Economy. Presently, she teaches Indian Banking & Financial System and Economic Journalism at SSE. She is a key member of the Center for Academic Writing at SSE, which assists students in writing essays, editorials, columns, commentaries, papers and reports of varied types. Furthermore, she guides students toward making of Annual Art Magazine and Yearbook. She also works towards the making of Oeconomica Scriptis — an in-house publication of SSE — which is a compilation of outstanding academic writings by students. Presently, she is serving as one of the Editors for a book titled Maintaining Financial Stability in Times of Risk and Uncertainty, anticipated to be released in 2018 by IGI Global. She is also serving as a Guest Editor to a Special Issue for the International Journal of Comparative Management, Inderscience Publishers Ltd. Her subjects of interest include Population Studies, Agricultural Economics, Banking & Finance, Managerial Economics, Public Finance, Theory of Political Economy and Supply Chain Management.

Sumitra Purushottam Pundlik, Prof., completed ME in CSE-IT form Savitribai Phule Pune University. He is working as Assistant Professor in MIT College of Engineering, Pune. She has more than 10 years of teaching Experience. Her area of interest is big data analytics and high-performance computing. She is a member of the IEEE, ACM, IAENG (H.K), LMISTE, and LMCSI.

Abirami R completed a B.E in the field of Computer Science and Engineering at Sathyabama University. Completed a M.E in the field of Computer Science at Rajalakshmi Engineering College.

Ch Rupa, Dr., has 13 years' work experience and has presented her work in several national and international conferences and published in many journals indexing DBLP, Scopus, SCI, Springer IEEE. She received many awards from various reputed organizations like the Young Engineer awards from the Institute of Engineers of India, once at the national level and again at the state level and also from JNT University, Kakinada. Dr. Rupa got a computer discipline award from the National Congress from 2011 to 2015, and again in 2017. She was a member of the board of studies to NEC (A). She has delivered a number of guest lectures in the workshops, FDPs, seminars, and conferences as a keynote speaker in both national and international events. She was also invited by the Global Research and Development Society, Linton University, Malaysia as a keynote speaker for its 12th Int. Conf. She has been handling DST funded projects on real time security systems. Dr. Rupa has filed intellectual property rights (IPRs) for her research work with the association of Prof. P. S. Avadhani, Andhra University. She is a Senior Member of the IEEE. Dr. Rupa has a lifetime membership with many professional societies like the Computer society of India (CSI); the Institution of Engineers, India; and the Cryptology Research Society of India (CRSI).

Md. Golam Sarowar is currently pursuing a bachelor of science in computer science and engineering at East West University, Bangladesh. his research interest includes Bioinformatics, Internet of things and big data mining. Profile can be accessed at https://www.researchgate.net/profile/Md_Sarowar3.

Siddhesh Shirsekar completed a M.F.A. (typography, lettering and calligraphy) at the esteemed Jamshtji Jijiboy (J.J) Institute of Applied Art. Institution Degree / certificate Class Year Mumbai University M.F.A. Applied Art 1st Class with Distinction 2016 Mumbai University 4th yr B.F.A. Applied Art 1st Class 2014. Academic awards: paper selected for International Conference on Intelligent Systems and Information (ICISIM), Aurangabad 2017; paper online published in international Journal of Engineering Research & Technology (IJERT 2017), Volume 6, Issue 05,

May 2017; Best in Typography in M.F.A. 2017 (SJJIAA); speaker presentation and poster display at 'ICoRD'17-IIT Guwahati; paper published in 'Smart Innovation, System & Technologies' Vol. 66, Research into Design for Communities- Volume 2 Proceedings of ICoRD 2017' (International Conference of Research in Design) by Springer Publication under the category (Design Creativity, Synthesis, Evaluation & Optimisation at IIT (Guwahati, Assam, India). Student Speaker at International Typographic Conference for Typoday 2016, Bangalore. Best in Communication Design (Art Direction) final year B.F.A 2014 (SJJIAA).

Vijender Kumar Solanki, Ph.D. is an Associate Professor in Computer Science & Engineering, CMR Institute of Technology (Autonomous), Hyderabad, TS, India. He has more than 10 years of academic experience in network security, IoT, Big Data, Smart City and IT. Prior to his current role, he was associated with Apeejay Institute of Technology, Greater Noida, UP, KSRCE (Autonomous) Institution, Tamilnadu, India & Institute of Technology & Science, Ghaziabad, UP, India. He has attended an orientation program at UGC-Academic Staff College, University of Kerala, Thiruvananthapuram, Kerala & Refresher course at Indian Institute of Information Technology, Allahabad, UP, India. He has authored or co-authored more than 25 research articles that are published in journals, books and conference proceedings. He has edited or co-edited 2 books in the area of Information Technology. He teaches graduate & post graduate level courses in IT. He received PhD in Computer Science and Engineering from Anna University, Chennai, India in 2017 and ME, MCA from Maharishi Dayanand University, Rohtak, Haryana, India in 2007 and 2004, respectively and a bachelor's degree in Science from JLN Government College, Faridabad Haryana, India in 2001. He is Editor of the International Journal of Machine Learning and Networked Collaborative Engineering (IJMLNCE) ISSN 2581-3242, Associate Editor of the International Journal of Information Retrieval Research (IJIRR), IGI-GLOBAL, USA, ISSN: 2155-6377 | E-ISSN: 2155-6385. He is a guest editor with IGI-Global, USA, InderScience and many more publishers. He can be contacted at spesinfo@yahoo.com.

Ankit Songara has completed his Bachelor of Technology (B.Tech.) with Computer Engineering from VIET, Jodhpur, India, in 2015. He is presently pursuing his M.Tech. in Mobile Computing from National Institute of Technology, Hamirpur, Himachal Pradesh, India.

Thomas Specht, Prof. Dr., is a professor of the Faculty for Informatics and leads the Institute for Enterprise Computing. His research focus is on distributed systems, cloud computing and component-based systems.

Subbian Umamaheswari, Dr., received her MCA degree from Bharathidasan University, Trichy, India, and a M.Phil degree and a Ph.D. degree from Bharathiar University, Coimbatore, India. She is currently working as an Associate Professor, Department of Computer Science, Dr. G. R. Damodaran College of Science, Coimbatore, India. Her current research interests include web programming, cloud computing, mobile databases, Internet of Things, data analytics and mobile ad hoc networks. She has publications in peer reviewed journals and International Conferences. She is a Member of IEEE. She has been offered with the Summer Research Fellowship by Indian Academy of Sciences, Bangalore at University of Hyderabad May and June 2015. She also got the fellowship from Internet Society to attend the IETF 101 Meeting at London during March 2018. She has published research papers in reputed journals and conferences.

Krishnan Umachandran has over 30 years of experience as a shop floor engineer in various opportunities, Krishnan became a consultant for a spectrum of industries to include beverages, batteries, chemicals, ceramic tiles, leather, electricity, manufacturing, facilities, IT, and ITES. Backed with corporate, consulting, and academic (CCA) expertise, his quest is for escalating human potential with proven strategic initiatives of multidisciplinary interests such as pedagogy; market promotions, customer retention; and leadership-management-organization development.

Index

E

G

H

I

L

M

N

P

Q

R

S

T

U

V

W

www.ingramcontent.com/pod-product-compliance
Lightning Source LLC
LaVergne TN
LVHW080600300926
849162LV00005B/1006